Introduction to Supply Chain Management Technologies

Second Edition

Introduction to Supply Chain Management Technologies

Second Edition

David Frederick Ross

CRC Press
Taylor & Francis Group
Boca Raton   London   New York

CRC Press is an imprint of the
Taylor & Francis Group, an **informa** business

CRC Press
Taylor & Francis Group
6000 Broken Sound Parkway NW, Suite 300
Boca Raton, FL 33487-2742

© 2011 by Taylor and Francis Group, LLC
CRC Press is an imprint of Taylor & Francis Group, an Informa business

No claim to original U.S. Government works

International Standard Book Number: 978-1-4398-3752-8 (Hardback)

Library of Congress Cataloging-in-Publication Data

Ross, David Frederick, 1948-
  Introduction to supply chain management technologies / David F. Ross.
    p. cm. -- (Resource management series)
  Rev. ed. of: Introduction to e-supply chain management : engaging technology to build market-winning business partnerships. 2003.
  Includes bibliographical references and index.
  ISBN 978-1-4398-3752-8 (alk. paper)
  1. Business logistics--Management. 2. Business logistics--Data processing. 3. Business logistics--Information technology. 4. Internet. I. Ross, David Frederick, 1948- Introduction to e-supply chain management. II. Title. III. Series.

HD38.5.R6753 2011
658.7--dc22   2010036525

**Visit the Taylor & Francis Web site at
http://www.taylorandfrancis.com**

**and the CRC Press Web site at
http://www.crcpress.com**

Contents

Preface

At the opening of the second decade of the twenty-first century, the revolutionary changes brought about by the expansion of the Internet into all facets of business had become distinctly mainstream. When work on the first edition of this book, aptly entitled *Introduction to e-Supply Chain Management,* was begun in late 2001, the world was in the process of financial retrenchment and growing doubt about the ability of governments and businesses to prevent another financial "tech bubble" debacle, control the despicable greed of leaders of companies like Enron and WorldCom, and mediate away the senseless violence of global terrorism. Almost a decade later, the cycle of human error and folly seems to have repeated itself as the "housing bubble," the bankruptcy of once rock-solid financial institutions and the awarding of outlandish bonuses for failure, and the wars in Iraq and Afghanistan threaten global stability and cast a long specter over the confidence and trust people once had in their public and business leaders. However, one thing has emerged pointing to continuous progress against this backdrop of a seemingly cyclical view of human affairs, and that is the explosive power of the Internet.

As the world moved past 9/11, the potential of the Internet, despite the false starts as dot-com after dot-com company crashed, seemed to only grow stronger and point to an entirely new way of conducting both business and personal life. Over the ensuing years a "new economy," despite all the hype, was indeed forming based on the Internet's ability not only to connect and informate but also to network communities of users who increasingly were taking responsibility for driving intelligence about what companies should sell, what they wanted from their buying experiences, how brand and company images were to be projected, and even the direction enhancements and innovations were to take.

Something else was also emerging about how effortlessly and seamlessly the use of the Internet was permeating the marketplace as well as the home. In 2003, businesses on the cutting edge of Web technologies were prefixing everything with an "e-": there was e-business, e-commerce, e-procurement, e-sales, and so forth. Software companies were quick to follow suit by linking applications to the Internet with powerful new technology tools and marketing them as a means for customers to gain access to the exciting new world of the Internet. In addition, new business management concepts centered on Web-driven connectivity, messaging, and collaboration became the foremost buzzwords capturing the imagination of management gurus, consultants, and scholarly journals.

By 2010, however, the novelty and hype surrounding Internet-based applications had passed. The revolutionary aura had disappeared by mid-decade: just about everyone had become so accustomed to using Web-based business functions in their daily jobs and personal lives that the little "e-" prefix simply disappeared from software company marketing brochures, scholarly and popular periodicals, and common parlance. This did not mean, however, that the capabilities of Internet-driven technology had bottomed out. Advancements in connectivity and networking architectures continued to advance Internet capabilities, making them more powerful and pervasive as well as easier to implement and use. In the end there was no denying it: e-business had simply become just "business."

The decision to revise the first edition had a lot to do with the demise of the "e-" prefix. Organizations and universities using the book began to complain that the stories about the environment of the early 2000s as well as the pervasive use of the "e-" prefix were severely detracting from the effectiveness of what was still a solid text. The decision to undertake a revision started first as an effort at removing what had become an anachronism. As work began, however, it had become abundantly clear that subtle but deep changes had been occurring to the application of the Internet to supply chain functions. The result was a wholesale restructuring of chapters and updating of content. The overall mission of the book remained the same: to provide a window into the concepts, techniques, and vocabulary of the convergence of supply chain management (SCM) and the Internet, thereby empowering executives involved in restarting a stalled economic environment and returning public confidence in an era every bit as chaotic and destructive as post-9/11 world of the first edition. However, the continuous advancement into what has become known as 2.0 Web technologies required a deep review of the relevance of what had been formerly said and the inclusion of new material on the cutting edge of today's SCM environment.

Each chapter in this book attempts to explore and elaborate on the many different components of the combination of SCM and today's Internet technologies. The first chapter focuses on defining Web-enabled SCM and detailing its essential elements. The argument that unfolds is that technology-driven SCM is a management model that conceives of individual enterprises as nodes in a supply chain web, digitally connected, and collectively focused on the continuous evolution of new forms of customer value. The chapter offers a new supply chain model. Instead of a monolithic pipeline for the flow of products and information, SCM is described as consisting of three separate elements: the *demand channel,* the *process value chain*, and the *value delivery network*. In addition, the chapter also introduces a concise view of the six competencies constituting this new view of SCM. Among the topics discussed are customer management, collaboration, operations excellence, integrative technologies, channel alignment, and supplier management.

Chapter 2 is concerned with exploring the foundations of supply chain technologies and is entirely new to the second edition. This chapter attempts to describe in easily understood terms the technical foundation of today's SCM systems. The chapter begins by exploring how computer technologies have reshaped the way companies utilize information to plan and control internal functions and create interactive, collaborative relationships with their customers and trading partners out in the supply channel network. Among the topics defined are the principles of information processing, integration, and networking. Following, the chapter focuses on an in-depth exploration of the basic architectural elements of today's enterprise information business systems. Included is a review of the three main system components: enterprise technology architecture, enterprise business architecture, and inter-enterprise business architecture. The chapter continues with a discussion of the elements for effective technology acquisition and current gaps inhibiting the effective implementation of collaborative technologies in today's business environment. The chapter concludes with an analysis of today's top new technologies, including *Software-as-a-Service* (SaaS), wireless and *radio frequency identification* (RFID), and *global trade management* (GTM).

Chapter 3 strives to remain close to the exposition of the e-business focus of the first edition while greatly expanding on the coverage of the architecture of today's modern enterprise system. The chapter begins with an overview of the five basic components constituting a business system. Among the elements discussed are the utilization of the database, transaction management, management control, decision analysis/simulation, and strategic planning. Following, the chapter details the principles of effective system management. The goal is to see how an *enterprise business system* (EBS) capable of effectively running an enterprise must have accountability, transparency, accessibility, data integrity, and control. Once these basics are completed, the chapter proceeds to discuss the historical evolution and components of today's EBS. During the review, a detail of the repository of applications constituting today's *enterprize resource planning* (ERP) and SCM software packages are considered. The chapter concludes with a discussion of the various forms of connectivity made available by the Internet. Topics detailed are Web-based marketing, e-commerce, e-business, and e-collaboration.

Chapter 4 is an entirely new chapter that seeks to explore the application of information technologies to execute Lean, Adaptive, Demand-Driven Supply Networks. The chapter explores how integrative technologies can provide organizations and their supply chains with access to demand and supply signals that assist them to reduce the latency that grows in a supply channel from the point a business, environmental, transactional, or out-of-bounds metric occurs to the point that it reaches the final downstream node in the channel network. The speed of receipt, processing, simulation, decision-making, and communication of event-resolution and then the ability of businesses and their supply channels to react to optimize resources and reduce the threat of disruption to channel flows is at the core of today's SCM.

The concepts and computerized toolsets associated with *customer relationship management* (CRM) are explored in Chapter 5. For the most part, chapter content remains close to the first volume. It begins with an attempt to define CRM, detail its prominent characteristics, and outline its primary mission. Next, the discussion shifts to outlining a portrait of today's customer. The profile that emerges shows that customers are *value driven*, that they are looking for strong partnerships with their suppliers, and that they want to be treated as unique individuals. Effectively responding to today's customer requires a *customer-centric* organization. The middle part of the chapter attempts to detail the steps for creating and nurturing such an organization. The balance of the chapter is then focused on e-CRM technology applications, such as Internet sales, sales force automation, service, partnership relationship management, electronic billing and payment, and CRM analytics. The chapter concludes with new sections introducing two of today's most important customer management technologies: *customer experience management* (CEM) and social networking.

Chapter 6 remains fairly close to the first edition, with minor updates to new technologies introduced over the past decade. The chapter is concerned with exploring the application of SCM practices and Web-based tools to the management of manufacturing. The discussion begins by reviewing the role of manufacturing in the "age of the global enterprise." Of particular interest is the discussion on the availability of a bewildering array of technology tools to assist in the management of almost every aspect of manufacturing from transaction control to Internet-enabled product life cycle management. The chapter discusses one of today's most important drivers of productivity—the ability of manufacturing firms to architect collaborative relationships with business partners to synchronize through the Internet all aspects of product design and time-to-market. The chapter concludes with an analysis of today's advanced manufacturing planning functions that seek to apply the latest optimization and Web-based applications to interconnect and make visible the demand and replenishment needs of whole supply network systems.

In Chapter 7, the functions of technology-driven purchasing and *supplier relationship management* (SRM) are explored. After a definition of purchasing has been outlined, a possible definition of SRM is attempted. Similar to the CRM concept, the strategic importance of SRM is to be found in the nurturing of continuously evolving, value-enriching business relationships and is focused on the buy rather than the sell side. The application of Web-based functions have opened an entirely new range of SRM toolsets enabling companies to dramatically cut costs, automate functions such as sourcing, *request for quotation* (RFQ), and order generation and monitoring, and optimize supply chain partners to achieve the best products and the best prices from anywhere in the supply network. The chapter concludes with a full discussion of the anatomy of today's e-sourcing system followed by an exploration of the e-sourcing exchange environment, today's e-marketplace models, and the steps necessary to execute a successful e-sourcing software implementation.

Chapter 8 is concerned with detailing the elements of logistics management in the Internet Age. The discussion begins with a review of the function of logistics and its evolution to what can be called *logistics resource management* (LRM). After a detailed definition of the structure and key capabilities of LRM, the chapter proceeds to describe the different categories of LRM available today and the array of possible Web-based toolsets driving logistics performance measurement and warehouse and transportation management. New to the discussion is a comprehensive review of warehouse management and transportation management software. Afterward, strategies for the use of third party logistics services are reviewed. The different types of logistics service providers, the growth of Internet-enabled providers, and the challenges of choosing a logistics partner that matches, if not facilitates, overall company business strategies are explored in depth.

The final chapter is concerned with SCM technology strategies and implementation approaches. The bulk of this chapter constituted Chapter 4 in the first edition. This chapter seeks to explore how companies can build effective market-winning business strategies by actualizing the opportunities to be found in today's technology-driven supply chain. The discussion begins with an investigation of how today's dependence on supply chains has dramatically altered business strategy development. Structuring effective business strategies require companies to closely integrate the physical capabilities, knowledge competencies, and technology connectivity of their supply chain networks along side company-centric product, service, and infrastructure architectures. Building such a powerful technology-driven SCM strategy requires that companies first of all energize and inform their organizations about the opportunities for competitive advantage available through the convergence of SCM enablers and networking applications. As the chapter points out, strategists must be careful to craft a comprehensive business vision, assess the depth of current supply channel trading partner connectivity, and identify and prioritize what initiatives must be undertaken to actualize new value-chain partnerships. The chapter concludes with a detailed discussion of a proposed SCM technology strategy development model. The model consists of five critical steps, ranging from the architecting of purposeful supply chain value propositions to assembling performance metrics that can be used to ensure the proposed SCM technology strategy is capable of achieving the desired marketplace advantage.

Acknowledgments

In the first edition of this book I wrote in the preface that writing a book on the science of supply chain management technologies is like trying to hold quicksilver. This lament proved to be much closer to the truth than I imagined. The novelty that was e-business in 2002 had become distinctly "old hat" by the end of the decade. Advancements in Web technologies and software architectures had simply made the application of the Internet so easy, so seamless, that it had become invisible to users on all levels of the supply chain: "e-business" had simply evolved to just "business." Doubtless, many of the ideas and relevancy of the resources used to create this book are destined with alarming quickness to be out-of-date as the book moves past its publication date. Still, to begin with, I would sincerely like to express my sincere thanks to the many students, professionals, and companies that I have worked with over the past several years who have contributed their ideas and experience.

I would especially like to thank the executive and editorial staff at CRC Press for so eagerly welcoming the project. My sincerest sympathy goes out to the family, friends, and colleagues of my friend and editor Ray O'Connell, who unexpectedly passed away in the middle of the project. Ray was always so supportive of my projects and effortlessly guided them through to publication. Ray will be seriously missed.

I would also like to thank the entire staff at the University of Chicago Library for their help. Finally, I would like to express my thanks to my wife, Colleen, and my son, Jonathan, for their support, encouragement, and understanding during the many months this book was written.

Author

A distinguished educator and consultant, David F. Ross, PhD, CFPIM has spent over thirty-five years in the fields of production, logistics, and supply chain management. During his thirteen years as a practitioner, he held several line and staff positions. For the past twenty-two years, Dr. Ross has been involved in ERP and supply chain management education and consulting for several software companies. He has also been active teaching supply chain management courses on the university level. He has taught for several years in the Master of Science in Supply Chain Management program at Elmhurst College. He has also taught supply chain management at Northwestern University's Kellogg School of Management. Currently, Dr. Ross is Senior Manager—Professional Development for APICS: The Association for Operations Management where he is involved in courseware design and operations management education.

Besides many articles and white papers, Dr. Ross has published several books in logistics and supply chain management. His first book, *Distribution Planning and Control* (Spencer, 1996), is used by many universities in their logistics and supply chain management curriculums and is a foundation book for APICS's Certified in Production and Inventory Management (CPIM) and Certified Supply Chain Professional (CSCP) programs. A second edition was published in January 2004 and is one of six books on the current APICS's "best seller" list. His second book, *Competing Through Supply Chain Management* (Spencer, 1998), was one of the first critical texts on the science of supply chain management. His third book, *Introduction to e-Supply Chain Management* (CRC Press, 2003), merged the concepts of e-business and supply chain management. This book has also been adopted as a cornerstone text for APICS's Certified Supply Chain Professional (CSCP) program. His fourth book, *The Intimate Supply Chain: Leveraging the Supply Chain to Manage the Customer Experience* (CRC Press, 2008), explores how supply chains can be constructed to provide total value to the customer.

Chapter 1

Supply Chain Management: Architecting the Supply Chain for Competitive Advantage

Over the past dozen years, a wide spectrum of manufacturing and distribution companies have come to view the concept and practice of supply chain management (SCM) as perhaps their most important strategic discipline for corporate survival and competitive advantage. This is not to say that companies have been unmindful of the tremendous breakthroughs in globalization, information technologies, communications networking, e-commerce, and the Internet exploding all around them. It is widely recognized that these management practices and technology toolsets possess immense transformational power and that their ability to continuously innovate the very foundations of today's business structures have by no means reached it catharsis.

SCM is important because companies have come to recognize that their capacity to continuously reinvent competitive advantage depends less on internal capabilities and more on their ability to look *outward* to their networks of business partners in search of the resources to assemble the right blend of competencies that will resonate with their own organizations and core product and process strategies.

Today, no corporate leader believes that organizations can survive and prosper isolated from their channels of suppliers and customers. In fact, perhaps the *ultimate* core competency an enterprise may possess is not to be found in a temporary advantage it may hold in a product or process, but rather in the ability to continuously assemble and implement market-winning capabilities arising from collaborative alliances with their supply chain partners.

Of course, companies have always known that leveraging the strengths of business partners could compensate for their own operational deficiencies, thereby enabling them to expand their marketplace footprint without expanding their costs. Still, there were limits to how robust these alliances could be due to their resistance to share market and product data, limitations in communication mechanisms, and inability to network the many independent channel nodes that constituted their business channels. In addition, companies were often reluctant to form closer dependences for fear of losing leverage when it came to working and negotiating with channel players.

Today, three major changes have enabled companies to actualize the power of supply chains to a degree impossible in the past. To begin with, today's technologies have enabled the convergence of SCM and computerized networking toolsets capable of linking all channel partners into a single trading community. Second, new SCM management concepts and practices have emerged that continually cross-fertilize technologies and their practical application. Finally, the requirements of operating in a global business environment have made working in supply chains a requirement. Simply, those companies that can master technology-enabled SCM are those businesses that are winning in today's highly competitive, global marketplace.

This opening chapter is focused on defining today's technology-driven SCM and exploring the competitive challenges and marketplace opportunities that have shaped and continue to drive its development. The chapter begins with an examination of why SCM has risen to be today's perhaps most critical business strategic paradigm. Next, a short review of the evolution of SCM has been explored. Once the contours of SCM have been detailed, the chapter offers a full definition of SCM. The argument that unfolds is that the merger of today's integrative information technologies and SCM has enabled companies to conceive of themselves as linked entities in a virtual supply chain web, digitally connected and collectively focused on the continuous evolution of competitive advantage. Having established a working definition of technology-enabled SCM, the balance of the chapter explores a new way of looking at the supply chain by breaking it down into three separate components: the *demand channel*, the *process value chain*, and the *value delivery network*. The chapter concludes with a detailed review of the six critical competencies constituting the theory and practice of SCM.

The Foundations of Supply Chain Management

In today's business environment, no enterprise can expect to build competitive advantage without integrating their strategies with those of the supply chain systems in which they are entwined. In the past, what occurred outside the four walls of the business was of secondary importance in comparison to the effective management of internal engineering, marketing, sales, manufacturing, distribution, and finance activities. Today, a company's ability to look *outward* to its channel alliances to gain access to sources of unique competencies, physical resources, and marketplace value is considered a critical requirement. Once a backwater of business management, creating "chains" of business partners has become one of today's most powerful competitive strategies.

What has caused this awareness of the "interconnectiveness" of once isolated and often adversarial businesses occupying the same supply chain? What forces have rendered obsolete long-practiced methods of ensuring corporate governance, structuring businesses, and determining the relationship of channel partners? What will be the long-term impact on the fabric of business ecosystems of an increasing dependence on channel partnerships? What are the possible opportunities as well as the liabilities of channel alliances? How can information technology tools like the Internet be integrated into SCM and what new sources of market-winning product and service value will be identified?

The supply chain focus of today's enterprise has arisen in response to several critical business requirements that have arisen over the past two decades. To begin with, companies have begun to look to their supplier and customer channels as sources of cost reduction and process improvement. Computerized techniques and management methods, such as enterprise resource planning (ERP), business process management (BPM), Six-Sigma, and Lean process management, have been extended to the management of the supply chain in an effort to optimize and activate highly agile, scalable manufacturing and distribution functions across a network of supply and delivery partners. The goal is to relentlessly eradicate all forms of waste where supply chain entities touch while enabling the creation of a linked, customer-centric, "virtual" supply channel capable of superlative quality and service.

Second, in the twenty-first century, companies have all but abandoned strategies based on the vertical integration of resources. On the one side, businesses have continued to divest themselves of functions that were either not profitable or for which they had weak competencies. On the other side, companies have found that by closely collaborating with their supply chain partners, new avenues for competitive advantage can be uncovered. Achieving these advantages can only occur when entire supply chains work seamlessly to leverage complementary competencies. Collaboration can take the form of outsourcing noncore operations to channel specialists or leveraging complimentary partner capabilities to facilitate the creation of new products or speed delivery to the marketplace.

Third, globalization has opened up new markets and new forms of competition virtually inaccessible just a decade ago. Globalization is transforming businesses and, therefore, supply chains, strategically, tactically, and operationally. As they expand worldwide in the search of new markets, profit from location economies and efficiencies, establish a presence in emerging markets, and leverage global communications and media, companies have had to develop channel structures that provide them with the ability to sell and source beyond their own national boundaries. Integrating these supply channels has been facilitated by leveraging the power of today's communications technologies, the ubiquitous presence of the Internet, and breakthroughs in international logistics.

Fourth, today's marketplace requirement that companies be agile as well as efficient has spawned the engineering of virtual organizations and interoperable processes impossible without supply chain collaboration. The conventional business paradigms assumed that each company was an island and that collaboration with other organizations, even direct customers and suppliers, was undesirable. In contrast, market-leading enterprises depend on the creation of panchannel integrated processes that require the generation of organizational structures capable of merging similar capabilities, designing teams for the joint development of new products, productive processes, and information technologies, and structuring radically new forms of vertical integration. Today's most successful and revolutionary companies, such as Dell, Amazon.com, Intel, W. W. Granger, and others, know that continued market dominance will go to those who know how to harness the evolutionary processes taking place within their supply chains.

Fifth, as the world becomes increasingly "flat" and the philosophies of lean and continuous improvement seek to reduce costs and optimize channel connections, the element of risk has grown proportionally. Companies have become acutely aware that they need agile, yet robust connections with their supply chain partners to withstand any disruption, whether a terrorist attack, a catastrophe at a key port, a financial recession, or a devastating natural event like Hurricane Katrina. SCM provides them with the ability to be both *responsive* (i.e., able to meet changes in customer needs for alternate delivery quantities, transport modes, returns, etc.) and *flexible* (i.e., able to manipulate productive assets, outsource, deploy dynamic pricing, promotions, etc.). SCM enables whole channel ecosystems to proactively reconfigure themselves in response to market events, such as introduction of a disruptive product or service, regulatory and environmental policies, financial uncertainty, and massive market restructuring, without compromising on operational efficiencies and customer service.

Finally, the application of breakthrough information technologies has enabled companies to look at their supply chains as a revolutionary source of competitive advantage. Before the advent of integrative technologies, businesses used their supply chain partners to realize tactical advantages, such as linking logistics functions or leveraging a special competency. With the advent of integrative technologies, these tactical advantages have been dramatically enhanced with the addition of strategic capabilities that enable whole supply chains to create radically new regions of marketplace value virtually impossible in the past. As companies implement increasingly integrative technologies that connect all channel information, transactions, and decisions, whole channel systems will be able to continuously generate new sources of competitive advantage through cybercollaboration, enabling joint product innovation, online buying markets, networked planning and operations management, and customer fulfillment.

For over a decade, marketplace leaders have been learning how to leverage the competitive strengths to be found in their business supply chains. Enterprises such as Dell Computers, Microsoft, Siemens, and Amazon.com have been able to tap into the tremendous enabling power of SCM to tear down internal functional boundaries, leverage channelwide human and technological capacities, and engineer "virtual" organizations capable of responding to new marketplace opportunities. With the application of integrative information technologies to SCM, these and other visionary companies are now generating the agile, scalable organizations capable of delivering to their customers revolutionary levels of convenience, delivery reliability, speed to market, and product/service customization. Without a doubt, the merger of the SCM management concept and the enabling power of integrated information technologies is providing the basis for a profound transformation of the marketplace and the way business will be conducted in the twenty-first century.

The Rise of Supply Chain Management

While the concept of SCM can be traced to new thinking about channel management that is little more than a decade old, its emergence has its roots in the age-old struggle of producers and distributors to overcome the barriers of space and time in their effort to match products and services as closely as possible with the needs and desires of the customer. Most texts attribute the foundations of SCM to the historical evolution of the logistics function. Logistics had always been about synchronizing product and service availability with the time and place requirements of the customer. Over the past 50 years, the science of logistics management has advanced from a purely operational function concerned with transportation and warehousing optimization to a business strategy for linking supply chains together in the pursuit competitive advantage. The SCM concept, enhanced by the power of integrative information technologies, is the maturation and extension of the logistics function. This section seeks to explore briefly the origins of SCM and sets the stage for a full definition of SCM to follow.

Historical Beginnings

For centuries, enterprises have been faced with the fundamental problem that the acquisition of materials and the demand for goods and services often extended far beyond the location where products were made. It had been the role of logistics to fill this gap in manufacturing and distribution systems by providing for the efficient and speedy movement of goods from the point of supply to the point of need. Companies that could effectively leverage their logistics functions were able to more profitably operate and focus their productive capabilities while extending their reach to capture marketplaces and generate demand beyond the compass of their physical locations. When viewed from this perspective, the use of the metaphor of linking customers and suppliers into a "chain" is most appropriate and describes how networks of supply and demand partners could be integrated not only to provide for the product and service needs of the entire channel system but also to stimulate demand and facilitate the synchronization of channel partner competencies and resources unattainable by each business operating on its own.

Historically, synchronizing chains of demand and supply had always occupied an important position in the management of the enterprise. As far back as the beginning of the twentieth century, economists considered the activities associated with effectively managing business channels to be a crucial mechanism by which goods and services were exchanged through the economic system. However, despite its importance, this concept, first termed *logistics* (a term originally used to describe the management of military supplies), was slow to develop. Most business executives considered channel management to be of only tactical importance and, because of the lack of integration among supply network nodes, virtually impossible to manage as an integrated function. In fact, it was not until the late 1960s, when cost pressures and the availability of computerized information tools enabled forward-looking companies to start the process of revamping the nature and function of channel management, that the strategic opportunities afforded by logistics began to emerge.

Stages of SCM Development

The emergence of the SCM concept can be said to have occurred in five distinct stages [1], as illustrated in Table 1.1. The first can be described as the era of decentralized logistics management. In this period, logistics functions were often segmented and attached to sales, production, and accounting departments. Not only were activities that were naturally supportive, such as procurement, inbound/outbound transportation, and inventory management, separated from one another, but narrow departmental performance measurements actually pitted fragmented logistics functions against each other. The result was a rather disjointed, relatively uncoordinated, and costly management of logistics. By the mid-1960s, however, growing business complexities and lack of a unified logistics planning and execution strategy were forcing companies to reassemble logistics functions into a single department. Academics and practitioners alike had begun to systematize these ideas around a new term, Physical Distribution Management.

The second stage in the evolution of SCM can be said to have revolved around two critical focal points: logistics centralization and the application of the *total cost concept* to logistics. The movement toward centralization was driven by three converging factors. First, the economic and energy crises of the mid-1970s had dramatically driven up logistics costs. Second, explosions in product lines and increased competition required everyone in the supply channel to deliver products on time,

Table 1.1   SCM Management Stages

SCM Stage	Management Focus	Organizational Design
Stage 1 to 1960s Decentralized Logistics Management	• Operations performance • Support for sales/marketing • Warehousing • Inventory control • Transportation efficiencies • Physical Distribution Management concept	• Decentralized logistics functions • Weak internal linkages between logistics functions • Little logistics management authority
Stage 2 to 1980 Total Cost Management	• Logistics centralization • Total cost management • Optimizing operations • Customer service • Logistics as a competitive advantage	• Centralized logistics functions • Growing power of logistics management authority • Application of computer
Stage 3 to 1990 Integrated Logistics Management	• Logistics concept founded • Support for JIT, quality, and continuous improvement • Use of logistics partners for competency acquisition	• Closer integration of logistics and other departments • Closer integration of logistics with supply partners • Logistics channel planning • Logistics as a strategy
Stage 4 to 2000 Supply Chain Management	• Concept of SCM • Use of extranet technologies • Growth of coevolutionary channel alliances • Collaboration to leverage channel competencies	• Trading partner networking • Virtual organizations • Market coevolution • Benchmarking and reengineering • Integration with ERP
Stage 5 2000 + Technology-Enabled Supply Chain Management	• Application of the Internet to the SCM concept • Low-cost networking of channel databases • e-Business • SCM synchronization	• Networked, multi-enterprise supply chain • .coms, e-tailers, and market exchanges • Organizational agility and scaleability

avert obsolescence, and prevent channel inventory imbalances. Finally, new concepts of marketing, pricing, and promotion facilitated by the computer necessitated a thorough change in the cumbersome, fragmented methods of traditional channel management. These organizational changes, in turn, fed the need to ascertain the total costs involved in logistics, not simply the individual cost of transportation or warehousing. Complementary to these drivers was the application of new computerized technologies and management methods that enabled logistics to effectively participate in efforts directed at company just-in-time (JIT), quality management, and customer service initiatives.

The third stage of SCM development can be said to revolve around the two quintessential catchwords that epitomized the 1980s: competition and quality management. Competition came in the form of tremendous pressure from foreign competitors often deploying radically new management philosophies and organizational structures that realized unheard of levels of productivity, quality, and profitability. The threat also came from the power of new total quality management philosophies and how they could enable more flexible and "lean" processes, tap into the creative powers of the workforce, and generate entirely new forms of competitive advantage. Companies also began to understand that supply channels could be leveraged as a dynamic force for competitive advantage. By enabling trading partners not only to integrate their logistics functions but also to converge supporting efforts occurring in marketing, product development, inventory, manufacturing, and quality management, companies could tap into reservoirs of "virtual" resources and competencies unattainable by even the largest of corporations acting independently. The importance of these breakthroughs was highlighted by the adoption of a new term, Integrated Logistics, in place of the old Physical Distribution Management title.

Stage 4 of the evolution of SCM began in the mid-1990s. Reacting to the acceleration of globalization, the increasing power of the customer, organizational reengineering, outsourcing, and the growing pervasiveness of information technologies, companies began to move beyond the integrated logistics paradigm to a broader and more integrated view of the chain of product and delivery suppliers. At the core of this conceptual migration was a distinct recognition that sustainable competitive advantage could be built by optimizing and synchronizing the productive competencies of each channel trading partner to realize entirely new levels of customer value. Using the supply chain operations reference (SCOR) model as a benchmark, the differences between Stage 3 logistics and Stage 4 SCM can be clearly illustrated.

- ■ *Plan.* In Stage 3 logistics, most business functions are still inward-looking. Firms focus their energies on internal company scenario planning, business modeling, and corporate resource optimization and allocation management. ERP systems and sequential process management tools assist managers to execute channel-level inventory flows, transportation, and customer fulfillment.

In contrast, Stage 4 SCM companies begin to perceive themselves as belonging to "value chains." Optimizing the customer-winning velocity of collective supply channel competencies becomes the central focus. Companies begin to deploy channel optimization software and communications technologies like EDI to network their ERP systems in order to provide visibility to requirements needs across the entire network.

■ *Source.* Companies with Stage 3 sourcing functions utilize the integrated logistics concept to work with their suppliers to reduce costs and lead times, share critical planning data, assure quality and delivery reliability, and develop win–win partnerships. In contrast, Stage 4 SCM sourcing functions perceive their suppliers as extensions of a single supply chain system. SCM procurement seeks to build strong collaborative partnerships based on a small core of approved suppliers with contracts focused on long-term quality targets and mutual benefits. SCM purchasing utilizes technology-driven toolsets that enable the collaborative sharing of real-time designs and specifications as well as for replenishment planning and delivery.

■ *Make.* Stage 3 organizations resist sharing product design and process technologies. Normally, collaboration in this area is undertaken in response to quality management certification or when it is found to be more economical to outsource manufacturing. In contrast, Stage 4 companies seek to make collaborative design planning and scheduling with their supply chains a fundamental issue. When possible, they seek to closely integrate their ERP and product design systems to eliminate time and cost up and down the supply channel. SCM firms also understand that speedy product design-to-market occurs when they seek to leverage the competencies and resources of channel partners to generate "virtual" manufacturing environments capable of being as agile and scaleable as necessary to take advantage of marketplace opportunities.

■ *Deliver.* Customer management in Stage 3 companies is squarely focused on making internal sales functions more efficient. While there is some limited sharing of specific information on market segments and customers, databases are considered proprietary and pricing data are rarely shared. Stage 4 SCM firms, on the other hand, are focused on reducing logistics costs and channel redundancies by converging channel partner warehouses, transportation equipment, and delivery capabilities. Customer management functions look toward automation tools to facilitate field sales, capability to promise tools, customer relationship management (CRM) software, mass customization, and availability of general supply chain repositories of joint trading partner market and customer data.

Stage 4 SCM organizations possess the power to move beyond a narrow focus on channel logistics optimization to one where channel partners strive to identify the best core competencies and collaborative relationships among their trading partners in the search for new capabilities to realize continuous breakthroughs in product design, manufacturing, delivery, customer service, cost management, and value-added services before the competition. Through the application of connectivity technologies linking trading partners, enterprises have the capability to pass information up and down the supply chain so that critical information regarding customer demand and inventory is visible to the entire channel ecosystem.

Today, SCM has progresses to Stage 5 in its development. Through the application of integrative information technologies like the Internet, SCM has evolved into a powerful strategy capable of generating *digital* sources of competitive value based on the real-time convergence of networks of suppliers and customers into collaborative supply chain systems. Actualizing technology-integrated SCM is a three-step process. Companies begin first with the integration of supply channel facing functions within the enterprise through technology solutions like ERP. An example would be integrating sales and logistics so that the customer rather than departmental measurements would receive top attention. The next step would be to integrate across trading partners channel operations functions such as transportation, channel inventories, and forecasting. Finally, the highest level would be achieved by utilizing the power of technologies such as the Internet and e-business to synchronize the entire supply network into a single, scalable "virtual" enterprise capable of optimizing core competencies and resources from anywhere at any time in the supply chain to meet market opportunities.

Although the remainder of this book will concentrate in detail on the enormous changes to SCM brought about by the application of integrative information technologies, the high points of these changes are as follows:

1. *Product and Process Design.* As product life cycles continue to decline and development costs soar, firms have been quick to utilize technology enablers to link-in customers to the design process, promote collaborative, cross-company design teams, and integrate physical and intellectual assets and competencies in an effort to increase speed-to-market and time-to-profit. In the past, efforts utilizing traditional product data management systems and exchange of design data had been expensive, cumbersome, and inefficient. Internet technologies, on the other hand, now provide interoperable, low cost, real-time linkages between trading partners.

2. *e-Marketplaces and Exchanges.* Buyers and suppliers have traditionally been concerned with proprietary channels characterized by long-term relationships, negotiation over lengthy contracts, long-lead times, and fixed margins. Today, technologies like the Internet are completely reshaping this environment. Companies can now buy and sell across a wide variety of Internet-enabled marketplaces ranging from independent and private exchanges to auction sites.

3. *Collaborative Planning.* Historically, enterprises were averse to sharing critical planning information concerning forecasts, sales demand, supply

requirements, and new product introduction. Today, as many organizations increasingly outsource noncore functions to network partners, the ability to transfer planning information real-time to what is rapidly becoming a "virtual" supply chain has become a necessity. Today, integrative technologies provide the backbone to transfer product and planning information across the business network to achieve the two-way collaboration necessary for joint decision-making.

4. *Fulfillment Management.* The collapse of the dot-com era at the beginning of the century revealed one of the great weaknesses of e-business. Customers may have access to product information and can place orders at the speed of light, but actual fulfillment is still a complex affair that occurs in the physical world of materials handling and transportation. Solving this crucial problem requires the highest level of supply chain collaboration and takes the form of substituting as much as possible information for inventory. Some of the methods incorporate traditional tools, such as product postponement, while others utilize Web-based network functions providing logistics partners with the capability to consolidate and ship inventories from anywhere in the supply network and generate the physical infrastructures to traverse the "last mile" to the customer.

As this section concludes, it is clear that channel management is no longer the loose combination of business functions characteristic of Stages 1 and 2 logistics. New integrative technologies and management models have not only obscured internal company functions, they have also blurred the boundaries that separate trading partners, transforming once isolated channel players into unified, "virtual" supply chain systems. Today's top companies are using technology-driven connectivity to reassemble and energize SCM processes that span trading partners to activate core competencies and accelerate cross-enterprise processes. They are also using technology to enable new methods of providing customer value by opening new sales channels as they migrate from "bricks-and-mortar" to "click-and-mortar" business architectures. The next section continues this discussion by offering a detailed definition of SCM that will serve as the cornerstone for the rest of the book.

Defining Supply Chain Management

From the very beginnings of industrialized economies, businesses have been faced with the fundamental problem of how to optimize the dispersion of their goods and services to the marketplace. When producers and customers are in close proximity to each other, demand signals can be quickly communicated by the customer and products and services in turn promptly delivered by suppliers. As the time and distance separating production and the point of consumption increases, however, the ability of companies to deliver to national or even international markets rapidly diminishes. Without the means to effectively move product rapidly from the supply source to the customer, producers find their ability to expand their businesses restricted and markets themselves limited to a narrow array of goods and services.

Bridging this gap between demand and supply has required companies to employ two critical functions. The first is the application of the concept and practice of *logistics*. The role of logistics is to execute efficient and cost-effective warehousing and transportation capabilities that enables companies to satisfy the day-to-day product and service requirements of customers. The second function is the utilization of other companies located in geographically dispersed markets who are willing to assume responsibility for the distribution of goods and services as representatives of producers. This function is termed the *supply value chain,* and the relationships governing the rights, duties, and behavior of producers and partners termed SCM.

Defining Logistics Management

Over the centuries, logistics has been associated with the planning and coordination of the physical storage and movement of raw materials, components, and finished goods. There are many definitions of logistics. The Council of Supply Chain Management Professionals (CSCMP) defines logistics as "that part of SCM that plans, implements, and controls the efficient, effective forward and reverses flow and storage of goods, services, and related information between the point of origin and the point of consumption in order to meet customers' requirements" [2]. In a similar vein, APICS—The Association for Operations Management defines logistics as "the art and science of obtaining, producing, and distributing material and product in the proper place and in proper quantities" [3].

A very simple, yet comprehensive description, defines logistics as consisting of the *Seven Rs*: that is, having the *right product*, in the *right quantity* and the *right condition*, at the *right place*, at the *right time*, for the *right customer*, at the *right price*. Logistics provides *place* utility by providing for the transfer of goods from the producer through the *delivery value network* to the origin of demand. Logistics provides *time* utility by ensuring that goods are at the proper place to meet the occasion of customer demand. Logistics provides *possession* utility by facilitating the exchange of goods. Finally, logistics provides *form* utility by performing postponement processing that converts products into the final configurations desired by the customer [4]. Together, these utilities constitute the value-added role of logistics.

Understanding the organizational boundaries and functional relationships of logistics can be seen by separating it into two separate, yet closely integrated spheres of processes, as illustrated in Figure 1.1. The *Materials Management* sphere is concerned with the *incoming flow* of materials, components, and finished products into the enterprise. This sphere comprises the flow of materials and components as they

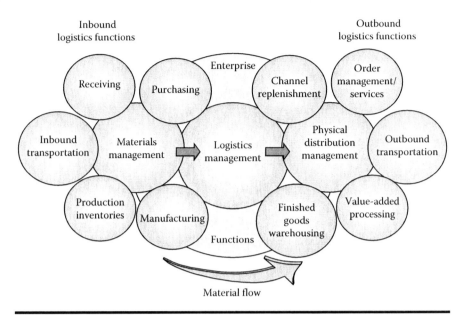

Figure 1.1   Logistics management functions.

move from purchasing through inbound transportation, receipt, warehousing and production, and presentation of finished goods to the delivery channel system. The sphere of *Physical Distribution* is concerned with the *outbound flow* of goods from the place of production to the customer. Functions in this sphere encompass warehouse management, transportation, value-added processing, and customer order administration. Finally, logistics management is a connective function, which coordinates and optimizes all logistics activities as well as links logistics activities with other business functions, including marketing, sales, manufacturing, finance, and information technology.

Defining Supply Chain Management

As they sought to penetrate deeper into the marketplace and rationalize and accelerate product supply and distribution, businesses have always known that by using the capabilities and resources of partners in the supply chain, they could dramatically enhance the footprint of their own core competencies. As far back as the late 1970s, firms had recognized that by linking internal logistics functions, such as transportation and warehousing, with those of channel partners, they could reach new markets, increase pipeline velocities, and cut costs far better than they could by acting in isolation. By the final decades of the twentieth century, however, it had become abundantly clear that the prevailing use of logistics partners needed to be dramatically expanded and elevated to a strategic level. In place of the opportunistic, tactical use of channel partners to achieve a short-term objective, strategists began advocating the transformation of these transient relationships into integrated, mutually enriching partnerships. Logistics channels were to be replaced by "value networks."

Although it an be said that logistics remains at the core of what supply chains actually do, the concept of SCM encompasses much more than simply the transfer of products and services through the supply pipeline. SCM is about a company *integrating* its process capabilities with those of its suppliers and customers on a strategic level. Integrative supply chains consist of many trading partners participating simultaneously in a collaborative network containing multiple levels of competencies and driven by various types of relationships. SCM enables companies to activate the synergy to be found when a community of firms utilizes the strengths of each other to build superlative supply and delivery processes that provide total customer value.

SCM can be viewed from several perspectives. Like most management philosophies, definitions of SCM must take into account a wide spectrum of applications incorporating both strategic and tactical objectives. For example, APICS—The Association for Operations Management defines SCM as

> the design, planning, execution, control, and monitoring of supply chain activities with the objective of creating net value, building a competitive infrastructure, leveraging worldwide logistics, synchronizing supply with demand, and measuring performance globally. [5]

In their text *Designing and Managing the Supply Chain,* Simchi-Levi and Kaminsky define SCM as

> a set of approaches utilized to efficiently integrate suppliers, manufacturers, warehouses, and stores, so that merchandise is produced and distributed at the right quantities, to the right locations, and at the right time, in order to minimize systemwide costs while satisfying service level requirements. [6]

Finally, the CSCMP defines SCM very broadly as encompassing the

> planning and management of all activities involved in sourcing and procurement, conversion, and all logistics management activities. Importantly, it also includes coordination and collaboration with channel partners, which can be suppliers, intermediaries, third-party service providers, and customers. In essence, supply chain management integrates supply and demand management within and across companies. [7]

While these definitions provide for a generalized understanding, they are inadequate to describe the depth and breadth of both the theory and practice of

SCM. Gaining a more comprehensive understanding begins by approaching SCM from three perspectives: one tactical, one strategic, and finally, one technology-enabled. The *tactical* perspective considers SCM as an operations management technique that seeks first to optimize the capabilities of the enterprise's operations functions and then to direct them to continuously search for opportunities for cost reduction and increased channel throughput by working with matching functions to be found in supply chain customers and suppliers. The mission of SCM at this level is focused on synchronizing day-to-day operations activities with those of channel partners in an effort to streamline process flows, reduce network costs, and optimize productivity and delivery resources centered on conventional channel relationships. Finally, SCM in this area is dominated by a *sequential* view of the flow of materials and information as it is handed-off from one channel to another.

Tactical SCM can be broken down into four key value-enhancing activities. The first set of activities, *channel supplier management,* involves optimizing the inbound acquisition and movement of inventories and includes supplier management, sourcing and negotiation, forecasting, purchasing, transportation, and stores receipt and disposition. After inventory receipt, companies can begin executing the second major channel activity, *product and service processing.* In this group of functions can be found product engineering, product manufacturing, and product costing. The third group of activities, *channel customer management,* includes finished goods warehousing, value-added processing, customer order management, channel fulfillment, and transportation. The final group of activities, *channel support activities,* focuses on utilizing channel partners to facilitate financial transactions, marketing information flows, electronic information transfer, and integrated logistics. The objective of operational SCM is to engineer the continuous alignment of internal enterprise departments with the identical functions to be found in supply chain partners. Channel operations synchronization will accelerate the flow of inventory and marketing information, optimize channel resources, and facilitate continuous channelwide cost reduction efforts and increased productivity.

The second perspective of a comprehensive definition is associated with SCM as a *strategy.* The principle characteristic of strategic SCM can be found in the transition from a supply chain model that is interfaced, sequential, and linear to one centered on functional and strategic *interoperability.* The mission of SCM at this level is to propel channel trading partners beyond a concern with logistics optimization to the establishment of collaborative partnerships characterized by the integrating of cross-channel correlative processes that create unique sources of value by unifying the resources, capabilities, and competencies of the entire network ecosystem to enhance the competitive power of the network as a whole and not just an individual company.

While the *tactical* and *strategic* perspectives of SCM have constituted a revolution in business management in and of themselves, the power unleashed by the merger of SCM and today's integrative information technologies has added a third and radically new perspective, changing completely the nature of supply chain theory

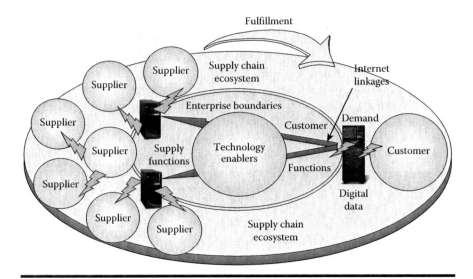

Figure 1.2 The technology-enabled supply chain.

and practice. In fact, it is very difficult to define SCM without acknowledging that at the heart of its development stands the tremendous integrating and networking capabilities of today's information systems. The driver of this dramatic change can be attributed to the application of the Internet. As illustrated in Figure 1.2, the deployment of technology toolsets based on the Internet has enabled companies to develop new methods of integrating with their customers, suppliers, and support partners. Technology-enabled SCM extends the reach of channel management systems beyond enterprise boundaries to integrate in real-time the customer/product information and productive competencies to be found in customers' customers and suppliers' suppliers channel systems. The synergy created enables companies to dramatically improve revenues, costs, and asset utilization beyond a dependence on internal capabilities and resources.

Finally, technology-enabled SCM provides today's supply chains with the means to realize the strategic possibilities of the original SCM model. At the dawn of the Internet Age, companies had come to realize that they were not simply isolated competitors struggling on their own for survival but were in reality part of a much larger matrix of intersecting business systems composed of intricate, mutually supporting webs of customers, products, and productive capacities played out on a global scale. What had been missing in the past was an effective mechanism to enable the intense networking of commonly shared strategic visions and mutually supportive competencies among channel partners. The merger of the Internet and SCM, on the other hand, offers whole supply chains the opportunity to create value for their customers through the design of agile, flexible systems built around dynamic, high-performance networks of Web-enabled customer and supplier partnerships and critical information flows. As detailed in the next section these radical improvements to SCM have required theorists and practitioners alike to revisit the current definitions of SCM.

Redefining Supply Chain Management

The emerging strategic capabilities of SCM and its fusion with today's integrative information technologies necessitate a revision of existing definitions of SCM. The goal is to demonstrate how technology tools are transforming SCM from an operational strategy for the optimization of internal logistics functions to a connected network of associated business partners focused around the common goal of exception customer satisfaction.

A New SCM Definition

Based on these factors, SCM can be redefined as a

> strategic channel management philosophy composed of the continuous regeneration of networks of businesses integrated together through information technologies and empowered to execute superlative, customer-winning value at the lowest cost through the digital, real-time synchronization of products and services, vital marketplace information, and logistics delivery capabilities with demand priorities.

The elements constituting this definition are revealing. The concept *continuous regeneration of networks of businesses* implies that successful supply channels are always in a state of mutating to respond to the dynamic nature of today's ceaseless demand for new forms of customer/supplier collaboration and scaleable product and information delivery flows. This element defines how supply channels will organize to compete. *Customer-winning value* refers to the requirement that supply chains need to continuously reinvent unique product and service configuration and agile delivery capabilities to meet the demands of an evolving marketplace. This element defines the mission of the supply chain. And finally, *digital, real-time synchronization* refers to the application of technology enablers that network disparate channel databases and transaction programs for decision-making on a supply chain level. Internet and e-commerce technologies have now made it possible to transcend extranet enterprise integrators with low-cost, inclusive Web-enabled tools that merge, optimize, and effectively direct supply channel competencies.

Redefining Supply Chain Components

Since its inception, the concept of SCM has been portrayed (Figure 1.3) as a pipelinelike mechanism for the movement of products, information, and financial

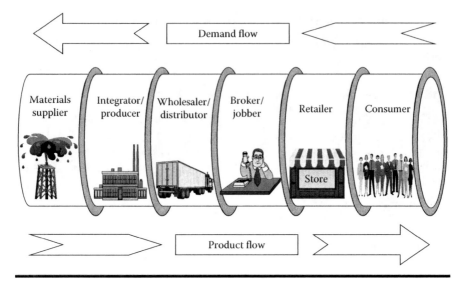

Figure 1.3 The SCM pipeline.

settlement up and down the channel of supply. As detailed, the supply chain pipeline can be composed of a variety of players, each providing a range of specialized functions. What the actual structure of a given supply chain will look like depends on the following of factors:

- Level of supply channel complexity and intensity of buyer/supplier collaboration
- The intensity of market penetration (number of tiers or levels needed to move the product from the supply point to the end customer)
- The extent of integrative intensity (the level of backward or forward control over adjacent channel nodes)
- The intensity of distribution (the number of intermediaries used by the supplier in the delivery process)

For example, a company like Coca-Cola that markets a commodity product will utilize a very complex, global distribution channel composed of several tiers of intermediaries who will provide a variety of value-added services as the product makes its way to the consumer.

While useful, the description of the supply chain as a monolithic pipeline is inadequate for today's technology-enabled supply networks. Actually, the supply chain pipeline should be understood as consisting of three separate, yet integrated channel components, as illustrated in Figure 1.4 [8]. The first can be described as the *demand channel*. The function of this component is to provide a conduit where information in the form of intelligence about marketplace wants and needs, actual orders for products and services, and surrounding ancillary support requirements is

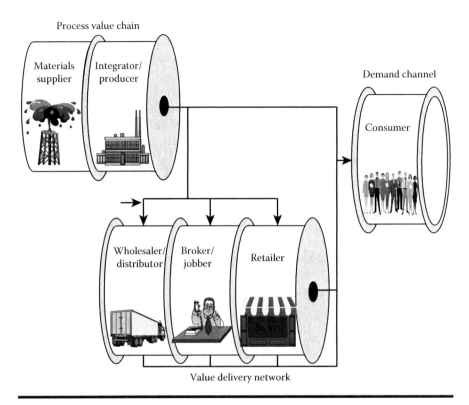

Figure 1.4 Supply chain components.

communicated up the pipeline, passing through a possible variety of intermediaries, and eventually ending with the producer. In addition, in the demand channel can be found payment for goods and services transacted by channel intermediaries and final customers.

In today's business environment, effectively managing demand has become absolutely essential. Supply channel strategists have called for the use of a "demand-pull" originating directly with the customer in place of forecasts, to drive not only the organization's marketing and sales efforts but also the operations side of the business, including production, logistics, distribution, planning, and finance. Outside the organization, as channel partners become so tightly networked that they can respond as a single "virtual" supply entity, channel demand management has challenged business nodes to coevolve closely synchronized, collaborative strategies and processes. The goal is to collectively enhance customer value by increasing the velocity of inventory, services, information, order management, and fulfillment at the "moment of truth" when the customer encounters a supply partner in the delivery network.

The second channel component illustrated in Figure 1.4 can be termed the *process value chain*. This component is composed of networks of materials, component, and resource suppliers that are used by channel integrators to manufacture or assemble the product. The role of the process value chain is to receive demand information in the form of marketing intelligence and actual orders, and then translate that demand into the products and services demanded by the customer. The ability of supply chains to respond to the demand channel begins, therefore, with the capacity of channel integrators to optimize their productive resources and leverage effectively their materials suppliers to make and have available their products and services when the customer wants them at the least cost and with the minimum of effort.

Once the anticipated portfolio of goods and services has been produced, they then enter the third component of the supply chain, the *value delivery network*. The goal of this component is the structuring of delivery channels that facilitate the effective distribution of the product/service composite that reflects as closely as possible the values demanded by the customer. The actual structure of the delivery channel is the responsibility of the integrators and intermediaries that constitute the channel and will be dictated by the nature of demand and the respective capabilities of the channel network constituents. For example, a producer shipping direct to the end-customer will shoulder the responsibility for delivery without the use of intermediaries. On the other hand, other companies will utilize intricate combinations of wholesalers, brokers/agents, export/import intermediaries, and retailers who buy, warehouse, negotiate, and sell. Finally, there is a host of delivery network *facilitators*, such as third party logistics services, financial institutions, and marketing and advertising agencies, that expedite the distribution process.

Channel Configuration

Dividing the supply chain into demand, production, and delivery components enables companies to assemble an effective supply channel configuration. As illustrated in the *supply chain attribute matrix* (Table 1.2), the depth and complexity of the supply network will be determined by several critical attributes [9]. The first, and most critical, is the choice of *process method*. Depending on the nature of the product and competitive strategy, producers can choose between project, jobbing, batch, line, or continuous processing (or a combination). A short definition of the process choices can be seen in Table 1.3. Once the process choice has been made, the most critical decision in the process value chain resides in determining *supply channel complexity*. This attribute is concerned with how the process choice impacts the scope and depth of the integrative partnerships existing between primary and secondary suppliers and the producer. Supply chain partnerships can be described as cooperative alliances formed to plan for and access channel capabilities that will enable the smooth flow of components and materials across the process value network.

The actual intensity of supplier collaboration will vary by the type of process deployed. For example, project producers normally form relatively loose, intermittent

Table 1.2 Supply Chain Attribute Matrix

		Supply Chain					
		Process Value Chain			*Value Delivery Network*		
	Process Method	*Process Intensity*	*Product Variety*	*Supply Channel Complexity*	*Market Penetration Intensity*	*Integrative Intensity*	*Distribution Intensity*
	Project	High	High	Very low	Low	High	Low
	Jobber	High	High	Low	Low	High	Low
	Batch	Medium	Medium	Medium	Medium	High/medium	Low/medium
	Line	Medium/low	Medium/low	High	High	Medium/Low	Medium/high
	Continuous	Low	Low	Very high	Very high	Low	Very high

(Left margin, bottom-to-top arrow: Standard/Intense Distribution; Custom/Single Source)

partnerships, depend on a wide variety of suppliers, negotiate contracts focused on short-term objectives such as price and delivery, and clearly demarcate boundaries of responsibility for quality and cost management. In contrast, at the other end of the matrix, continuous producers develop deep collaborative relationships with a small number of suppliers centered on contracts specifying quality, cost reduction, mutual benefit sharing, and joint responsibility. Continuous producers want supply organizations that are highly networked and synchronized to provide agile responses to rapid changes in production requirements.

The selection of process method will also impact choices in the *value delivery network*. As detailed in Table 1.2, three attributes can be identified. The first, *market penetration intensity,* describes the depth/tiers in the delivery network producers and/or intermediaries must pursue to satisfy market objectives. The desired level of penetration will require the formulation of strategies associated with capabilities and costs (executing delivery activities), channel power (distribution of influence among channel players), and competitive actions (presence of competitors and buying alternatives). As indicated in the matrix map, since project producers create highly customized, unique products, they normally have fairly shallow delivery chains. Line and continuous producers, on the other hand, are constantly searching to expand penetration of commodity-type product/service offerings in an effort to gain an ever-increasing market share.

The second delivery network attribute can be described as *integrative intensity.* This characteristic determines the level of integration backward, forward, or horizontally a producer chooses to pursue in the delivery channel. Backward/forward integration identifies how much control a company seeks over predecessor supply (backward) or downstream customer-facing (forward) delivery agents.

Table 1.3 Process Choices

Process Type	Description
Project	Products produced in a project environment are unique and one of a kind, matched to an individual customer's requirements. This process produces a wide range of custom products very high in cost and price.
Jobbing	Jobbers are normally focused on the processing of a range of product models characterized by complex, often non-repeated configurations of features and options. These products tend to be low in volume while high in cost and price.
Batch	Batch production consists in processing products in large lot sizes. While a range of products is offered, economies of scale in conversion usually lead producers to offer several basic models with a limited variety of features. Batch production enables producers to migrate from a job shop to a flow pattern where lot sizes of a desired model proceed irregularly through a series of processing points or even possibly a low volume assemble line.
Line	In a line environment a limited number of products with no or an extremely limited number of features are produced. Line production is often characterized by connected, mechanized processes, such as a moving assembly line. Key attributes of this process model are volume manufacturing, delivery, and concern with cost. Product design and quality are determined before the sale occurs.
Continuous	Producers utilizing this method sell a very narrow range of highly standardized, commodity products produced in high volumes utilizing dedicated equipment. Price and low cost are the key determining factors for this type of production.

Horizontal integration refers to a level of control whereby a company acquires a linked channel intermediary. According to the matrix map, the more a product/service is customized, the more intense will be a company's control over channel delivery functions. For example, a producer of highly configurable plant heating and cooling equipment will most likely control the entire channel from marketing, to production, to actual delivery. On the other hand, an electronics commodity producer like Sony will depend on a complex delivery network that assumes the responsibility for a wide range of functions, from promotion and pricing to delivery, merchandizing, and retail.

The last delivery network attribute, *distribution intensity*, is concerned with the number of intermediaries to use at each channel level. There are four strategies that can be deployed. In the first, *single source distribution*, a producer performs all delivery network activities. This strategy is chosen by producers who wish to retain exclusive control over elements such as brand, delivery, price, promotion, and service. In the second strategy, *intensive distribution*, a company seeks to utilize a broad and deep distribution network to reach as many customers as possible. The third strategy, *exclusive distribution*, is pursued by companies who want to limit the number of intermediaries who deal with customers. This strategy can take two forms: *exclusive distribution*, in which a producer authorizes exclusive distribution rights to a select group of intermediaries, and *exclusive dealing*, where a producer requires intermediaries to sell only its products. The final strategy, *selective distribution* permits select groups, but not all possible intermediaries, to handle and sell the producer's products.

The *supply chain attribute matrix* provides an easy format to use in determining the structure of a supply chain. For example, line and continuous process-based producers (such as foods, low-cost electronics, and basic services) normally produce commodity goods or raw materials and seek as many outlets to sell to as possible. On the other hand, batch and line producers, who produce, assemble, and make-to-order goods (such as major appliances, automobiles, and designer apparel) often seek exclusive or selective distribution in order to retain control over pricing, promotions, and market image as well as to prevent the inroads of competitors. Finally, capital goods producers will normally use the single source distribution strategy. Makers of highly customized project-oriented products either purposefully (to eliminate any form of competition) or unintentionally (due to the unique nature of the product/service) will pursue this strategy.

Supply Chain Competencies

The supply and delivery network-building attributes of SCM have revolutionized the role of the supply chain and infused channel constituents with radically new ways of providing total customer value. Instead of a focus just on purchasing, warehouse, and transportation management, today's SCM practices require entire supply chains to collectively work to activate an array of competencies, as illustrated in Figure 1.5 [10]. A detailed discussion of each competency is as follows.

Customer Management

In the past, customer management strategies focused on optimizing economies of scale and scope by pushing standardized goods and services into the marketplace regardless of actual customer wants and needs. Today, companies can no longer

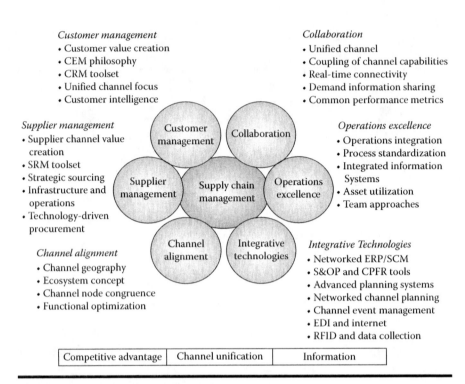

Customer management
- Customer value creation
- CEM philosophy
- CRM toolset
- Unified channel focus
- Customer intelligence

Collaboration
- Unified channel
- Coupling of channel capabilities
- Real-time connectivity
- Demand information sharing
- Common performance metrics

Supplier management
- Supplier channel value creation
- SRM toolset
- Strategic sourcing
- Infrastructure and operations
- Technology-driven procurement

Operations excellence
- Operations integration
- Process standardization
- Integrated information Systems
- Asset utilization
- Team approaches

Channel alignment
- Channel geography
- Ecosystem concept
- Channel node congruence
- Functional optimization

Integrative Technologies
- Networked ERP/SCM
- S&OP and CPFR tools
- Advanced planning systems
- Networked channel planning
- Channel event management
- EDI and internet
- RFID and data collection

Competitive advantage	Channel unification	Information

Figure 1.5   Supply chain management competencies.

compete by pursuing strategies built solely on volume and throughput, but instead must migrate to a supply chain perspective where the collective competencies and resources of their channel ecosystems can be leveraged in the pursuit of unique avenues of customer value and superior service. As customers demand to be more involved in product/service design, pricing, and configuration of their own buying solutions, companies have come to understand that creating customer value rests on establishing enriching customer relationships.

In today's era of the customer-centric marketplace, businesses can not hope to survive without the activation of an array of enabling techniques that have come to coalesce around CRM. As will be discussed in much greater detail in Chapter 5, the goal of CRM is to provide complete visibility to all aspects of the customer, from facilitating the service process, to collecting data concerning customer buying history, to optimizing the buying experience. As a toolbox of technology enablers and a management philosophy, CRM attempts to realize these objectives through the following points:

- CRM is supportive of the strategic missions of all channel network players by providing the mechanisms necessary to align their resources and capabilities in the search to build profitable, sustainable relationships with customers.

- The customer experience management (CEM) side of CRM is focused on optimizing the customer experience wherever it occurs in the supply chain. Such an objective requires the use of metrics and analytical tools capable of providing a comprehensive, cohesive customer portrait that in turn can drive acquisition and retention initiatives.
- CRM assists in gathering intelligence as to which customers are profitable, what products and services will spawn differentiating customer value propositions, and how each channel member can architect sales and delivery processes that consistently deliver to each customer an exception buying experience.
- CRM is about nurturing mutually beneficial, long-term customer relationships resilient enough to sustain continuous improvement opportunities and tailored solutions beyond immediate product and service delivery.
- CRM is a major facilitator of supply chain collaboration. Customer value delivery must be considered as a single, unified event that ripples up and down the supply chain. Firms participating in integrated, synchronized delivery networks capable of providing collectively a seamless response to the customer will be those that will achieve sustainable competitive advantage.

CRM provides business with the mechanisms necessary not only to quantify customers' behaviors and preferences but also to invite customers to be value chain collaborators. CEM provides business with a process to strategically manage a customer's entire experience, both emotionally and physically, with a product or a company. Together, CRM and CEM provides companies with the ability to architect a mosaic of processes and toolsets for the generation of fast-flow, flexible, synchronized delivery systems that enable customer self-service, configuration of customized, individualized value-solutions, and fulfillment functions providing the highest level of service and value.

Supplier Management

Effective management of the sourcing function resides at the very core of competitive supply chains. Managing procurement, however, is more than just the acquisition of products and services. In fact, for several decades, businesses have known that the relationship between buyer and seller, and not just the price and quality of the goods, determined the real value-add component of purchasing. With the application of today's enabling technologies, this viewpoint has spawned a new concept and set of business practices termed supplier relationship management (SRM).

The mission of SRM is to activate the real-time synchronization of inventory and service requirements of buyers with the supply capabilities of channel partners. The goal is to actualize a customized, unique customer buying experience while simultaneously pursuing cost reduction and continuous improvement performance objectives. SRM seeks to fuse supplier management functions—information systems, logistics, resources, skills, cost management, and improvement—found across the entire supply chain into an efficient, seamless process driven by relationships founded on trust, shared risk, and mutual benefit.

Attaining these objectives requires realization of several key components. To begin with, companies must look beyond the everyday purchase of materials to *strategic sourcing.* Focus on a strategic view not only provides for technology-enabled product-cost search from a universe of suppliers that best fit requirements but also reveals the depth of supplier competencies, availability of value-added services, level of desired product quality, capacity for innovative thinking, and willingness to collaborate. Among the benchmarks can be found expected cost savings, enhanced process efficiencies, reductions in cycle times, channel inventory minimization, and increased process optimization as a result of a closer matching of channel demand with network capabilities and resources.

Secondly, the procurement process has always been driven by technology toolsets that facilitate the communication of requirements, negotiation of quality, pricing, and delivery objectives, and accounts payable processes. Today, the application of the Internet to SRM has enabled purchasers to leverage new forms of procurement functions, such as online catalogs, interactive auction sites, spend analytics, and trading exchanges. These integrative technologies have provided purchasers with tools for the real-time, simultaneous synchronization of demand and supply from anywhere, anytime, across a global network of suppliers.

The third key component of SRM is the activation of procurement infrastructures linked directly to the customer. Procurement functions unable to respond in a timely fashion to changes in the marketplace with complimentary organizational, technical, and performance management objectives will consistently result in suboptimal customer performance. At the heart of the SRM-driven organization is the ability to widen traditional purchasing functions to include new Internet-driven players, such as trading exchanges, consortiums, and other e-commerce services providing for payment, logistics, credit, and shipping.

Channel Alignment

The geography of any supply chain is composed of its supply and delivery nodes and the links connecting them. In the past, this network was characterized as a series of trading dyads. In this model, channel partners created trading relationships one partner at a time without consideration of the actual extended chain of customers and suppliers constituting the entire supply chain ecosystem. In reality, a company like Wal-Mart deals with literally thousands of suppliers, and their supply chain resembles more a networked grid of business partners (Figure 1.6) than the pipelinelike structure portrayed in Figure 1.3. Maintaining a strategy of trading partner dyads as the supply chain expands risks decay of cost management objectives, leveraging resource synergies, and maintaining overall marketplace competitiveness.

To counter the inertia of the trading partner dyads strategy, effective SCM requires a continuous focus on network node congruence. This means that each

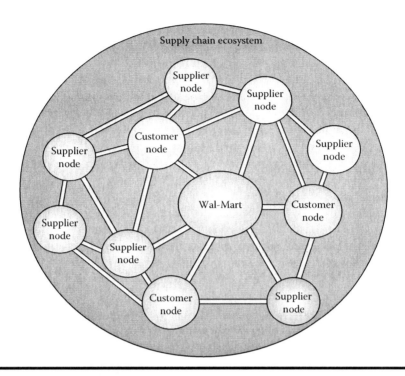

Figure 1.6   The supply chain as a network grid.

channel constituent must construct an individual business strategy and set of operational objectives that simultaneously provides for competitive advantage for both the firm and the collective channel network. This step should also reveal the gaps and regions of potential conflict existing between the strategies and metrics of individual partners. Without strategic and operational alignment, the chain will have weak links that will easily break as the pressure of demand variability and missing partner capabilities appears at times of channel stress.

As supply chain convergence matures, the number of nodes occupying peripheral positions can be more closely integrated into the direct channel. The goal is to increase the length of the contiguous supply chain, thereby expanding opportunities for collaboration, customer value, and operations excellence while minimizing conflict and increasing compromise over costs, performance metrics, service value propositions, and delivery velocity targets. Achieving supply chain congruence can be even contentious as companies find themselves working with several separate channel networks as their business ecosystems evolve in new directions.

Supply Chain Collaboration

The keystone of SCM can, perhaps, be found in the willingness of supply network partners to engage in and constantly enhance collaborative relationships with each other. *Collaboration* can be defined as an activity pursued jointly by two or more entities to achieve a common objective. It can mean anything from exchanging raw data by the most basic means, to the periodic sharing of information through technology-based tools, to the structuring of real-time architectures capable of leveraging highly interdependent infrastructures in the pursuit of complex, tightly integrated functions ensuring planning, execution, and information synchronization.

As displayed in Figure 1.7, the intensity of the collaborative content can vary. It can be internally driven and focused on the achievement of local objectives. On the other hand, it could seek to use technology to deepen interchannel operations linkages, drive shared processes and codevelopment, and even foster a common competitive vision for the whole channel. The *value* of collaboration is gauged by how effectively firms are leveraging the competencies of the distributed knowledge of the channel base, reducing redundant functions and wastes, sharing a common vision of the supply chain, and constructing the technical and social architectures, thereby enabling whole channel networks to achieve marketplace leadership.

According to Finley and Srikanth [11], there are several critical imperatives that must be pursued if collaboration is to reach its potential. First, it is essential that individual company strategies and goals be aligned across the supply chain. Convergence of objectives ensures network partners are pursuing a common

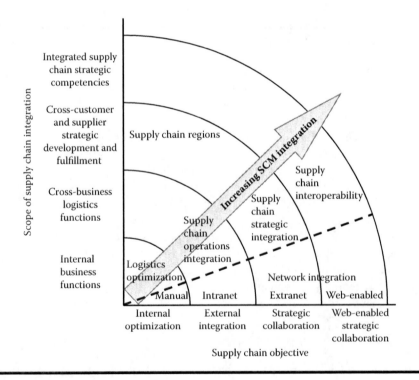

Figure 1.7   Span of SCM collaboration.

channel vision, planning and execution processes, and performance metrics that benefit both individual businesses and the whole supply chain. Next, it is crucial that companies segregate the various channels in which they participate. Because each channel (i.e., wholesale, export, retail, mass merchandize, etc.) possesses a range of different processes, delivery dynamics, and performance indicators, a tailored strategy for each channel, rather than a forced global approach, would allow channel planners a way to effectively design an integrated approach despite the presence of local variations. Finally, unified channel collaboration makes it easier to ensure connectivity of all channel nodes, availability of the proper technology tools for information visibility and real-time transfer, acceptance of common performance metrics and benefits, and access to demand patterns and expectations as they stream across the supply chain.

While there can be little doubt collaboration is one of the key foundations of SCM and is recognized as a high-corporate priority, the path to successful collaboration is blocked by many barriers. According to a recent survey [12], five barriers were singled out by respondents as inhibiting collaboration in their companies. Seventy-five percent agreed that organizational structures propagating turf protection was the most pervasive barrier. The four other barriers were singled out as resistance to change (58%), conflicting measures (55%), lack of trust (42%), and weak managerial support (42%). To make matters worse, the same study conducted a decade earlier yielded almost the same measurements. Despite that fact that such numbers indicate that truly cohesive and collaborative supply chain teams are the exception and not the rule, today's advanced supply chain leaders—Apple, Dell, Proctor & Gamble, Wal-Mart, and Cisco—are winning and outdistancing the competition because they understand their success rests on seeing themselves as the drivers of value chain collaboration.

Operations Excellence

All organizations seek to optimize productive functions while removing costs, and this objective becomes even more critical in SCM. Ideally, operations excellence compels every firm in the channel network to optimize both their own performance and, by extension, the performance of the entire supply chain. Simply, as the competency of each individual channel node is increasingly integrated with other participants, collectively the supply chain will have access to a range of processes and benefits individual companies would be incapable of achieving acting on their own. By the collaborative nature of SCM, supply network participants are compelled to look beyond the performance of their own organizations to the supply chain in its entirety.

The use of the metaphor of an ecosystem best describes this sense of mutual dependence of channel entities on each other if they are to survive and prosper [13]. Instead of the biological environment, supply chain ecosystems consist of intertwined matrices of customers, market intermediaries, and suppliers as well as other stakeholders, such as government, unions, and associations, that involve cooperation as well as conflict. While product and service competition remains at the core of business, the ability to generate value, engage in innovation, and repel challenges from competitors requires today's organization to invest in expanding its community of allies. It is therefore through the powerful mechanism of coevolution that companies in the twenty-first century will be able to architect centers of innovation where, by orchestrating the competencies and vision of a network of partners, they can build unbeatable competitive advantage.

Implicit in the business ecosystem concept is the need for the joint acceptance of several key partnership attributes. Effective SCM relationships, whether informal or formal, are broadened and matured by building the following [14]:

- *Trust.* This attribute is perhaps the foundation of SCM. While trust is hard to quantify, it means that channel partners can have faith in the intentions and actions of each other, that individual company strategies are formulated with the good of the entire network in mind, and that companies will not use positions of power to abuse more dependent members.
- *Reliability.* Being reliable means that a company can count on its partners to exhibit consistent, predictable, and honest behavior over the long run. Lack of congruence between commitments and behavior erode supply chain relationships. In addition, the use of coercion to force partners to act in a prescribed way often results in less-than-reliable behavior and is antithetical to the establishment of strong channel relationships.
- *Competence.* This attribute is concerned with the capability of a partner to support and perform its channel the role as initially promised. Competence refers to the ability of the partner's organization to provide the people, processes, knowledge, experience, technology, and resources that will ensure the viability of the channel relationships.
- *Risk Sharing.* Risk is part of every business endeavor. Risk in a supply chain relationship, however, often extends beyond normal uncertainties, because there is an implied external vulnerability arising from dependence on partners to perform their agreed upon roles.
- *Loyalty.* Trust, reliability, and a willingness to risk all contribute to a sense of loyalty between channel partners. Loyalty is a two-way street: each partner not only performs predictably but is also willing to assist each other to resolve problems or ameliorate risk. Loyalty enables parties to engage in deeper commitment to the relationship and by extension enriches the entire supply chain ecosystem.

Integrative Technologies

The convergence of integrative information technologies and SCM constitutes the very core of this book. It has been argued earlier that it is virtually impossible to think of SCM without the power of the enabling technologies that have shaped and driven its development into a management science. In today's highly competitive global marketplace, having the best product or service is simply not enough: now, having the best *information* has become the decisive differentiator between market leaders and followers. Supply channel transparency requires a single version of data encompassing demand, logistics, demand-capability alignment, production and processes, delivery, and supplier intelligence across the firms comprising the end-to-end supply chain. Generating a single view of the supply chain requires information technologies that enable collection, processing, access, and manipulation of complex views of data necessary for determining optimal supply chain design and execution configurations.

Technology applications to SCM can be divided into three broad areas. To begin with, the most complex technology toolsets can be found in the utilization of major business systems, such as ERP, that seek to cover the entire enterprise. On the next level can be found point technology solutions, such as advanced planning systems (APS) and transportation management systems (TMS), which seek to optimize specific functions or provide for cross-channel visibility. In the third and final level of technology tools can be found execution solutions, such as electronic data interchange (EDI), the Internet, or radio frequency identification (RFID).

The Impact of SCM

The intense focus on today's technology-driven SCM is the result of the realization that companies can not hope to gather on their own without closely linked channel partners the necessary resources and core competencies that permit them to efficiently and profitably design, manufacture, market, and distribute goods and services to what has become a global marketplace. The six SCM competencies illustrated in Figure 1.6 enable companies to realize the three essential success factors shown at the bottom of the figure. To begin with, by architecting highly integrated supply networks through the application of connective technologies like the Internet and social networking, companies can achieve continuous *competitive advantage* by jointly developing and delivering winning products and services before the competition. Second, by converging the collective resources and innovative capabilities found among network partners, SCM enables whole supply chains to act as if they were a single *unified channel* capable of delivering customer value seamlessly across intersecting supply chains. And finally, by deploying connective technologies, companies can leverage *information* to lead marketplace change, preserve brand integrity, effectively identify customers, and provide total customer value as well as overall profitability.

Summary and Transition

Today, even the best-run organizations have begun to feel the strain of effectively responding to the realities of the twenty-first century marketplace. New concepts of what constitutes customer value, requirements for more agile, flexible product design and manufacturing processes, the growth of "virtual" companies and collaborative processes, and the continuous fragmenting of marketplaces demanding customized products and services to satisfy a "marketplace of one" have rendered obsolete past organizational models. In addition, the ubiquitous presence of technology, the ultimate enabler promising the real-time synchronization of information with anyone, anytime, anywhere, has reshaped the very nature of business and altered fundamental thinking about the way companies market, buy, sell, and communicate with their customers and suppliers.

Perhaps the most salient of these changes has been the realization that to survive and prosper, companies must perceive themselves not as self-contained competitive entities but rather as collaborative nodes in a much larger network of intersecting business systems composed of intricate, mutually supportive webs of customers, products, competencies, resources, and information played out on a global scale. The best enterprises understand that to thrive in today's marketplace, it is imperative that they continuously activate the synergies that occur when they leverage resources and human capabilities wherever they can be found in the supply chain to establish virtual, scalable organizations capable of responding decisively to any marketplace challenge.

Over the past half decade, the principles supporting this nascent management paradigm has gathered around the term SCM. SCM provides today's enterprise with the strategic vision as well as the operational principles necessary to integrate once isolated companies into unified value-generating networks. Although SCM constitutes a virtual revolution in business management in and of itself, the merger of SCM with integrative information technologies has propelled SCM to a new level by linking the concept with powerful networking tools enabling the real-time integration and synchronization of channelwide processes and databases. These enhancements to SCM have rendered obsolete prior interpretations and called for the following redefinition:

> SCM is a tactical and strategic management philosophy that seeks to network the collective productive resources of intersecting supply channel systems through the application of integrative business technologies in the search for innovative solutions and the synchronization of channel capabilities dedicated to the creation of unique, individualized sources of customer value.

SCM has become today's most important management concept because it enables enterprises to exploit the explosion of technology tools that are transforming the realities of the twenty-first century marketplace. Investigating in greater detail the foundation of current technology is the subject of the next chapter. The discussion focuses first on a basic understanding of today's technology architectures and why technologies possess such powerful transformational capabilities.

Following, a full analysis is given of why technology is so important to the effective functioning of SCM and how companies can leverage their convergence to build exceptional, highly flexible supply chains capable of mastering today's complex business environments.

Notes

1. For an interesting insight into the early stages of logistics management see Donald J. Bowersox, Bernard J. LaLonde, and Edward W. Smykay, *Readings in Physical Distribution Management: The Logistics of Marketing* (London: Macmillan, 1969). Also of interest is Bowersox's retrospective on the history of SCM as a discipline in "SCM: Past as Prologue," *Supply Chain Quarterly,* (2nd Quarter, 2007).
2. This definition can be found at accessed June 22, http://cscmp.org/aboutcscmp/definitions.asp.
3. APICS—The Association of Operations Management, *APICS Dictionary*, 12th ed. (APICS, 2008), p. 74.
4. This definition can be found in David F. Ross, *Distribution Planning and Control: Managing in the Era of Supply Chain Management,* 2nd ed. (Norwell, MA: Kluwer Academic, 2004), p. 37.
5. *APICS Dictionary*, p. 134.
6. David Simchi-Levi, Philip Kaminsky, and Edith Simchi-Levi, *Designing and Managing the Supply Chain: Concepts, Strategies, and Case Studies,* 2nd ed. (Boston, MA: McGraw-Hill Irwin, 2003), p. 1.
7. This definition can be found at accessed June 22, http://cscmp.org/aboutcscmp/definitions.asp.
8. This three-part division of the supply chain was first introduced in David Frederick Ross, *The Intimate Supply Chain* (Boca Raton, FL: CRC Press, 2009), pp. 54–58.
9. A simplified version of the *supply chain attribute matrix* first appeared in ibid., pp. 66–68.
10. This figure has been adapted from ibid., p. 103.
11. Foster Finley and Sanjay Srikanth, "7 Imperatives for Successful Collaboration," *Supply Chain Management Review* 9, no. 1 (January/February 2005), pp. 30–37.
12. Stanley E. Fawcett, Joseph Andraski, Amydee M. Fawcett, and Gregory M. Magnan, "The Art of Supply Change Management," *Supply Chain Management Review* 13, no. 8 (November 2009), pp. 18–25.
13. This metaphor was coined by James F. Moore in his book *The Death of Competition: Leadership and Strategy in the Age of Business Ecosystems* (New York: Harper Business Systems, 1996).
14. These points have been summarized from Robert B. Handfield and Ernest L. Nichols, *Introduction to Supply Chain Management* (Upper Saddle River, NJ: Prentice Hall, 1999), pp. 83–89.

Chapter 2

Supply Chain Technology Foundations: Exploring the Basics

As it has in all areas of today's business environment, the application of integrative information technology has caused a revolution in the concept and practice of supply chain management (SCM). In fact, the transformative topics discussed throughout this book—the real-time networking of channel trading partners, collaborative design, social networking, e-commerce, radio frequency identification (RFID), and others—would be impossible without the connective power of today's array of information technologies. In addition, without them, metrics relating to customer demand, base information like item masters and pricing, transaction databases, visibility to disruptions impacting execution activities such as delivery, availability, and quality, and a myriad of additional supply channel events would remain tightly locked within organizational departments and invisible to the suppliers and customers located outside the business.

The implementation of information technology has become an absolute requirement for success in today's business environment. Several factors are driving this movement. To begin with, companies have had to turn to computerized applications in order to deal with the increasingly complex requirements of doing business in a fast-paced global environment. Second, customers and suppliers have continued to demand instantaneous response and full information visibility, online, in real-time. Third, the rapidly changing contours of dealing with shrinking life cycles for everything have made the ability to closely link customers, producers, and suppliers a requirement for competitive survival. Fourth, the integrative power of the Internet is requiring supply chains to have the ability to rapidly transfer transaction information and marketplace intelligence from buyer to supplier. And finally, information technology has become a competitive advantage. Wal-Mart's satellite-connected systems, American Airline's Sabre System, Federal Express's package-tracking systems, and Cisco's "virtual manufacturing environment" are examples of how technology automates complex tasks, generates useable information, and networks supply chain nodes tightly together to manage complexity and expand competitive advantage.

Successfully leveraging technology requires organizations, if not whole supply chains, to find answers to the following critical questions. What are the goals of information technology from the perspective of the business? What computerized technology components (hardware, software, peripherals, etc.) are necessary to realize information goals? What technology toolsets need to be implemented across the supply chain if channel partners are to be closely linked to form a virtual supply network? What are the trends in today's information technologies and how do they impact the supply chain? What are the methodologies and tasks necessary to create a sustainable supply chain information technology environment?

If the purpose of the first chapter is to define the concept of SCM, the goal of this chapter is to discuss in a nontechnical manner the nature of the computerized technologies that lay behind SCM and how they be leveraged to find answers to the above and other questions. Talking about technologies such as enterprise resource planning (ERP), the Internet, software as a service (SaaS), and cloud computing is one thing; actually understanding from a practical perspective the nuts-and-bolts of the computerized architectures and the networking and communications toolsets that activate them is another. The good news is that the computer architecture is not scary—it just that it keeps morphing into radically new capabilities that quickly obsolete current toolsets while revealing exciting new possibilities for linking customers and suppliers. Which is what, after all, SCM is all about.

The chapter begins by exploring how computer technologies have reshaped the way companies utilize information to plan and control internal functions and create interactive, collaborative relationships with their customers and trading partners out in the supply channel network. Among the topics defined are the principles of information processing, integration, and networking. Following, the chapter focuses on an in-depth exploration of the basic architectural elements of today's enterprise information business systems. Included is a review of the three main system components: enterprise technology architecture, enterprise business architecture, and inter-enterprise business architecture. The chapter continues with a discussion of the elements for effective technology acquisition and current gaps inhibiting the effective implementation of collaborative technologies in today's business environment. The chapter concludes with an analysis of today's top new technologies, including SaaS, wireless and RFID, and global trade management (GTM).

The Importance of Information Technology

The concept of *technology* is connected with the creation and management of human knowledge. The word "technology" stems from the ancient Greek *techne*. We have come to understand technical as meaning having special knowledge of a mechanical or scientific nature. However, the original meaning of *techne* arises from the possession of neither art nor skill. In the authentic sense of *techne,* knowledge is about bringing forth or uncovering the existence of things, the relation of things to each other. Unlike knowledge that is about data, *techne* is concerned with the actualization of possibilities, about the creation of new things. Furthermore, this knowledge, unlike calculative thinking, is cumulative and builds on itself. Therefore, any agent, including the human imagination, that provides the opportunity to generate new forms of knowledge can be said to be a *technology*.

From the very beginning of mankind, people have used physical objects to provide them with capabilities to do things that were beyond the physical constraints imposed by the body. Whether it was the use of simple tools made from flint and wood or styluses and brushes to create works of art, physical devices provide an enriching knowledge that opens new vistas beyond the simple recognition of the repetitiveness of diurnal events. The more powerful the capability of an instrument that generates a new *techne,* the more fundamentally the infrastructure of the physical world is reorganized. The new knowledge quickly renders obsolete existing practices and patterns of behavior. It enables thinking that lies hidden and evokes the awareness of fresh choices.

Basics of Information Technology

Information technology utilizes computer-based tools [1] to assist people in the management of information arising from processing activities in support of the knowledge requirements of their organizations. While today's computers have grown exponentially more powerful and ubiquitous, by themselves they provide little use outside of their calculative power. Information technology is most useful when the data it contains and its ability to network the thoughts and purposes of people is directed at activating capacities for both task management and innovation. In essence, information technology succeeds when people, the computer systems, and information are tightly integrated.

It can be said that the application of information technology to any human enterprise consists of three separate yet inextricably linked concepts of knowledge. The first type is centered on the use of computerized technology to *automate* knowledge. The second type of knowledge arises when technologies are used to perform productive and administrative processes. An interesting word that was coined by Zuboff to describe this category of knowledge is *informate*. New sources of knowledge about activities, events, and objects are made visible when a technology *informates* as well as *automates* [2]. The third and most sophisticated type of information arises when people use technology to *network* their tasks, ideas, and aspirations to produce a form of collective, opt-in/opt-out fusion of opened-ended knowing and experiencing.

Using Technology to Automate Knowledge

Accompanying the rise of civilization can be found the use of physical devices created to reproduce and extend the capacities of the human body to perform work. With the advent of the Industrial Revolution at the end of the eighteenth century, the complexity and sophistication of machines to perform work grew exponentially. In place of human hands and skills making the totally of a product, machines were increasing deployed that could be used to *automate* productive processes. The advantages of machine automation are dramatic. The human body is fragile, limited in range and speed, and, when used in production, prone to imperfection, fatigue, and variation. Machines, on the other hand, could be programed and rationalized to respond with unwavering precision, accuracy, and speed. The benefits of automation are substantial: labor reduction, elimination of repetitive tasks, improvements in worker productivity, increased order accuracy, working nonstop 24/7, protection from hazardous products, and standardization of processes to name a few.

The success of automation resides in the ability of designers to incorporate the human knowledge necessary to perform the productive task directly into the machine. In a "manual" work environment the knowledge and skills to perform a task are invested in the worker. The act of work is action-centered, based upon sentient information derived from physical cues, and marked with a level of personalism where there is a strong linkage between competence and the produced object. The goal of automation is the transfer of these human sources of productive knowledge directly into the machine. While there has grown an enormous debate on the implication of the impact of automation on the worker that is beyond the scope of this book, it is true that the linkage of human knowledge and machine programability and predictability has generated tremendous wealth and determined the living standards of the modern world.

The very foundations of computerization can be traced to this concept of information automation. When computers were first introduced into business, their job was to facilitate calculative tasks such as forecasting, mathematical modeling (inventory, queuing method, etc.), and payroll. As the concept of software developed, businesses were began to utilize the concept of databases containing fixed records such as customer and supplier masters and variable records such as sales and purchasing history. Developers were also quick to see that by transferring and then standardizing process knowledge, such as order taking and inquiry previously performed by humans, work could be rendered more accurate, controllable, visible, and accessible. In today's world, it would be hard to function without the automation to our daily tasks provided by personal computers, personal data assistants, cell phones, and Internet connectivity.

Using Technology to Create Knowledge

When computerized technology was first used to perform tasks, a powerful by-product was identified: automated functions produced a new form of information *about* the activities performed. For example, when a purchase order is launched from a software program an array of new data is created that could be used to monitor supplier performance, view accounts payable detail, gauge inventory values, schedule deliveries, and other activities. Thus, the use of a computerized technology not only utilizes work information in the form of a programed instruction, it also produces usable information.

> It both accomplishes tasks and translates them into information. The action of a machine is entirely invested in its object, the product. Information technology, on the other hand, introduces an additional dimension of reflexivity: it makes its contribution to the product, but it also reflects back on its activities and on the system of activities to which it is related. Information technology not only produces action but also produces a voice that symbolically renders events, objects, and processes so that they become visible, knowable and shareable in a new way. [3]

According to Zuboff, this dynamic of information technology both *automates* and *informates* [4]. Automation is the result of the transfer of knowledge from people to machines and, in the process of performance, has enabled the generation of new knowledge that illuminates the value of further automation.

The impact of information technology on people is dramatic. Where work is performed with very simple or no technology, the worker is individualized, with the skill or knowledge used to produce self-contained, sentient, and centered on manipulating "things." This dynamic is significantly changed with the advent of information technology. As process knowledge is absorbed into the automation, work becomes much more abstract and workers are required to master new skills associated with understanding and manipulating information. The result of this technologically based transformation also fundamentally alters the relationships between people, which now become more collaborative, interactive, and intimate, with a strong sense of mutual responsibility. As information becomes more integrative across time and space, traditional roles in organizations evolve into cross-functional, cross-knowledge teams capable of further leveraging the data emerging from the performance of technology-driven work into new value-adding possibilities.

As we move into the second decade of the twenty-first century, the debate that raged in the concluding decades of the previous century about the alienation of the worker in the face of automation seems now irrelevant. Much of the reason is that the fear that automation would destroy worker knowledge and self-worth and subject them to increased rationalization, control, isolation, and eventual impotence simply has not occurred. Instead of stunting people's creativity, today's information technology has had a liberating, empower impact. Business have implemented an array information technologies that permit their employees to have access to and manipulate vast storehouses of data and to activate new opportunities to generate radically new processes and products and then to communicate them to the marketplace.

Using Technology to Integrate and Network Knowledge

The third type of information residing at the core of today's computerized technologies consists of two fundamental principles. The first is the availability of a technical infrastructure that links computer systems and people. The word used to describe this process is *integration*. Unfortunately, there has been a great deal of confusion concerning the definition of this principle. It is often erroneously used synonymously with *connectivity* and *interfacing*. Connectivity means connecting processes together, such as when a telephone system connects customers with order management functions. Interfacing means bringing information from one system and presenting it for input to another, such as occurs in an electronic data interchange (EDI) transaction. Although both assemble and transmit information to the enterprise, neither connectivity nor interfacing has the capability to integrate people and processes. In fact both processes treat information as proceeding serially through the information channel: they consider transacting parties simply as independent business functions transmitting and receiving data as separate entities according to local operations requirements and performance measurements.

In contrast, *integration* calls for the elimination of the barriers separating functions within or between an organization and its business partners. Integration means to come in touch with, to form, coordinate, or blend into a functioning whole. Organizationally, integration means leveraging information technologies that enable physically remote pools of people, data, and processes to function together as if they were a single, merged set of capabilities directed at the same performance goals. Integration accomplishes this objective by activating the creative thinking within and between organizations and then enables them to work as a single, virtual enterprise. Integration attempts to bring into alignment the challenges and opportunities offered by information technologies and the cultures and capabilities of the modern organization.

The second technology dimension at the core of information technology is *networking*. In the past, computer system architecture permitted only hierarchical communication. As each processor completed its task, the output was then available for the next processing task, which, in turn, passed its output to the next downstream processor. With the advent of client/server architectures and Internet browsers, the process of communicating information has shifted from processing hierarchies to connecting different computers and their databases horizontally in a network. The growing availability of open source operating systems like Linux and programing language like Java are targeted at solving the problem of the dissimilarity of hardware operating systems.

An advantage of networked systems is that information can be stored and retrieved from a central location in the network from any connected computer. In addition the network provides a collaborative medium where people can now communicate directly to other people in the network regardless of physical location. This peer-to-peer networking enables companies to combine the capabilities, skills, and experience of their people by integrating and directing their talents so they can work more efficiently and productively. What is more, the establishment of cross-enterprise teams can occur not only inside the organization but also can be extended to suppliers, customers, and trading partners constituting the entire value chain.

Defining Integrative Information Technology

In the above discussion, information technology was said to consist of three separate yet inextricably linked concepts of knowledge that arise through automation, informated output, and integrative networking. Throughout this text this three part division of knowledge has been grouped into the term *integrative information technology*. When broken apart, the term gives reference to each knowledge type. The word *technology* refers to the automation capabilities of technology. The word *information* refers to the capability of technology tools to informate, to generate information arising from technology driven tasks that can be used to deepen, expand, and reconfigure the context of work. Finally, the work *integrative* refers to the ability of today's computerized connectivity to activate real-time, peer-to-peer networks that enable people to cut across functional barriers and interweave common and specialized knowledge to explore new business opportunities. In the next section the content of integrative information technologies will be explored.

Enterprise Information Technology Basics

The challenges of driving business profitability and connecting and synchronizing supply chains continues to demand as never before creative thought and purposeful action. *Internally,* these challenges require companies to development integrative information technologies that can rapidly collect, analyze, and generate information about customers, processes, products, services, markets, and company and partner distribution channels to guide purposeful decision-making. Building a real-time knowledge bank enables the creation of seamless links and the coordination of the capabilities of individual companies, their customers, and their trading partners. *Externally,* these challenges require companies to deploy information technologies that enable them to architect supply channels that are collaborative, agile, scalable, fast-flow, and Web-enabled. The goal is to present customers with a single, seamless response to their wants and needs by creating a unique network of value-creating relationships. Connectivity and synchronization at this level require the end of channel information silos and the construction of collaborative,

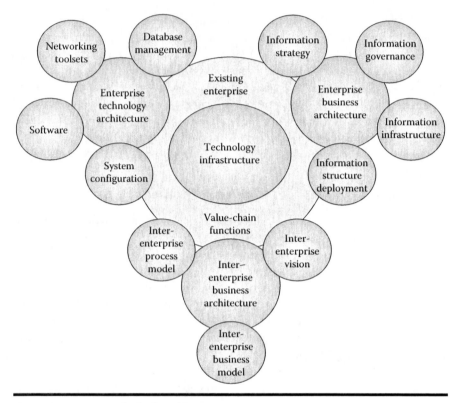

Figure 2.1 SCM technical architecture.

channelwide communication and information management directed at a single point—total customer satisfaction. The impact of information connectivity and integration, however, extend beyond merely facilitating internal and operations functions: they provide a launch point for the generation of new sources of products and services, whole new businesses and marketplaces, and radically new forms of competitive advantage.

Today's robust information systems enable companies to view internal business, information system, and interorganizational channel architectures as a fully integrated set of components as portrayed in Figure 2.1. A detailed discussion of each component is as follows.

Enterprise Technology Architecture

As a principle, it can be stated that the capabilities of a business are directly related to the enabling power of its technology architecture. Simply put, the ability of a company to effectively execute productive functions is in direct proportion to the velocity by which the organization can create, collect, assimilate, access, and transfer information. Before the computer, information processes could move only as fast as human efforts, assisted by crude forms of automation, could manage it. With the advent of the computer, capable of handling information and communications data in volumes and at speeds previously thought unimaginable, the limitations on information management were dissolved, revealing whole new avenues for business management and rendering obsolete the former role of information management.

When seen from a high level, the information technology infrastructure consists of four essential components as illustrated in Figure 2.1. The first and central core of computerized information management is the *database*. As discussed briefly above, the database is the repository of all data that has been captured by users and consists of various categories of information from numerical vales to words to graphics. There are several different types of database. A hierarchical database stores related information according to predefined categories in a "treelike" arrangement. A network database is used by a network installation tool to allocate and track network resources. And finally, the most common type is the relational database [5].

The relational database is the model most used in today's software systems. The key feature of this model can be found in the centralization of key master records such as items, customers, and suppliers that can be referenced by multiple software applications without duplication. In addition, besides increased security, flexibility, scalability, and performance, relational databases enable the use of sophisticated data mining toolsets such as structured query language (SQL) for data reporting and analysis. Since databases store an enormous amount of data, they require a database management system (DBMS) that allows users to create, access, and query that data. These components of information technology can be seen in Figure 2.2.

The second component of today's technology architecture is *networking*. A computer network can be defined as two or more computerized devises linked together over a geographical area. Networking permits users to exchange data with each other and to generally share hardware or software connected to the network. The computerized communication devices are connected to either a local area network (LAN) used for internal networking in a private company, or devices can be connected to a wide area network (WAN) that is used in a dispersed networking environment such as the Internet. There are several possible variations. An *Intranet* is a proprietary network that only permits people like employees to access information and application software. An *Extranet* is a private network utilizes the Internet protocol and the public telecommunications system to permit selected outside business partners to enter parts of a proprietary network. And finally, a virtual private network (VPN) utilizes the public telecommunications infrastructure (e.g., Internet) to access a network through secured entry.

The third component of technology architecture is *software*. Software is a general term used to describe the various kinds of programs that are used to enter, maintain, display, and access the information resident in databases. Without effective software companies are unable to utilize the knowledge both within their databases and through networking with the business partners. In today's Internet and

Figure 2.2 Computer components.

social networking age, software success, or failure, can have a dramatic impact on the well-being of the organization. In addition, information management personnel must be ever vigilant to ensure that the usability of their software systems keeps up with the changes in programing languages and design necessary to ensure compatibility of their information systems with their strategy in dealing with the marketplace.

There are basically two types of software: *operating system software* and *application software.* The former controls the application software and determines how hardware devices and peripherals will work together. The most famous are Microsoft Windows, Mac OS X, and Linux that are used for personal computer environments. The second type of software is application software. This type is used to perform processing activities, such as accounts payables and receivables, payroll, and order entry. Software also provides tools for displaying the database and generating reports.

In the past, the development and maintenance of software was a time consuming: difficult, and expensive process. Today, almost all software is created using a tool called service-oriented architecture (SOA). Basically, SOA enables developers to pick the parts—or services—they need from installed systems and to assemble them into an application that corresponds to the needs of a specific business process. The principle is simple: imagine that a programer has access to every application subroutine not only in own their system, but to any application available say through the Internet. After importation into a new application, when the program ran only the subroutine would be used and not the original program. For example, if a designer wanted to use Microsoft Word's spell check, it could be imported without using all of Word. SOA is about breaking up monolithic applications into reusable services and, via Web standards, flexibly composing those services to create and keep in synchronization the software and the company's real world business processes.

The final component of an information technology infrastructure is the actual *configuration* of hardware, system software, and application software. In the past, system architectures consisted of dumb terminals and peripherals hardwired directly into the computer. Today, most system architectures utilize powerful personal computers (PCs), middleware applications (linking PCs and servers), and communications enablers, like the Internet, to build networks capable of integrating people and knowledge both internally and externally. A sample view of a WAN configuration can be seen in Figure 2.3. In this example, company ABC has two warehouses in Chicago and Los Angles that sell finished goods. Each has their own servers that contain their SCM software. Both are also networked via the Internet to each other. They sell to same customer. In this architecture the customer is networked to both warehouses that sell products via e-commerce to the marketplace. In this setup the customer can place orders, review order status, and change orders in either warehouse directly through ABC Company's marketing Web site.

Enterprise Business Architecture

In today's business environment enterprises and their supply chains must be considered as fragile, yet dynamic connections of customers and trading partners constantly maneuvering for competitive advantage. In turn, these smaller galaxies of business relationships are evolving within a much wider context of the industries and economies that comprise the totality of global business. Far from being isolated, self-contained systems, each is both simultaneously growing in internal complexity while at the same time drawn increasingly into positions of greater dependence on other systems. Unlike the physical universe where constellations are racing away from a common point, today's corporations are finding the space separating each other shrinking and the need for communion expanding.

The response of businesses to these twin principles of internal evolution and increasing channel dependence is found in the continuous deconstruction and

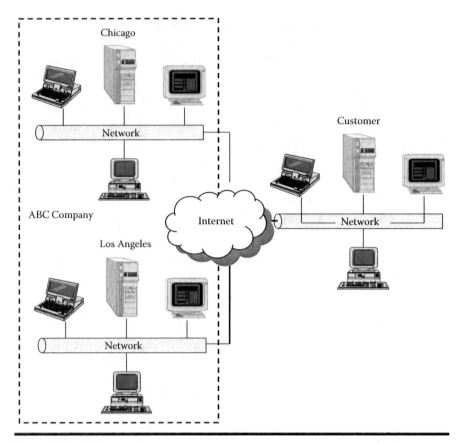

Figure 2.3   WAN.

reinvention of the enterprise business architecture. The term *enterprise business architecture* can have a very wide meaning. It encompasses the components of the firm that are responsible for the performance of ongoing transactions involved in buying, making, and selling products. It also refers to the corporate cultural that has evolved over time and drives current and future attitudes, expectations, and value judgments about what is the mission of the firm. It consists of the particular configuration of human and computerized resources that accumulate, analyze, and utilize the enterprise's repository of information. An finally, it consists of the core competencies of its human resources that breath life into and serve as the directing instruments of the totality of the firm's business components. Without an effective architecture an enterprise's evolution to more successful models would cease and its ability to adapt to change in the face of new business paradigms and technologies would rapidly disintegrate.

The foundation of an effective enterprise business architecture resides in the creation of a comprehensive information strategy that directs the organization towards a coherent, integrated environment for managing and delivering information in support of both short-term objectives and long-term the business goals. The effectiveness of a business architecture is the ability of its processes and systems to respond to changing conditions and customer interactions as they occur. Operational responsiveness means capitalizing on opportunities like reduced cycles times, new sales, lower spend, and reduced risk by identifying and handling problems quickly and accurately. Strategic responsiveness means capitalizing on opportunities by quickly adjusting business processes and channel configurations to pursue new marketplaces, drive greater efficiencies by engaging in options like outsourcing or partnering, and reducing risk by mastering complex challenges emerging from a changing business and regulatory landscape.

Today's business architectures will, of course, have varying degrees of operational and strategic responsiveness. Regardless of the actual structure, the goal is to improve responsiveness over time and increase the value and effectiveness of the business systems comprising the company's architecture, transforming them into competitive weapons. Managing this evolution requires the utilization of the following four-fold agenda to guide information strategy development and definition as well as the governance of supporting information infrastructure [6].

1. *Information Strategy Definition.* The information strategy establishes the principles that will guide the organization's efforts to create and exploit their information system databases and application capabilities. The information strategy provides an end-to-end vision for all components constituting the information system structure and is driven by an organization's business strategy and operating framework. The information strategy establishes the culture, nomenclature, and common objectives to assist all levels of users realize the information system vision. Organizationally, the strategy sets the ongoing framework and guiding set of principles to ensure that current and future investments in people, processes and technologies align and support an agile and flexible information environment. The information strategy is a critical component of the enterprise business architecture. It represents a commitment, across the enterprise, to recognize and treat the information system and its contents as strategic assets.

2. *Information Definition and Governance.* Defining the content of a business information repository and its governance is a critical component of an enterprise business architecture. This is a nuts-and-bolts issue that requires answers to basic questions as to the nature and accuracy of system information, where it is located and how long it is kept, and how does the business access and use it for decision-making. Information definition and governance can enhance the quality, availability, and integrity of database information by fostering cross-organizational collaboration and policymaking.

   Effective governance of system data requires establishment of a specific corporate organization whose mission is to define the policies and managing practices of information assets over their life cycle. Some of the objectives of information definition and governance include the following:

- Defining governance infrastructure and the toolsets deployed to ensure ongoing database and application excellence
- Defining ongoing governance processes
- Developing technology and business system architecture standards and practices
- Providing for initiatives aimed at monitoring and improving database and process quality
- Establishing necessary organizational policies and cross-organizational oversight
- Establishing an effective training program aimed at all levels of the organization

Information definition and governance is easily overlooked. However, if the business architecture is to provide the level of information capable of revealing competitive advantage and expanding business performance definition and governance must be seriously embraced as a critical pillar of overall information system accountability.

3. *Information Infrastructure.* If a business is to leverage their information systems as a competitive advantage, it is essential that they construct and maintain an effective enterprise-level information infrastructure. Anything less results in significant operational inefficiencies associated with data duplication, inaccuracy, and inaccessibility. As detailed in the discussion of the technology architecture, an enterprise information infrastructure framework identifies the technology necessary to establish the DBMS, software, networking toolsets, and the actual configuration of these components into a technology solution governing the functioning of the enterprise.

According to an IBM study [7], enterprise information infrastructure frameworks include the following elements:

- *Information integration management systems,* such as enterprise intelligence (data warehouse) solutions, ERP, and customer relationship management (CRM), integrates transaction data for decision-making.
- *Data master file management* enables master data—such as customer, supplier, items, and employee data—to be easily accessed to provide accurate information for any business application.
- *Dynamic data warehousing* provides the capabilities to turn historical data into relevant, real-time predictive analytics that enable timelier, more insightful business decisions.
- *Enterprise content management* (ECM) provides content management, discovery, and business process management to guide enterprise transformation.
- *Operations management* encompasses servers and data management tools supporting various platforms.

- *Business intelligence and performance management* provides decision makers across the organization with information they need to understand, oversee, and drive the business so they can align their actions with organizational objectives.
- *Metadata management* enables designers to organize and define the meaning of data within an organization's system. The goal is to assure consistency, completeness, and visibility through service directories, data directories, content directories, translation, retrieval, and navigation processes.

4. *Information Structure Deployment.* The enterprise information architecture must provide a deployment roadmap that defines how the various technology toolsets governing systems management will be configured, maintained, and improved over time. Among the critical tasks constituting the action plan can be found leveraging existing investment in hardware and software, prioritization of technology projects, dates for technology rollout, capabilities required to support and access relevant information, reference to forthcoming business and governance practices, and discussion of future investments in emerging technologies.

Most organizations today have some form of information technology strategy, set of governance guidelines, hardware and software infrastructure, and program for continuous information structure improvement and deployment. What differentiates first-in-class from the herd is the ability to address each of the above four components of enterprise business technology simultaneously while ensuring integration with the third component of the enterprise information business system framework portrayed in Figure 2.1—the inter-enterprise business architecture.

Inter-Enterprise Business Architecture

Much has already been said about using information technology to integrate and network knowledge not only within the enterprise, but also between business partners found along the supply chain continuum. While it is absolutely critical for today's enterprise to architect tightly integrated internal business organizations empowered by information technology, simply reengineering functions, rationalizing processes, and removing redundancies will barely provide the necessary fuel the engines of competitive advantage in a global economy. The really significant gains in productivity are about integrating and normalizing the processes that link companies to their channel trading partners.

The process change described here is not about simply outsourcing a peripheral function: it is about architecting a collaborative community of trading partners collectively driven by a common mission to deliver the highest level of customer service possible at any node in the supply chain. This movement from a company and product-centric strategy to a customer-centric, channel network strategy will require the constant aligning and optimizing of the value delivered by the entire supply chain and not just an individual company. Realizing this strategy requires the adoption of new technologies and new roles forcing management and the workforce to adapt, expand, and transition to meet the requirements of an effective inter-enterprise architecture.

Although this book looks at the application of integrative information technologies to each component of today's business environment—sales and service management, manufacturing and supply chain planning, procurement, warehousing, and logistics—as separate entities, in reality, the boundaries are artificial. The same can be said of the distinctions between enterprise and supply partnerships, between demand chain and supply chain. In fact, the basis of business today is, in realty, the application of repositories of integrated, networked information and resources, whether its source be customers, trading partners, market information, or products, to establish a collaborative medium where people can now communicate directly to other people in the network regardless of physical location so they can work more efficiently and productively. The promise of today's integrative information technologies is that it provides for the cross-enterprise unification of databases that can be used in an almost infinite variety of ways to provide a different rich-context to every accessing entity. In such a unified inter-enterprise environment there is no distinction between employee, partner, and customer portals. They are all windows into the same repository of information, differing only in the roles that are defined for each user, the interactions that are associated with each role, and the level of access permitted for each role.

Such a vision of inter-enterprise unity requires that all members of the supply chain be closely integrated and their databases and information flows closely synchronized to eliminate distortions and the "bullwhip effect" in the communication of information. Architecting inter-enterprise structures capable of synchronous information flows requires channel planners to develop and constantly attended to a joint strategy that seeks to utilize the best technology toolsets to realize targeted individual company and supply chain objectives. Such a program involves the following critical processes.

1. *Architecting a Shared Inter-enterprise Vision.* The development of an information technology strategy in isolation from the challenges of integrating with channel trading partners is destined to failure. While it is true that the shared vision emanating from inside the organization provides companies with a cohesive force to drive a common direction, focus, and personal and team motivation, strategists must be careful to include a vision of how the internal functions of the company are to fit into a much wider supply chain vision. Such an undertaking requires a comprehensive knowledge of internal and partner core competencies, technology capabilities, and commitment to supply chain collaboration. The goal is not only to structure a functional

system of cross-channel business, but also to use the framework as method to leverage the entire business ecosystem to discover breakthrough propositions made possible by the common shared information platform. The process [8] used to guide this analysis is straightforward: an optimized or "green field" architecture is envisioned and documented. Next, the existing "as is" structure is matched against the optimized architecture. Strategists then perform a gap analysis, uncovering where resources and competencies in the existing supply chain occur. Finally, a model of the inter-enterprise architecture should emerge that can be used as the basis for all subsequent e-SCM strategy enhancements.

2. *Inter-enterprise Business Modeling.* Once the inter-enterprise (supply chain) vision and strategy has been completed, companies can then begin the task of establishing the enterprise-facing portion of the business model. The inter-enterprise business model provides a high-level description of the technology integration points connecting each business in the supply chain network. The objective of the model is to detail the for the supply chain as a whole the following marketplace dynamics:

   ■ Target market/market segment, including expected share, profitability, service goals, customer retention, and new customer acquisition

   ■ Products and services, including product line profitability, life cycle management, new product/service introduction, and manufacturing strategies

   ■ Financial elements, including return on assets (ROA) management, return on investment (ROI) management, potential revenue growth, and internal productivity cost measurements

   ■ Product distribution, including logistics management, depth of channel integration, cost structure, levels of automation, and cost management

   A critical element in the inter-enterprise model is the information architecture. It is the responsibility of the each channel partner's information architects to configure and maintain a repository of computerized process components that enable the desired levels of communication and networking among channel members. By effectively designing these technology components, solution developers can integrate internal business work rules, roles, tasks, and policies with those of their channel partners as well as rapidly respond to supply channel business environmental changes.

4. *Inter-enterprise Process Modeling.* Once the inter-enterprise vision and business model have been formulated, the next step is to detail the process model that describes the external processes that govern daily supply chain functions. Developing the process map requires strategists to know precisely which business functions are going to be inter-enterprise processes, what technology infrastructures must be in place, and how the organizational structure should be constructed. Constructing effective inter-enterprise processes is a critical project consisting of the following steps [9].

■ *Engineer Trading Partner Processes.* The task of generating the desired inter-enterprise linkages must involve the full participation of customers and suppliers regardless of the desired complexity of the proposed connectivity. For example, the interaction of information driven through an EDI system will be different than the use of a Web-based system. While a significant degree of process standardization is the target, architects must be prepared to fashion process components that are customizable to meet individual buying service requirements. For example, Internet buying process components, such as the work flows associated with internal request for quotation (RFQ), sourcing, approval, and order management, must be able to interact with complementary work components resident in the supplier's technology infrastructure architecture.

■ *Degree of Process Interaction.* A robust architected system will provide companies with multiple levels of connectivity depending upon the business requirements of their trading partners. At the lowest level can be found the loose coupled model. This level of connectivity between trading partners utilizes information technologies, such as EDI and the Web, simply as a medium to replace paper-based information documents. At the next level, *process handoff,* the technology connectivity has been architected to permit transactions to automatically trigger processes in the systems of trading partners. For example, a sales transaction posted in a retailer's system will trigger a replenishment notice in the planning system of a first-tier supplier. At the highest level, *virtual enterprise,* the process components are used jointly and operate in real-time. This level of architecture provides each linked node in the supply chain full access to information across the channel network galaxy. For instance, information concerning a customer would be available to every trading partner, thereby removing unnecessary database redundancies and suboptimizations.

■ *Internal Infrastructure and System Reengineering.* Regardless of the level of supply chain connectivity deployed, the work force will be required to function in a cross-enterprise mode, and not simply according to the needs of a single company. Technology-wise, supply chain interoperability will require enhancements to existing systems or the purchase of point solutions, such as business-to-business (B2B), to supplement legacy system deficiencies. Normally, this process will require customization of the "wrapper programs" of existing ERP/SCM systems to accommodate the linkages needed to work with new packaged software, portals, and other interoperability solutions.

■ *e-Application Architecture.* The technology architecture that emerges should support the inter-enterprise business strategy. According to Hoque [10],

- e-application architecture involves determining individual integration points between the application and data sources, the application and back-end installed software, and between multiple back-end systems.

> Every decision should be made with a good deal of attention toward not only the functionality of the application the day it is rolled out, but also the ability of the platform to scale up to support heavy usage, added business attributes, new users, and additional functionality in later revisions.

The e-application architecture should determine which inter-enterprise processes components will constitute the final platform. The completed architecture should contain the definition of what the networked infrastructure should look like, how networked resources will be accessed, the source and type of the data the networked resources will utilize, and resolution of data, hardware, process, and human resource ownership ambiguities.

■ *Pilot, Go Live, and Iterate.* The challenges of developing and implementing a comprehensive inter-enterprise technology solution are fraught with difficulties, potentially enormous expenses, and significant trauma to even the best of organizations. Most experts caution against trying to do too much in the first round. Implementers should view the process as iterative: utilize a minimalist approach and begin with the easiest processes or the ones that provide the biggest payback. The goal is to keep expectations, costs, and trauma to the organization realistic and doable while ensuring the company is pursuing a path that keeps it at the forefront of the competition.

New Technologies

The most attractive element about today's integrative information technologies is not the wizardry of the technology, but the *information* it makes available. Information provides companies with a window to reality, and the more accurate and timely, the better businesses are able to make the decisions that will assist them to better service their customers, develop their work forces, invest their resources, control their costs, and remain competitive. Today, business concepts and applications like SCM, the Internet, open computing environments, and the connectivity with channel trading partners should be viewed first and foremost as methods to increase information availability. While business applications such as ERP, CRM, SRM, Internet-enabled business, and other computerized toolsets remain as the basis for information management, several new developments are reshaping the face of business computing today.

New Generation of Technology Enablers

Today's technology marketplace has undergone a dramatic sea change. Just over a decade or so, technology purchasers were presented with functionality rich, often proprietary mega systems that were also difficult and costly to implement, difficult to integrate with other technology components, and difficult to enhance with non-native functionality. A perfect example is CRM. High-powered systems from vendors like Onyx, Oracle, PeopleSoft, SAP, and Siebel bombarded customers with the sheer power of their robust functionality. Unfortunately, by 2005, customers had wearied of the enormous effort in time, money, organizational stress, and the often large portions of the software that lie unimplemented. As the inevitable collapse occurred and vendors disappeared, companies demanded a better return for their investment and software that was easily adaptable to fit their business strategies, workflows, work force capabilities, and their native system environments. Instead of bundles of prepackaged functionality, CRM users were asking for smaller, more easily managed components accompanied by more robust toolsets.

Whether it is ERP, CRM, or other software suite, software companies of all flavors can expect today to experience the following range of requirements from prospects before they will purchase any form of software.

■ *Technology Architecture.* As companies look to their application suites to expand interoperability and inter-enterprise integration capabilities, the utilization of technology tools sets such as SOA, *Java*, and Microsoft's *.NET* framework for Web services development is a critical litmus test for today's integrative information technologies. These platforms make it easy to develop powerful Web application servers that facilitate integration and interoperability with add-on application point solutions. Such standards also permit companies to publish their Web components as services and access components from Internet partners without writing custom interfaces. Finally, these platforms make database access more flexible by converting data to XML documents that can be easily passed to enterprice business systems (EBS) backbones, wireless devices, or other applications.

■ *Adaptability.* Past software developers focused on building systems that provided users with robust toolsets that enabled them to reconfigure or build new functionality. However, a focus on building giant systems meant that there was also a lot of built-in structure that made the components difficult to change. While today's marketplace considers robust toolboxes a requirement, it wants less built-in structure and a greater focus on company specific, native business processes.

■ *Cost.* Today's systems often require huge expenditures for hardware, software, and training. In addition, a substantial budget will be necessary to accommodate modification or enhancement requests. In contrast, the newer wave of software is built on open architectures and, therefore, requires less expense for hardware, training, and modification.

■ *Implementation.* Today's application systems have become so function-rich that implementers often have a hard time embarking on system configuration. Users can get lost in functionality that adds minimal value, while crucial functions are given insufficient attention. In addition, because of functionality robustness, it is difficult to modify to meet native requirements. While software today is no less rich in functions, system architects have built in flexible configuration toolsets, portal-type capabilities that enable merger of outside applications, and even preconfigured solutions targeted at specific industries that facilitate adapting standard functions to closely mirror existing business processes.

■ *User Adoption.* A common complaint is that today's business systems are too expansive and too complex for users. Today's new wave of software is easier to use, has less interface clutter, is more focused on native functions, and is more likely to gain rapid user acceptance and validation.

Software-as-a-Service (SaaS)

The robustness of today's networking technologies have spawned a new industry directed at relieving companies of the cost and organizational burden of implementing cutting-edge software and network-building by providing major business system applications and technical infrastructure for a subscription fee. Computer systems, application software, networks, the Internet—all are fundamental tools that every business must utilize. Maintaining these tools, however, are normally outside the core competencies of most businesses. What SaaS basically does is provide companies with the ability to lease software over a secure Internet-accessed network. Often there is little or no up-front investment required by participants. The subscriber does not own, license, or even keep a copy of the software on its in-house systems. Rather, the SaaS company hosts the software and the related IT services, including upgrades and maintenance. In addition, the SaaS company can provide the network linking the client's offices, homes, and operating locations to the data center. Finally, the SaaS company provides the licenses, implementation, training, system management, and user support necessary to ensure the subscriber receives the anticipated value. In return, the SaaS company determines payment on a subscription basis for an agreed upon period of time [11].

A new development in SaaS is the concept of *cloud computing*. While the ultimate definition of cloud computing has yet to be determined, there is little doubt that it represents the next major business shift in the IT industry and business computing. In addition, according to the Aberdeen Group, cloud computing is emerging in conjunction with green IT and sustainability as major corporate strategies [12]. The idea, basically, is to realize the possibility of enterprise IT organizations meshing seamlessly an internal network "cloud" of applications with a public "cloud," hosted by an integrator that charges on a pay-as-you-go basis [13]. This concept has been defined as

> a model for enabling convenient, on-demand network access to a shared pool of configurable computing resources (e.g., networks, servers, storage, applications, and services) that can be rapidly provisioned and released with minimal management effort or service provider interaction. [14]

Salesforce.com, a billion-dollar-a-year company, provides an excellent example of SaaS cloud computing capability. At the center of its cloud strategy is Force.com, the platform where Salesforce hosts its own sales force automation, customer service, and other applications, but also where subscribers can build their own applications using Salesforce's programing language, APIs, and custom interface framework offered through the Salesforce online store, AppExchange. Since so much of the basic code for a needed application is already running on Force.com, it has become easier for developers to customize existing apps and write small, limited-function apps of their own. [15]

ERP: Going Mobile
Today's ERP systems are leveraging communications technologies to allow users access to their ERP systems from anyplace, at any time. CDC sofware, a supplier of enterprise software applications sells a new mobile application enabling users to perform real-time ERP inquiries via their Blackberry devices. The application helps increase the productivity of sales professionals and others by allowing them to obtain up-to-the-minute information regarding the status of customer accounts. It also allows executives to monitor the progress of key accounts remotely. Information that can be accessed via the Ross Mobile application includes: customer details, sales orders, sales pricing, accounts receivable aged summary, accounts receivable detail inquiry, current inventory, projected inventory, and order and shipping status. *Source*: Renee Robbins, "Real-time ERP inquiries available via Blackberry devices." http://www.mbtmag.com/article/356982-real_time_ERP_inquiries_available_via_Blackberry_devices.php ?rssid_20213

The range of EBS applications offered by SaaS companies varies and can range from companies like Salesforce.com (see Figure 2.4), which specializes in sales and customer relationship applications, to SAP and Oracle, which offer their entire suites of software products via the Internet. The decision to choose using a SaaS supplier requires companies to seriously think through issues such as business applicability, security, and cost. Historically, according to Murphy [16], transportation management has been a significant user of SaaS, where companies need to stay connected to a complex network of carriers. However, other supply chain functions, such as demand planning, forecasting, collaborative planning, forecasting, and replenishment (CPFR), and GTM are being transferred to SaaS. Still, there are risks involved when using a SaaS company. One is the fear that vital company data will be lost or compromised. Then there is the issue of loss of control and possible system downtime. Still another is the IT effort necessary to integrating SaaS applications into the company's existing system model.

Companies looking to implement a cloud solution are confronted with several challenges. Foremost is understanding the scope of the business requirements, followed by assessing project ROI, overcoming a lack of internal expertise, and simply understanding with cloud computing actually does. In selecting a SaaS solution companies should use the following touch points:

Figure 2.4   Salesforce.com Web site.

- *Determine Requirements.* What are to be the outsourced application(s)? How many users and locations will there be and what level of performance and support is needed?
- *Determine Technical Environment.* What types of network connections and client devices are needed? Is there any integration required with existing systems and applications? What implementation and integration services will be necessary?
- *Evaluate SaaS Vendor.* Evaluate potential partners with regard to their strategic direction, experience, technology infrastructure, depth of customer base, cash flow, and investors.
- *Evaluate Level of SaaS Services.* Is the SaaS vendor certified, and does it have a track record in the desired applications? What types of services are provided, including remediation policies for downtime, quality of security, network and bandwidth options, and backup and recovery capabilities?

- *Determine Level of Technical Support.* Are there limits on the number and type of support calls? What is the SaaS's escalation policy? What are the days and hours of available service?
- *Evaluate Scalability and Control of Assets.* Can the SaaS provider respond to periods of peak load? Does the SaaS provider have control over its own assets, and if not, what guarantees does it offer? Does the SaaS have the potential to grow along with its clients?
- *Validate the SaaS provider's Strategy.* Assess the SaaS's overall sales and service strategy and commitment to the industry.

What's in a cloud?

Some definitions of commonly used cloud terms:

Public cloud computing: Infra-structure—including computers, storage and operating systems —and applications offered as services on a subscription basis. For example, Amazon.com Inc's Elastic compute cloud service offers virtualized serves that users rent to run work-loads over the internet. The public cloud provider controls where the workloads execute.

Private cloud computing: Infra-Infrastructure—including computers, storage and operating systems—and applications offered as vitualized services to user. The organization owns and controls its infrastucture and appli-cations, which run behind a corporate firewall. The IT organization managers the virtualization tools, policies and workload deployment.

Hybrid cloud computing: A mix of public and private virtualized infrastructure and applications where work loads shift in and out of each sphere, in an automated way or not based on policies.

Wireless Technology

Today, the presence of wireless information capture and transmission pervades just about every facet of the supply chain. For example [17],

- The problem of gathering accurate pallet dimensions (DIM) in high-demand, cross-docking warehouse environments without slowing pallet movement was solved at one company by calculating the weights right on the forklift with the pallet dimensions transmitted wirelessly to warehouse computers.
- A trash collection firm in San Diego responded to complaints that trash was not being collected by embedding sensors in RFID technology on the truck's back gate lift to record every barrel emptied at specific locations for wireless feedback to the central office.
- For haulers of frozen or refrigerated goods, temperature-monitoring equip-ment was installed that provided wireless feed of temperature records starting

from the distribution center to the end of the channel with the retailer or restaurant customer.

The use of RFID, Bluetooth, global positioning systems (GPS), and other technologies, working in tandem with cloud computing environments, Web Portals, and back-end systems to track everything from apparel, equipment, and appliances to animals and even people is transforming the information processing capabilities of the supply chain and promising to enable a new dimension of business management. In general these toolsets for monitoring and tracking can dramatically assist everyday business uses such as providing supply chains with tracking capabilities, solving or averting problems, and increasing efficiencies.

Wireless can be defined in its most rudimentary form as the transmission of data between devices that are not physically connected. A wireless device may be anything from a personal digital assistant (PDA), to a laptop, a two-way pager, a global positioning satellite antenna, or a remote sensor. The data communication can occur at short range using infrared technology, at a wider range using high-speed wireless LANs located on a fixed structure, or globally using satellites. The goal of wireless technologies is to provide mobile workers access and input to any database, any time. It enables collaborative information exchange where physical colocation is not feasible. It also assists in tracking, locating, and managing movable assets such as cargo, containers, laboratory equipment, and delivery trucks. By creating networks of objects, wireless technologies not only have the ability to exponentially improve the flow of information through the supply chain at the speed of light, they also can provide the intelligence to examine patterns and trends about channel business strengths and weaknesses [18].

RFID and American Apparel

American apparel, the largest U.S. clothing manufacturer, has used RFID since 2007 and is implementing it in 46 of its 300 stores. As the only supplier to its stores. data about the products never has to leave the system to be shared with outside companies.

Every garment is accompanied by an avery dennison RFID before arriving at a store. A motorola reader captures data from the tags at the store's receiving docks and posts into an inventory management system. The RFID tags are removed at the point of sale and another reader logs the transaction and alerts the employees that a replenishment action needs to be performed.

Sales at stores with the RFID systems are 14% higher on average than the non-RFID stores. In addition, staffing levels are 20% to 30% lower because employees don't have to spend time performing manual inventory checks. Finally, stockroom inventory in RFID enavled stores is down 15% as compared to non-RFID stores.

Source: Weier, Mary Hayes, "Slow and Steady Progress." *Information week* issue 1,248, (November 16, 2009), P. 33.

Probably the most talked about of the wireless devices is RFID. While the technology had been around since the turn of the millennium, it began to gain traction with the announcement by Wal-Mart that their top one hundred suppliers had to have RFID labels at the case and pallet level by January 2003. Wal-Mart hoped to use electronic product code (EPC) tags, which store details about products and transmit them to inventory systems using RFID chips and readers, to create a more efficient supply chain. However, by the end of 2009 only about 600 of Wal-Mart's 20,000 suppliers were participating in the effort. Despite the global recession of 2008–2009, ABI Research, a marketing intelligence company specializing in emerging technologies, estimated total revenues from RFID transponders, readers, software, and services would exceed $5.6 billion in 2009. Analysts expect the market to each more than $9.2 billion in 2014 [19].

Despite its benefits, however, the widespread use of RFID has yet to occur. Several reasons come to mind. To begin with the RFID tags are too expensive for universal use (7–15 cents a tag in 2009). Second, the technology is complex and costly to implement, requiring not only investment in the chips, but also readers, software, and the construction of new business processes. Third, no standard for tags and data formats has been formally adopted. Fourth, there have been problems with the technology itself. Some Wal-Mart suppliers have had problems with liquids and metal in their products that block the tag's data transmission. Fifth, the existence of networks sophisticated enough to make RFID data useful is not there. And finally, requirements by major companies mandating usage and a general realization of the inherent value in RFID by businesses have yet to materialize. Some companies simply are skeptical about realizing a reasonable return for the effort and expense [20].

There can be little doubt, however, that as wireless technology expands and matures, it will become an essential building block of SCM. Although the RFID-enabled supply chain has not materialized (tags for retail goods accounted for just 10% of the two billion tags sold worldwide in 2008), the technology has enjoyed a wider success in the public sector, particularly government, the military, and smart card projects for passports, ID cards, and prepaid transportation cards. Wireless enables a wider audience of participants within the organization and outside in the supply network to bring automation and efficiency to a new range of processes by making information ubiquitous and real-time. While there are still many issues to resolve, wireless technology is destined to have as much of an impact on business and technology architectures as the arrival of the Internet. In the meantime, the evolutionary nature of the wireless revolution will enable companies to implement and experiment with the technology by focusing on high-payback applications currently available to the marketplace.

Amazon's elastic compute cloud

Amazon is giving its elastic compute Cloud customer ,the ability to bid on unused computing capacity. The new acquisition model, called 'spot instances', lets Amazon web services users bid for computing power and pay by the hour and have their jobs processed if their bid exceeds Amazon's fluctuating spot price."

How it works is a customer posts a maximum bid. If the bid exceeds the spot price, their jobs will be run and price, their jobs will be run and priced at the current spot price. Customers with lower bids will have their jobs terminated and queued until the spot price falls below the original bid.

According to Amazon CTO Werner Vogels, this arrangement provides customers exact control over the maximum cost they would incur for their workloads, providing them with substantial savings. Amazon offers two other pricing models: on-demand instances (using a published rate) and reserved instances (pre-paid at discount rates for use up to three years).

Source: Claburn Thomas, "Amazon Auctions spare EC2 Server Capacity." *Information week*, issue 1,253, (December 21, 2009).

Global Trade Management Solutions

Companies going global today are beset a range of problems: unexpected transportation costs, higher inventory investment, and longer and more unpredictable cycle times, while experiencing demand for lower costs, more unique services, and improved responsiveness. As a result, businesses are searching for ways to render global supply chain processes more reliable, more flexible, and less expensive. As the increased risk and complexity associated with the effective management of export/import cost management, government compliance, and security regulations grow, companies have been turning for solutions to a relatively new software application termed GTM.

Historically, GTM was a small part of a typical company's business, run typically through stand-alone spread sheets. Today, managing the flow of goods, information, and money across borders (totaling more than $32.5 trillion in 2008) is a highly complex, regulated, and dynamic process. Companies can no longer rely on manual processes to manage their global trade operations, which is why the GTM systems market is one of the fastest-growing segments of the software industry. According to the Arc Advisory Group, GTM software in 2009 is a $673.5 million industry and is expected to reach $814 by 2012 [21].

GTM can be defined as a software solution that manages in global trade events by achieving efficiencies and excellence in four critical areas:

■ *Compliance.* Functionality in this area provides capabilities focused on automating customs and regulatory compliance activities, such as product classification, restricted party screenings and embargo checks, trade documents, calculation of total landed costs, assignment of export and import licenses, electronic communication with legal authorities, management of customs processes and transit procedures, and determination of preferential trade eligibility.

■ *Content.* This area is concerned with establishing accurate and complete trade content for every country a company trades with in order to successfully comply with trade regulations and prevent customs clearance delays. Such factors include denied parties and embargoed countries, harmonized system chapters and descriptions, license codes, descriptions and requirements, document templates, and duty, value added tax, excise tax, and seasonal tax.

■ *Connectivity.* Establishing and maintaining connectivity with a dynamic set of trading partners, as well as keeping up with customs modernization efforts around the world, is a critical component of a GTM system. While most companies depend on EDI, tech savvy executives have been increasingly turning to Web applications imbedded in their GTM systems. Some of the best GTM vendors, such as INTTRA, Descartes, and FreightGate, even offer SaaS solutions.

■ *Finance.* Global trade requires the effective management of transaction financing. A GTM must be able to create, present, and manage all documentation required for export and import letters of credit. For more advanced relationships, companies can use GTM functionality for open account payment methods to support international trade transactions. This method streamlines paperwork and documentation, automates invoice flow, and protects transactions from exceeding credit limits.

Companies can choose a GTM solution from one of the big ERP vendors, such as SAP, Oracle, QAD, and Infor who have incorporated GTM functionality into their packages, or they can acquire a "best-of-breed" solution. An emerging requirement is being able to link trade-compliance capabilities with international logistics execution and transportation management, supply chain visibility, and supply chain finance. In the end, GTM vendors will be able to thrive if they can improve the efficiency of their customers' global business processes, identify and implement best practices, and provide ongoing oversight.

Summary and Transition

The enabling power of integrative information technologies has caused a virtual revolution in the concept and practice of modern business management. In today's highly competitive, risk-prone economic environment the importance of timely, accurate, and complete information has become the mantra for competitive advantage. Inside the organization, information technologies enable companies to develop databases and implement applications that provide for the automation of repetitive tasks, the efficient management of transactions, and the timely collection, analysis, and generation of information about customers, processes, products, and markets necessary for effective decision-making. Out in the supply chain, information technologies enable strategists to architect channel networks that are collaborative, agile, scalable, fast-flow, and Web-enabled. Actualizing the potential of today's information technologies requires companies to move beyond viewing computers as purely a tool for *automating* business functions. Increasingly, the real value of information technology is to be found in its ability to enable integration and networking between channel trading partners providing for the robust capabilities and highly productive competencies necessary to generate new products and services, whole new businesses and marketplaces, and radically new forms of competitive advantage.

The structuring of integrative information technologies can be seen in the internal business, information system, and inter-organizational channel architectures characteristic of today's modern information systems. At the core of an effective business system can be found the *enterprise technology architecture*. This component consists of the configuration of the database management, networking, and software tools matching the specific needs of the enterprise. The next component, the *enterprise business architecture*, determines the strategy, governance, infrastructure, and infrastructure deployment activities based on the technology architecture. The third and final component, the *inter-enterprise architecture*, will determine the models to be used to create the networking, integration, and collaboration capabilities necessary to weld supply chain partners into a highly competitive value-added network.

The integrative technology tools discussed in this chapter form the backbone of today's modern business systems. Applications such as ERP, SCM, CRM, and a host of other software application are built around the database management and networking and integrative capabilities discussed in this chapter. Achieving information system objectives requires deploying these toolsets to construct business architectures that are efficient, configurable, scalable, purposeful, integrative, and collaborative. In the next chapter, the foundations of an effective supply chain software environment are explored.

Notes

1. The term *computer* is used here in its widest possible definition and includes everything from personal digital assistants (PDA), to notebooks, tablets, and desktops, to mini-computers, mainframe computers, and supercomputers.
2. See the discussion in Shoshana Zuboff, *In the Age of the Smart Machine* (New York: Basic Books, 1988), p. 10.

3. Ibid., p. 9.
4. Ibid., p. 10.
5. For a detailed discussion of database management concepts, see Stephen Haag, Paige Baltzan, and Amy Philips, *Business Driven Technology* (New York: McGraw-Hill-Irwin, 2006), pp. 61–66.
6. These four points have been adapted from the IBM Whitepaper "The Information Agenda: Rapidly Leveraging Information as a Trusted Strategic Asset for Competitive Advantage," found at accessed January 1, 2009, ftp://ftp.software.ibm.com/software/data/pubs/papers/info-agenda-wp.pdf.
7. Ibid.
8. This process has been detailed in Peter Fingar, Harsha Kumar, and Tarun Sharma, *Enterprise E-Commerce: The Software Component Breakthrough for Business-to-Business Commerce* (Tampa, FL: Meghan-Kiffer Press, 2000), pp. 228–229.
9. The three levels of inter-enterprise integration are detailed in ibid., pp. 235–236.
10. Faisal Hoque, *e-Enterprise: Business Models, Architecture, and Components* (Cambridge, UK: Cambridge University Press, 2000), p. 153.
11. For an excellent, detailed discussion of SaaS, see Gianpaolo Carraro and Fred Chong, "Software as a Service (SaaS): An Enterprise Perspective," at accessed October, 2006, http://msdn.microsoft. com/en-us/library/aa905332.aspx.
12. Bill Lesieur and Carol Baroudi, "Business Adoption of Cloud Computing," Aberdeen White Paper (September 2009), p. 5, at accessed September 1, 2009, http://www.aberdeen.com/launch/report/benchmark/6220-RA-cloud-computing-sustainability.asp. In this same report, a survey indicated that 45% of the respondents saw cloud computing as a way to reduce IT costs, 31% as a competitive weapon, 19% as a fix to currently inflexible infrastructures, and 13% as an easy way to support additional users or services.
13. See Charles Babcock, "Hybrid Clouds," *Information Week,* no. 1240 (Sept. 7, 2009), pp. 15–19.
14. National Institute of Standards and Technology, draft definition, version 14.
15. Mary Hayes Weier, "SaaS Leader Practices What It Preaches," *Information Week,* no. 1241 (Sept. 14, 2009), pp. 40–42.
16. Sean A. Murphy, "SaaS: Right for You?," *Supply Chain Management Review* 13, no. 6 (2009), pp. 43–48.
17. These examples were taken from Amy Zuckerman, "Managing with Mobility," *World Trade Magazine* (December 2008), pp. 30–32.
18. For a general treatment of wireless technologies, see the series of articles in Sameer Kumar, ed., *Connective Technologies in the Supply Chain* (Boca Raton, FL: CRC Press, 2007) and Amy Rogers Nazarov, "The Internet of Things Handbook," *Information Week—Special Supplement,* no. 1240 (Sept. 7, 2009), pp. HB4–HB14.
19. Bob Trebilcock, "Total RFID Revenue Expected to Exceed $5.6 Billion This Year," *Modern Materials Handling* 64, no. 8 (August 2009), pp. 9–10.
20. Mary Hayes Weier, "Slow and Steady Progress," *InformationWeek,* no. 1248 (Nov. 16, 2009), pp. 31–33.
21. Adrian Gonzales, "Beyond Software: The Role of Content and Connectivity in Global Trade Management," *ARC Advisory Group Whitepaper* (Dedham, MA: August 2009), p. 2.

Chapter 3

Supply Chain System Foundations: Understanding Today's Technology Solutions

The origins and continuous development of the supply chain management (SCM) concept is directly dependent on the information and networking technology foundations discussed in the previous chapter. While it is true that computers have been available for over 50 years, it has only been in the past 15 years or so that commercial computer systems have been able to expand beyond the boundaries of their own architectures to literally "talk" with one another. Until fairly recently, programs, data, and information were confined to the four walls of the business. Today, radical advancements in hardware architecture, programing languages, and communications devices have enabled enterprises to engineer systems that allow supply chain business partners to peer as if through a portal into once inaccessible databases, pass documents freely back and forth without concern for the constraints of time and distance, and interactively enter data, verify information, and assemble real-time networks unencumbered by proprietary systems and software.

In fact, the mechanics of the management and operational aspects of technology-enabled SCM discussed in this book—the ability to network geographically dispersed productive and information generating processes, the integration of supply chain strategies and operations, communications providing connectivity between enterprises, planning systems facilitating the transfer of demand and inventory data across the channel pipeline, and others—would be impossible without the enabling capabilities of today's integrative information technologies. SCM, coupled with the power of technology, is such a potent competitive and productive force because it is focused on networking the wants, needs, and capabilities of customers and suppliers and making them visible to all channel partners real-time and online. SCM and networking are technology correlatives: as SCM concepts and practices and integrative information tools expand, there can be little doubt that the connective and informating capabilities of each will be reciprocally enhanced, providing ever fresher application and strategic perspectives.

In this chapter, the application of integrative information technologies to create what is termed enterprise business systems (EBS) is discussed. The chapter begins with an overview of the five basic components constituting a business system. Among the elements discussed are the utilization of the database, transaction management, management control, decision analysis/simulation, and strategic planning. Following, the chapter details the principles of system management. The goal is to see how an EBS capable of effectively running an enterprise must have accountability, transparency, accessibility, data integrity, and control. Once these basics are completed, the chapter proceeds to discuss the historical evolution and components of today's EBS. During the review, a detail of the repository of applications constituting today's enterprise resource planning (ERP) and SCM software packages will be considered. The chapter concludes next, with the various forms of connectivity made available by the Internet examined. Topics detailed are Web-based marketing, e-commerce, e-business, and e-collaboration.

Business Information System Basics

The successful management of information is the key to business success. Managing information means that companies must find solutions to such questions as the following:

- What is the information that is needed to effectively automate, informate, and network the organization and its business partners?
- How is this information to be organized in a meaningful manner?
- What are the software applications to be deployed that enable people throughout the network to enter, access, and work with the information?
- How can the network of users ensure that the data is timely and accurate?

The answer to these and other questions is the ultimate goal of today's computerized business systems. Realizing this goal requires the application of technology tool-sets that automate tasks subject to human or environmental variance, activate new sources of information assisting people to effectively collect, access, and analyze system data, and provide a platform for the close networking of business partners across the supply chain.

The Five Basic Functions of Information Systems

At the foundation of every information system reside five common functions as portrayed in Figure 3.1 [1].

- *Enterprise Database.* The foundation data necessary to operate the business is located in the system database. These records are, for the most part, composed of two types of data. The first, *static data,* consists of core information elements, such as customer and supplier masters, item masters, product structures, product costing, warehouse geography, bills of material (BOMs) and process routings, manufacturing equipment, and channel structures that do not change during the performance of transactions. The second type, *variable data*, consists of databases, such as open order, transaction balance, and accounting, which are impacted through transaction management. Today's computing technologies enable companies to have a single database (relational database) that all departments can use jointly. Furthermore, the advent of connectivity tools, like the Internet, provide the capability for *database networking,* where key database files such as forecasts, channel inventories, and open order status, can be viewed by users both external as well as internal to the company.

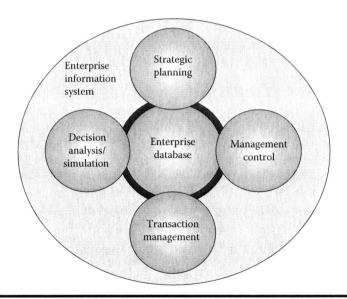

Figure 3.1   Enterprise system architecture.

- *Transaction Management.* Transaction functions enable users to enter and maintain database information and consist of such activities as order entry, inventory balance maintenance, order selection, allocation and shipment, and payables and receivables. Computerized transaction management programs serve several purposes. They provide for the accurate and timely entry and collection of transaction data; they automate difficult and time-consuming record-keeping; they enforce rules governing transaction data entry and maintenance; and, they enable companies to collect large volumes of information that can in turn be used for business analysis, management control, and strategic planning. The overall goal of transaction management functions is to ensure an accurate record of the firm's day-to-day operations.

- *Management Control.* The entry of transactions and the compilation of databases are meaningless without defined best practices and performance measurements to guide process-flow management and decision-making. This function focuses on toolsets that permit developers to create standardized applications ensuring uniformity across channel businesses and analysts to "mine" enterprise databases in an effort to uncover and detail operational and financial measurements relating to such issues as cost, asset management, customer service, productivities, and quality. Management control has the following goals: ensuring rationalization of business functions, providing the feedback necessary for the timely reformulation of operations plans and activities, ongoing measurement of the competitive capabilities of the enterprise, and development of plans providing for continuous improvement.

- *Decision Analysis/Simulation.* A critical function basic to an information system is the ability of planners to utilize a variety of modeling tools to assist in managing simple to increasingly complex processes during decision-making. Some of these applications, such as material requirements planning (MRP) and capacity management, provide mature and easy-to-use applications to simulate the impact of demand on inventories and productive capacities. Other computerized applications assist in the identification, evaluation, and comparison between alternative courses of action, such as vehicle routing and scheduling. Finally, in this area can be found powerful advanced planning systems (APS) and SCM applications driven by complex mathematical algorithms used to design supply chains, determine plant locations, and aggregate demand and supply data across channel networks. The goal of these functions is to provide planners with the capability to identify and evaluate the best choices from a range of competing alternatives.

- *Strategic Planning.* The role of strategic planning functions is to provide managers with the capability to construct long-term plans and forecasts used to determine enterprise financial goals, explore strategic business partner alliances, design marketing approaches, and define and develop productive capacities necessary to support product and service requirements. The plans developed in this function provide the basis for management control, decision analysis, and performance, and are used to drive the operations plans executed by transaction functions. The overall goal of strategic planning functions is to provide the enterprise with competitive advantage.

While the exact structure of the applications used by today's information systems varies by industry, a base architecture must provide the business with a common database, a suite of relevant transaction programs, reporting and display, some form of simulation, and tools for ongoing operations management and control as well as long-range planning.

Principles of System Management

Whether the capture of historical data for forecasting or the utilization of inventory transaction balance records, an effective information system requires databases that are of the highest integrity and are easily understood by the user. The following seven principles of information system management are fundamental to an efficiently run business system [2]:

■ *Accountability.* While computers provide the functions for the entry and maintenance of data, responsibility for the quality and integrity of that data resides squarely with the people who use the system. Without accountability, an information system will quickly spin out of control and lose its ability to provide meaningful information for planning and decision-making.

■ *Transparency.* One of the fundamental keys to effective information technology is that the mechanics of how the system works be simple, understandable, and apparent to the user. For the typical user, the elegance and sophistication of the technical architecture of the system is irrelevant. The issue is simply one of relevance and usability. If the mechanics of a particular application are easy to work with and conform to best practices, the system will be intelligently and competently used. Transparency means that the system provides the user with answers as to why and how the system requires a particular activity to be performed.

■ *Accessibility.* One of the fundamental problems of paper-based systems is that data are not readily accessible for decision-making. Computer systems remove the difficulties surrounding data retrieval by containing programs that provide for quick access and update of critical information, such as order status, that span company departments or even databases belonging to trading partners.

■ *Data Integrity.* The usefulness of information technology is directly dependent upon the accuracy and timeliness of its databases. *Accuracy* can be defined as the degree to which actual physical data corresponds to the same data recorded in the system. As a metric, the accuracy of most data should be 99% at a minimum. *Timeliness* can be defined as the length of the spatial delay between the moment a transaction occurs and the point in which it is recorded in the system. The speed of data update is critical in providing users with the ability to perform activities in a timely manner. Similar to accuracy, high levels of timeliness enable companies to remove uncertainties and increase the accuracy of decisions.

■ *Valid Process Simulation.* If an information system is to provide useful information, the transactional and maintenance programs in the system must work the way the business actually works. In effect, a business application is in actuality a *representation,* a simulation of the actual physical action performed during process execution. For example, if the act of physically receiving a product to a location is not mirrored by the transaction entered in the application, data integrity is diminished and the ability of users to keep data records accurately decreases exponentially as invalid data cascade through the system.

■ *Flexibility.* An information system must provide the users with the capability to perform transactions or manipulate data to meet the needs of both the business and customers and suppliers. For example, inventory allocation and shipping functions should permit interbranch transfer orders to be created restricting delivery to a single receiving location or alternatively allow multiple delivery points on a single order. In addition, the software itself should enable easy upgrade capabilities without causing the company undo cost or implementation time.

■ *Control.* A fundamental benefit of an information system is the ability to control business processes. Applications should provide users with reporting and exception messaging designed to alert them as early as possible to actual or pending out of control processes. For example, a forecasting module should be able to recognize changes in basic historical patterns or relationships at an early stage and provide forecasters with warning alarms to take preemptive action.

Without these basic disciplines in place, it would be virtually impossible to achieve the necessary levels of performance from an information system regardless of its sophistication and elegance. In the next section utilizing the basic functions and management attributes to build effective information system architectures guiding internal business, interorganizational, and technological structures will be discussed.

Enterprise Business Systems Foundations

Effectively managing today's multifaceted enterprise requires architecting business system solutions that facilitate automation of standardized tasks, the generation of information that can be used for decision-making, and the networking of knowledge of people within the organization and externally out in the supply chain. The goal of the process is not to create monolithic, rigid systems, but rather to configure scalable, highly flexible information enablers that provide the business with the capability to respond effectively with value solutions and collaborative relationships at all points in the supply channel network. The software solutions that emerge should be able to effectively respond to the various levels of information management required by the supply chain. As will be discussed, there are three possible dimensions today's enterprise system must encompass. The first is concerned with solutions that integrate *internal* data and processes. The second is concerned with linking *external* parts of the company together. The third, and final, dimension examines technology tools that link external customers and suppliers to enterprise demand and supply planning functions.

Evolution of Enterprise Business Systems (EBS)

At the dawning of the computer age, computers were first used primarily as a mechanism to automate individual business processes. The first computerized applications for such areas as payroll, general ledger, customer billing, and inventory, were stand-alone systems each having their own databases, application logic, and user interfaces. While these systems enabled departments to standardize work and generate useful databases for their own particular function, their value dramatically decreased when information needed to be shared *across* enterprise departments. Nonintegrated systems simply added information processing cost to the company, fostered a lack of data integrity, and caused added frustration to its employees. For example, since databases were localized, information used by multiple departments had to be entered again separately in each local department's system. Often the data needed to be translated from the language of one department to that of another (a good example is part numbers); data needed by a department but not found in the database of a feeding department had to be constructed and manually fed into their system [3].

By the early 1980s, software companies began to offer business systems that increasingly focused on linking departmental functions around a common database. Beginning first as manufacturing resource planning (MRP II) and then ERP, these systems were an adaptation and refinement of earlier computerized applications. Today, the theory and practical use of ERP-type systems have so transcended their original architectures that it would be more appropriate to call them EBS. Whether "home-grown" or purchased from a software developer, applications originating in these robust and highly integrated systems support a wide spectrum of businesses from manufacturers to nonprofit organizations, from universities to government agencies.

The most significant aspect of an effective EBS can be found in its ability to not only to organize, codify, and standardize an enterprise's business processes and data, but more important to *integrate* all these pockets of information into a unified repository. The goal of today's EBS is to optimize an enterprise's *internal* value chain by integrating all aspects of the business, from purchasing and inventory management to sales and financial accounting. By providing a common database and the capability to integrate transaction management processes, data are made instantaneously available across business functions, enabling the visibility necessary for effective planning and decision-making. In addition, by providing for information commonality and integration, an EBS eliminates redundant or alternative information management systems and reduces non-value-added tasks, thereby dramatically impacting a company's productivity.

There are several other benefits gained by implementing an EBS. As companies continually grapple with engineering continuous change, many project strategists look to the suite of "best practice" process designs embedded in packaged EBS business function work flows to guide application process decisions. Often existing operating processes that contained bad or obsolete practices could be removed or standardized by reconstructing them around the best practices resident in the software's business applications. Finally, as the needs of business change to meet new challenges, companies with standardized processes driven by an EBS are more adaptable to change. Paradoxically, as Davenport points out [4], "Standardization can lead to increased flexibility." A single, logically structured, and common information system platform is far easier to adapt to changing circumstances than a hodgepodge of systems with complex interfaces linking them together.

EBS System Components

The fundamental objective of an EBS is to extend the structure and benefits of computer integration and networking to encompass the entire enterprise and the supply chains in which they are intertwined. In practical terms, this means that not just a portion but all business functions, from purchasing and inventory control to forecasting and general ledger, must be integrated together. The goal is to place all internal enterprise processing, decision-making, and performance measurement in a common database system capable of being maintained and referenced in real-time by all company users and selected supply chain partners. The EBS itself is composed of two elements. The first is the *hardware,* consisting of the computer, input/output devices, and data storage/warehousing. The central features of this element are discussed in detail in Chapter 2. The second element, the *software,* contains the application programs used for processing transactions, displaying the database, printing reports, management control, and operational and strategic planning. This element is the subject of this chapter.

Core Functional Areas of Operation

The EBS is the workhorse of a business and truthfully can be said to constitute the hub or "backbone" of the enterprise's information infrastructure. An effectively implemented and utilized EBS links the different functions of the business, drives continuous improvement and process efficiency and effectiveness, provides the mechanism to support company strategies, and enables the pursuit of e-business technologies.

As illustrated in Figure 3.2, an EBS can be described as having two major components: the technical architecture and eight tightly integrated business modules, one for each of core functions found in a typical business. The *EBS technical architecture* refers to the hardware configuration, programing languages, graphic presentation, document output capabilities, database designs, and a host of other enablers available in the system. In a wider sense, the technical architecture refers to the choice of how the EBS is assembled. It can be composed of a single, homogenous, fully integrated business software system or it might be the result of a company's decision to assemble a best-of-breed portfolio model linking third party point solutions to a home-grown or previously implemented packaged solution.

The second component of an EBS is the array of eight possible core business applications that an enterprise might deploy. These applications are described as follows: customer management, manufacturing, procurement, logistics, product data, finance, asset management, and human resources. Depending on the nature of the business, an enterprise may utilize some or all of the modules. For example, the EBS of a manufacturing company that performs distribution functions will

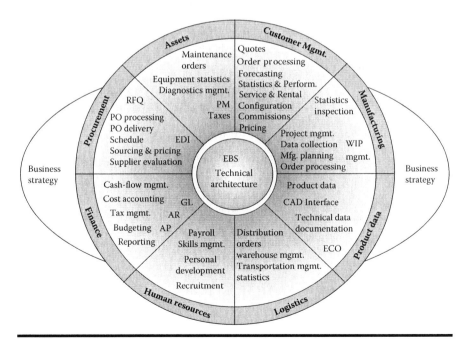

Figure 3.2   EBS backbone.

have all eight modules activated. A wholesaler would most likely have all except for manufacturing related functions. A dot-com catalog business would most likely have customer management, human resources, finance, and perhaps procurement and logistics functions. Regardless of the business environment, every company will at least have to install customer service, order management, and finance. These base functions would be difficult to outsource without loosing corporate integrity. A description of each of the eight modules is as follows:

- *Customer Management.* The primary role of this module in an EBS is to provide availability to the user interface screens that enable order entry, order promising, and open order status maintenance. Order entry and service maintenance are the gateway to the sales and marketing database. Second, this module should provide the data necessary to perform real-time profitability analysis to assist in calculating costs, revenues, and sales volumes necessary for effective quotation and ongoing customer maintenance. Third, a well-designed EBS will provide marketers with tools to design sophisticated pricing schemes and discount models. In addition, the software should permit the performance of miscellaneous functions such as order configuration, bonus and commissions, customer delivery schedules, tax management, currency conversion, customer returns, and service and rental. Finally, the customer database should be robust enough to permit the generation of sales budgets for forecast management and statistical reporting illustrating everything from profitability to contributing margins analysis.
- *Manufacturing.* Functions in this module comprise most of the foundation applications in the suite of a modern day EBS. Originating as a bill of material (BOM) processor, this module has been enhanced over the decades to include MRP processing, manufacturing order release, work in progress (WIP) management, cost reporting, and overall shop floor control. A critical integrative aspect is the real-time linkage of demand to supply management facilitating order-to-production and WIP modeling while promoting real-time available-to-promise (ATP) and capable-to-promise (CTP) to assist in customer order management. In addition to these basic tools, today's EBS manufacturing module also contains functionality for activities such as inspection, project management, capacity/resource management, and the compilation of production statistics. Finally, advances in technology have enabled the interface of EBS applications with "bolt-on" data collection devices and advanced planning and optimizing software.
- *Procurement.* In today's business climate the ability to effectively integrate procurement requirements with a variety of supplier management concepts and technology tools is one of the most important components of an effective EBS. Although much press has been given to Web-based business-to-business (B2B) technologies, basic management of procurement requires a close integration with internal MRP and maintenance, repair, and operations supplies

(MRO) systems. Today's EBS contains robust functionality to facilitate purchase order processing, delivery scheduling, open order tracking, receiving, inspection, and supplier statistics and performance reporting. In addition, detailed request for quotation (RFQ) must be available that ties back to customer demands and extends out to supplier management, negotiation, and pricing capabilities. Finally, the system architecture must include electronic data interchange (EDI) and Internet capabilities.

■ *Logistics*: The ability to link in real-time logistics functions to sales, manufacturing, and finance is fundamental to competitive advantage in the twenty-first century. Today's EBS must provide the mechanism to run the internal supply chain of the business as well as provide the necessary connectivity with remote trading partners located on the rim of the supply network. Critical tools in the module center on distribution channel configuration, warehouse activity management, channel replenishment planning and distribution order management, and the generation of distribution, asset, and profitability reporting. Also, of growing importance is the integration of EBS functions with "bolt-on" warehouse and transportation management systems, as well as applications supporting Web-based customer and SCM systems.

■ *Product Data.* At the core of manufacturing and distribution information systems reside the databases describing the products that a company builds and distributes. Often considered highly proprietary, these databases contain data ranging from engineering descriptions to details concerning cost, sources of acquisition, planning data, and product structure details. Besides obvious uses for inventory and manufacturing planning and shop floor management, these databases are critical for marketing product life cycle management analysis and costing, engineering product introduction, and financial reporting and analysis. As the speed of time-to-market and ever-shortening product life cycles accelerates, progressive companies have been looking to channel partners to implement collaborative technologies through the Internet that can link in real-time computer-aided design (CAD) and design documentation in an effort to compress development, introduction, and phase-out of products and services.

■ *Finance.* Without a doubt, one of the strong suits of an EBS is its ability to support effective management accounting. In fact, one of the criticisms leveled at EBS is that it is really an accounting system requiring everyone in the business to report on an ongoing basis each transaction they perform with 100% accuracy. Today's financial applications provide for the real-time reporting of all transaction information originating from inventory movement, accounts receivable, accounts payable, taxes, foreign currency, and journal entries occurring within the enterprise. The more timely and accurate the posting of data, the more effective are the output reports and budgets that can be used for financial analysis and decision-making at all levels in the business.

- *Assets.* Effective control of a company's fixed assets is essential to ensuring continuous planning of the productive resources necessary to meet competitive strategies. EBS databases in this module center on the establishment of equipment profiles, diagnostics and preventive maintenance activities, and depreciation tracking.
- *Human Resources.* The final module composing a modern EBS is the management of an enterprise's human capital. Functions in this area can be broken down into two main areas. The first is concerned with the performance of transaction activities, such as time and attendance reporting, payroll administration, compensation, reimbursable expenses, and recruitment. The second is focused on the creation of databases necessary to support employee profiles, skills and career planning, and employee evaluations and productivity statistics.

Secondary Functional Areas of Operation

Today's extended EBS is the product of the merger of the eight core business functions with the integrative networking capabilities of advanced Internet-enabled software applications. Realizing today's requirements for speed, agility, real-time control, and exceptional customer satisfaction means that enterprises must employ a variety of technology applications beyond the core business model that enable the construction of a business architecture that is effective, efficient, purposeful, integrative, and collaborative. The resulting system, consisting of a mixture of often disparate noncore EBS technology applications, data collection devices, B2B point solutions, and legacy systems, must be capable of being disassembled and rapidly reassembled into a single framework that matches the needs of the marketplace.

The merger of information technology suites as a business enabler is not a new idea. Historically, even in the days of silo-based organizations, the perpetual systems challenge centered on integration and collaboration issues. The advent of e-business simply heightened the gaps separating the "closed" environment of localized ERP and legacy systems and the requirement for "open" or networked operating systems, databases, and hardware platforms. Effectively meeting this challenge requires businesses to develop architectures and frameworks that permit enterprises to synthesize available integration software and existing solutions to address the ever-evolving structure of today's demanding business climate. Integrative information technology is about much more than just moving information back and forth: it is also about integrating into whole business processes the capabilities of enterprises that span the supply chain.

At first sight, the application suites comprising the universe of today's enterprise systems appear to be a maze of software products. ERP, customer relationship management (CRM), SCM, EDI, supplier relationship management (SRM) and other technology acronyms litter current business literature and impart an impression of

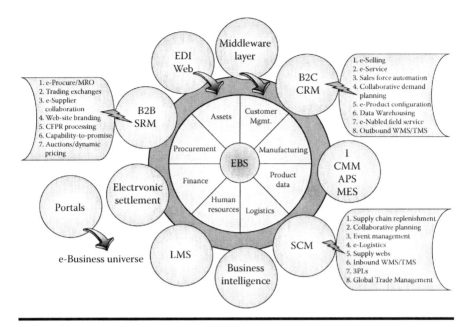

Figure 3.3 EBS universe.

incoherency, comprehendible only to a select club of industry analysts and "techies." Figure 3.3 is an attempt to assemble the core EBS functions and the possible components of today's extended business system into a coherent model. As can be seen, the model is divided into three distinct regions of software products: core EBS, Middleware, and connectivity-enabling applications. Each of the regions can be briefly explained as follows:

1. *EBS.* The eight core functions of EBS stand at the center of today's integrative information business solution. As described in detail at the beginning of this chapter, an EBS acts as the hub or "backbone" of an enterprise's transaction and information management functions. Using the core EBS functional software applications as a foundation, today's extended EBSs have been gradually converging with once stand-alone point solutions and Web-based applications to provide enhanced functionality and networking capabilities not available in the past. For example, supply chain planning (SCP) and APS have dramatically enlarged MRP planning functions. e-Procurement is encroaching on traditional purchasing functions, while CRM has radically expanded the one dimensional customer management functions of legacy systems. Also, the trend to outsource select business software to third parties, such as Software-as-a-Service (SaaS), is moving tomorrow's EBS toward being a mixture of proprietary backbone applications, on-demand add-on software suites, and e-business capabilities.

2. *EDI.* The definition, advantages, and drawbacks of EDI are discussed in the previous chapter. In summary, EDI refers to the structured transmission of data and documents between organizations by electronic means. Even in this era of technologies such as XML Web services and the Internet, EDI is still the data format used by the vast majority of companies for the transaction of electronic commerce. EDI documents generally contain the same information that would normally be found in a paper document used for the same business function. Trading partners are free to use any format for the transmission of information. While recently there has been a move toward using some of the many Internet protocols for transmission, most EDI is still transmitted using a VAN.

3. *Web-Based Applications.* The third major area of today's extended EBS is compromised of a variety of customer- and supplier-side software applications that are Web-enabled and directly integrated with EBS backbone applications. When they were first introduced, these applications were simply "bolted-on" to supporting EBS applications. Today, the former monolithic structure of EBS is being decoupled and merged with these supporting e-business tools. For example, as Figure 3.3 illustrates, APS that provide manufacturing functionality, such as finite loading, optimization, and production synchronization, has been integrated with traditional MRP to provide planning and control capabilities not possible within the standard ERP architecture. This merging of traditional ERP and specialized applications is expected to accelerate. A brief review of the applications illustrated in Figure 3.3 is as follows:

   – *B2C and CRM (Customer Relationship Management).* Utilizing the Internet to sell directly to the customer has become one of today's top avenues for sales management and is discussed in detail in the previous chapter. CRM applications enable companies to closely manage all aspects of a customer's relationship with an organization focused on increasing customer loyalty, retention, and profitability. CRM toolsets provide an integrated view of customer data and interactions allowing sales and marketing to be more responsive to customer needs. CRM components typically include the following: Internet-driven order management, sales force automation, promotions and event management, social networking, customer information storage, analytics for marketing research, and customer service.

   – *CPC (Collaborative Product Commerce).* This can be defined as the enablement through Web-based tools of virtual communities of manufacturing and product designers focused on new product development. Because it is Internet empowered, CPC permits companies to focus more intensely on their core capabilities while leveraging external design partnerships to dramatically decrease the time-to-market of product configuration and product rollout.

- *MES (Manufacturing Execution Systems)*. These systems manage and monitor work-in-process (WIP) data, including manual or automatic labor and production reporting, online inquiries, and links to tasks that take place on the production floor. Among the core functionalities of a MES can be found WIP tracking (labor, machines, materials), inventory management, production reporting (finished parts/goods, scrap, material issues, and returns), machine setup and downtime, online work orders, instructions, documentation, and drawings, quality management (sampling tests and employee certifications), automation integration (PLC/SCADA/HMI/SPC), and other functions. While most MES do nothing to connect the shop floor with the broader supply chain they do enable companies to better coordinate manufacturing with overall supply chain needs.

- *SCM (Supply Chain Management Applications)*. These applications enable not just companies, but whole supply networks to be closely integrated. SCM applications encompass all operations within the supply chain, including the sourcing, acquisition, and storage of inventories, the scheduling and management of WIP, and the warehousing and distribution of finished goods. SCM solutions enable businesses to streamline and automate the planning, execution, and control of these key activities. Additionally, since there are often many third parties involved in a supply chain, SCM software is designed to enhance communication, collaboration, and coordination with suppliers, transportation and shipping companies, intermediaries, and other partners. Internet-enabled SCM provides for the synchronization of the supply channel community, minimization of channel costs, optimization of network capabilities, real-time connectivity of channel supply and demand, and instantaneous visibility to demand and supply conditions within the entire supply chain network. Through the use of optimization software, whole supply networks can make better decisions concerning priorities, demand, inventory, and asset utilization.

- *BI (Business Intelligence)*. This comprises performance management and data warehouse software platforms. This software is used to access, transform, store, analyze, model, deliver, and track information to enable fact-based decision-making and extend accountability by providing all decision-makers with timely, relevant information. Additionally, business analytics is a framework that extends beyond software and systems to include culture, process, and performance strategies. The software tools considered part of business analytics span several areas, including analytics, data integration, query/reporting, and performance management.

- *B2B and SRM (Supplier Relationship Management)*. Much has already been discussed about the content of B2B Web-based applications. Tools in this part of the extended EBS application suite are focused on deploying independent, private, and consortia exchanges to facilitate materials and finished goods acquisition, requisitioning, sourcing, contracting, ordering, and payment utilizing online catalogs, contracts, POS, and shipping notices. Similar to CRM, SRM software ensures that effective collaborative relationships are being formed, the right suppliers are being used, supplier performance targets are being hit, and buyers are paying the right price for the quality and services contracted.

– *e-Finance and Human Resources.* The utilization of Internet technologies to facilitate financial and human resource functions is a developing element in assisting companies to build competitive advantage in today's complex, global business environment. e-Finance focuses on embedding core banking services, such as invoicing and payment, financing, and risk management within purchasing and logistics functions. A significant area is international commerce where Web applications are used to settle transactions in real-time regardless of currency. In the area of human resources, companies are expected to make much greater use of online tools to assist employees populate and access personnel databases, such as human skills repositories, recruiting, hiring, compensation, payroll, and knowledge management solutions housing the expertise and best practices of an entire corporate culture.

– *Portals.* Portal technologies are designed to provide company personnel and trading partners with secure, personalized access to data and self-service applications enabling them to react to changes in production schedules, forecasts, and other information. Usually information is accessed through Web browsers. The data retrieved is normally static and there is no direct access to the stored data. Not to be confused with PTXs, which require extensive system-to-system, application-to-application integration at the process level, portals provide people-to-system coordination where data from core functions such as ERP can be extracted, consolidated, and published on a portal for trading partners to see. Portals provide an effective medium to access information and knowledge across the supply network while at the same time reducing the costs of distributing and sharing content and applications.

EBS Networking and Integration Frameworks

The EBS application model suggested in Figure 3.3 implies that the range of EBS, resident-memory PCs, and Internet-enabled applications are integrated together so that information and business processes flow freely from any one place in the networked supply chain to any other place. In greater detail, the model assumes that *internally* companies have accessible synchronized repositories of data about customers, processes, and products that transcend the boundaries of departments, divisions, and geographical units. The model likewise assumes that *externally*, customers and suppliers have the ability to traverse company-centric silos to tap seamlessly into information or generate transactions any where in the supply chain. Such a vision requires cross-functional integration linking applications and data across business units, operating systems, and hardware platforms.

Networking Framework Tasks

While the benefits of networking are tremendous, actually executing an effective comprehensive strategy for both external B2B and internal EBS integration is a complex matter. According to a Forrester Consulting study [5], companies often spend more on integration-related projects than needed because they are implementing point solutions without understanding the full breadth and scope of their broader integration needs. At the time of the survey 80% of the 260 IT executive respondents rated their infrastructure as somewhat ineffective. Top among the causes were an overabundance of integration tools due to the implementation of single business unit solutions, toolsets based on older technology, and the fact that their main tools did not handle EDI and B2B solutions nor take advantage of SOA.

To meet this integration challenge, the following critical issues will need to be addressed:

- Evaluation of the capabilities of the existing integration toolset(s) and comparison to a list of current and future business needs must be performed. It is imperative that the following components are considered: enterprise application integration (EAI), B2B, business process management (BPM), and MFT.
- Integration standards must be available for document formats that enable the transfer of information between differing business systems.
- Automated and standard transformation and routing tools must be available to convert and transmit data in varying formats and be compatible with existing investments in systems, transports, and business documents formats.
- Tools needed to create and manage distributed business processes and the documents exchanged must be implemented.
- Strong security must be implemented that allows data transfer to be encrypted and digitally signed.
- The integration tools must leverage standard Internet transmission protocols as well as open data formats to facilitate data transfer between companies.
- The integration tools must be cost effective for small as well as large companies to enable mass market B2B transaction and trading.

In responding to such a formidable list of requirements it is critical that companies develop an integration framework strategy. Such a strategy enables companies to implement scalable integration capabilities that provide system designers with the ability to partition the business process logic found across application programs (*components*) and arrange the collaboration framework needed to meet functional requirements.

Middleware

At the heart of the integration architecture can be found a broad set of technology functions that has been termed *middleware*. The role of middleware is to coordinate and enable applications running on one computer to communicate with applications residing on another computer. Middleware provides an e-business structure to access legacy systems, EBSs, data structures, data warehouses, a repository containing a metamodel of the overall environment, and B2B interface points that constitute the range of network applications. Middleware provides the engine that enables internal and external business functions to pass data between each other. According to McWhirter [6], middleware can be said to fall into two categories:

■ *Data-Oriented Middleware*. In this category middleware facilitates the sharing of information between different applications, such as a CRM system and an EBS.
■ *Process-Oriented Middleware*. This category of middleware enables the processing and integrity of transactions and insures system integrity.

Both categories can reside on a server embedded in one of many application layers. Functionally, when a transaction occurs in an application like e-procurement, the data passes through to the middleware layer where it is translated into a common language by any one of a number of middleware tools, such as XML or Java. Following, the data are then passed are a readable format into another application layer, say the purchasing application inside of an EBS. An illustration of an e-business integration structure can be seen in Figure 3.4.

As e-business collaborative commerce has expanded, the requirement for more complex connectivity tools that can not only pass data, but also facilitate process-oriented business workflows, has increased. Organized around the term business process management (BPM), this subset of middleware is focused on integrating business processes across business units, applications, and enterprises. The goal is simple: how can companies align business processes to deliver key information and performance indicators as opposed to just moving data back and forth. Through modeling tools that enable the creation of processes via a dynamic graphical environment, BPM provides visibility to business processes that span many types of computer systems and architectures to assist managers monitor, measure, and resolve Web-enabled processes throughout an e-business environment. BPM also provides tools that automate and integrate processes by generating graphical user interfaces to trigger event management across intercompany and networked supply chains. Finally, BPM activity output provides managers with metrics so that process integration can be further fine-tuned [7].

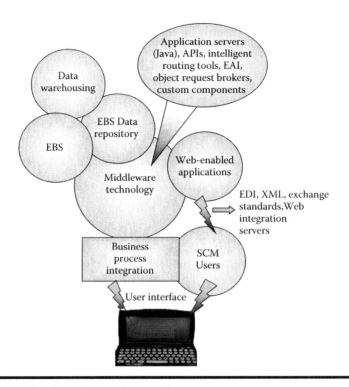

Figure 3.4 E-business integration architecture.

Standard EBS Systems

As detailed above, the array of possible core and secondary functions constitute the structure of today's EBS. In reality, not all companies utilize all of these functions. As such, today's EBS must be capable of being configured to meet the information solution requirements of various types of companies. In this section the two most common EBS types for manufacturers and distributors are explored.

Enterprise Resource Planning (ERP)

A common software system that has been applied by companies for many decades to manage their businesses is ERP. Historically, the value of ERP systems can be found in their ability to serve as a foundation for information and process integration and collaboration between company departments. Achieving competitive advantage requires all units of the organization to work together toward a common goal. ERP assists in achieving this objective by providing a central database that collects information from and feeds information into all of the functional applications areas so that effective transaction management, shared management reporting, and decision-making can be effectively performed.

Beyond these integrative characteristics, an ERP system must also posses the following architectural elements [8]:

■ *Flexible.* ERP system architecture should provide the user with the ability to enable those application functions that constitute its business. The capability for complex system configurability is a significant enhancement of today's ERP over past systems, which were both rigid and monolithic and difficult and expensive to customize.

■ *Modular and Open.* Today's ERP system should have an open architecture permitting any module to be interactive with any other module without affecting system functions. This also means that modules can be enabled or disabled with impairing system integrity. An "open system" means that the software can be run on multiple hardware platforms (IBM, Oracle, Microsoft, etc.). This feature is particularly important for businesses that have multiple systems and third party point solutions.

■ *Comprehensive.* The ERP system should support as wide a variety of organizational functions as is possible and support a wide range of business verticals.

■ *Networked.* The ERP system should possess functionality that allows users to escape beyond the four walls of the enterprise by possessing Internet connectivity to network them with supply chain partners.

An illustration of the functions of the typical ERP system can be found in Figure 3.5.

The architecture of an ERP system attempts to integrate all of the information processes of the enterprise and use the resulting synergy of planning and control enabled by this integration to continuously improve performance to the customer as well as establish uniform policies and practices across the business. An ERP system can be best understood by dividing Figure 3.5 into three regions. In *Region 1* can be found applications used for the development and disaggregation of top management plans governing the overall direction of the enterprise, products and market programs, sales and promotions campaigns, and aggregate product family priority and resource capacity planning. *Region 2* is composed of applications for demand management encompassing forecasting finished goods, order management, order promising, and managing intercompany channel demand and inventory replenishment priority and capacity planning. In *Region 3* can be found applications focused on operations execution activities such as purchasing, production activity control, transportation and shipping, fixed asset and human resource management, and financial accounting. A short description of these critical applications is as follows:

■ *Region 1: Strategic Planning.* Applications in this area provide corporate strategists with tools to develop the overall goals and objectives to be pursued by the enterprise. The output of these applications is focused on determining how the business is going to compete—product lines, financial investment

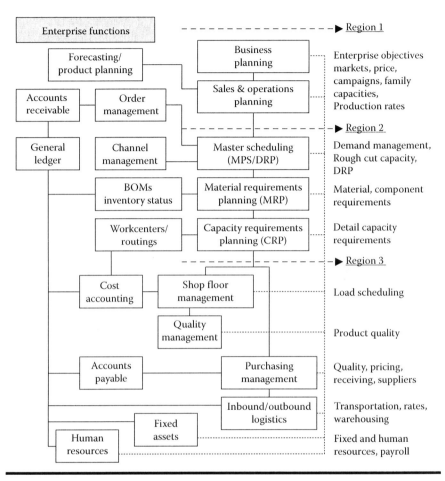

Figure 3.5   Basic ERP functions diagram.

and return, markets, research and development—in the future. Planning in this area begins with the utilization of detail transactional databases from marketing, finance, production, and distribution that in turn are used as the basis for long-range forecasting and decision-making. The strategic business plan developed from the process is then used as a framework determining the goals and objectives for the marketing and sales, finance, engineering, production, and logistics departments. An ERP system automatically integrates and translates these business plans into a common language that each department is able to utilize to define their own set of performance objectives. Sales & operations planning (S&OP) enables the business to build consensus teams in sales, operations, finance, and product development around the corporate strategy and to enable continuous strategy revision and realignment of departmental plans to meet changing circumstances. The output from strategic planning region is then feed into the next ERP region.

■ *Region 2: Demand and Supply Planning.* This region is divided into demand and supply planning management. The primary role of demand management is to enable quick and accurate order entry, order promising, and open order status maintenance. Order entry and ongoing service maintenance are the gateway to the sales and marketing database. Applications should also provide the data necessary to perform real-time profitability analysis to assist in calculating costs, revenues, and sales volumes necessary for effective quotation, ongoing customer maintenance, and accounts receivable. Next, these applications should provide marketers with tools to design sophisticated pricing schemes and discount models. In addition, the software should permit the performance of miscellaneous functions such as order configuration, bonus and commissions, customer delivery schedules, global tax management, customer returns, and service and rental. Finally, the customer database should be robust enough to permit the generation of sales budgets for forecast management and the generation of statistical reporting illustrating everything from profitability to contributing margins analysis.

The second component of *Region* 2 is supply planning management. At the heart of ERP can be found the oldest software applications resident in the system: the MRP processor used for planning finished goods, component and raw materials, and distribution channel inventories. This module contains functionality enabling the generation and maintenance of inventory database records and parameters, the efficient and timely replenishment of inventories, and inventory planning simulation. The master production schedule (MPS) receives demand on finished goods from the production plan, forecasts, actual customer orders, and interbranch demand and calculates the replenishment quantities necessary to maintain customer demand targets. Once completed, this application sends finished goods requirements down to the MRP by exploding product BOMs where component and raw materials are then calculated to provide the company's plan for materials and product replenishment.

Once MRP has generated the inventory priority plan, it must be reviewed against the available capacities within the firm and outside in the supply channel necessary to manage replenishment proposals. Capacity constraints can arise from a variety of sources including manufacturing capabilities, warehousing, transportation, and material handling equipment. The goal of the process is to identify bottlenecks that could constrain the efficient production, storage, and movement of goods. Today's ERP provides planners with the capability to view plant and channel capacities that match aggregate, finished goods, and detail inventory priority plans. If priority and capacity plans are not found to be in balance at any planning level, planners must resolve the imbalance by resource acquisition, outsourcing, or postponement of production or delivery. Finally, advances in technology have enabled the enhancement of these applications with the addition of "bolt-on" data collection devices, manufacturing execution systems, and advanced shop floor planning and optimizing software. Once the material and capacity plans have been authorized, they are then driven down into the third ERP region.

■ *Region 3: Manufacturing, Purchasing, Logistics, and Accounting.* Functions in this application comprise several plan execution activities. The manufacturing module controls order release, order scheduling and operations synchronization, inspection, general WIP management, and the compilation of production statistics. The purchasing module control the tasks associated with the acquisition of all purchased components and raw materials, RFQ, and maintenance, repair, and operations (MRO) needs. Today's ERP contains robust functionality to facilitate purchase order processing, delivery scheduling, open order tracking, receiving, inspection, supplier statistics, and performance reporting. The logistics applications are concerned with coordinating transportation, warehousing, labor, and material handling equipment with the place and time requirements of sales, manufacturing, and inventory management. Finally, ERP accounting modules enable budgeting, fixed asset, cost accounting, payables, receivables, and general ledger transactions and financial reporting.

EBS Architecture for SCM

One of the most crucial aspects of an EBS is its ability to adapt to different supply chain environments. The applications and processes detailed in the previous section can aptly be described as constituting an ERP system primarily for manufacturing companies. In this section the architecture of an EBS for running a distribution enterprise will be examined. For many software suppliers the actual difference between ERP and SCM systems is negligible. For example, software vendors SAP, Oracle, Epicor, JDA, and Infor include manufacturing planning and shop floor management as part of their "SCM" solutions. These major vendors have simply expanded their ERP packages to include SCM functions such as CRM, warehouse and transportation management, and e-business and then provided users with the ability to configure the exact EBS solution needed. There are, however, vendors like RedPrarie and Manhattan Associates whose software contains modules only for demand planning and forecasting, inventory optimization, order life cycle management, transportation, and distribution [9]. The standard module configuration of a SCM solution is illustrated in Figure 3.6.

This model illustrates the integrative, interactive, and closed-loop nature of the typical activities occurring within a distribution enterprise. Similar to the general ERP model, the SCM applications presents a systemic approach. The flow of information begins with business planning functions and proceeds through operations and financial execution and concludes with performance measurement. Besides

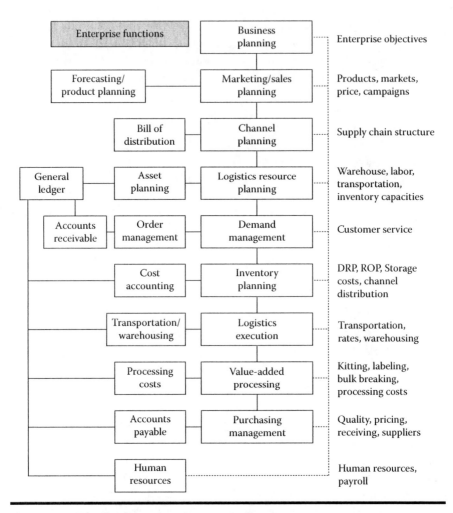

Figure 3.6 Distribution enterprise system diagram.

providing a systems approach to directing and measuring the entire enterprise, from internal functions to the management of supply channels, the SCM model provides a solution to the following problems plaguing the distributor:

1. The creation of business plans addressing the strategic decisions defining enterprise mission, market demand, and the allocation of financial, physical assets, and human resources and the mechanism to disaggregate those plans through logistics resource planning (supply chain capacity) and distribution requirements planning (DRP). The goal is the generation of time-phased plans where the demands of the entire supply chain are balanced against individual enterprise and total channel inventory, value-added processing, logistics, and purchasing capabilities.

2. The capability to be agile and flexible enough to respond to the many changes occurring in the business environment by providing visibility to customer demand and real-time status of the supply chain. Changes in market demand, supplier delivery problems, excess inventories, inventory channel imbalances, marketing promotions, and a host of other events are made visible along with the mechanism to adjust and reallocate critical resources.

3. By applying the time-phased logic of MRP and the ability to structure an integrated channel planning mechanism through the bill of distribution (BOD), the SCM model provides planners with a detailed workbench to keep channel inventories low and evenly distributed to meet the demands placed on each channel node, while providing the tools to resupply the supply chain as efficiently and cost effectively as possible. SCM architecture also enables the design of performance measurement programs that look beyond local agendas to how well each channel segment is supporting the overall business plan.

4. Finally, when integrated with today's newest supply chain event management and collaborative forecasting and planning toolsets, the SCM model provides a catalyst for the application of Internet-based technologies and lean principles targeted at removing barriers to time and space as well as to quality and excellence.

SCM software solutions offer distribution functions a fresh approach to running the supply chain that is fully compatible with today's newest Web-based technologies. Both as a management philosophy and as a suite of integrated business applications, SCM software permits distribution companies to synchronize demand and supply up and down the channel network and link their systems with the core and secondary EBS functions of supply chain partners. Such an approach enables entire supply chains to compete as if they were a single, seamless entity focused on channel cost reduction and superior customer service.

EBS Benefits and Risks

The decision to implement an EBS like MRP or SCM requires significant forethought as to the advantages as compared to the tremendous risks involved. All companies are agreed that the capability to automate out process error, decrease costs and increase productivities for a specific process is a significant benefit. However, the decision to use technology to informate, integrate, and network the entire business is an entirely different matter. The benefits can be substantial. As a planning and control mechanism, an EBS enables whole enterprises to organize, codify, and standardize business processes and data. Achieving such objectives in turn permits strategists and planners to optimize the business's *internal* value chain by integrating all aspects of the business, from purchasing and inventory management to sales and financial accounting. In addition, by providing a common database and the capability to integrate transaction management processes, data is made instantaneously available across business functions, enabling the visibility necessary for effective planning and decision-making.

An EBS provides other benefits. As companies grapple with managing continuous change to products, processes, and infrastructure, strategists are looking to the suite of "best practice" process designs embedded in today's EBS business functional work flows to assist in the removal of ill-defined or obsolete processes and the structuring of "best in class" processes by building them around the capabilities of EBS applications. Finally, as enterprises evolve to meet new challenges, companies with standardized processes driven by an EBS are more adaptable to change. A single, logically structured, and common information system platform is far easier to adapt to changing circumstances than a hodgepodge of systems with complex interfaces linking them together.

While the benefits can be dramatic, so can the risk companies take on during an implementation. The most direct is cost. Depending on the size of the company, an EBS implementation requires huge expenditures: according to the Panorama Consulting Group, in 2008 the average cost in the manufacturing and distribution industries was \$8.2 million [10]. Cost for an EBS implementation include both the hardware and software costs as well as education and consulting fees, customization, integrating and testing, data warehouse integration, and data conversion. What is more, according to a 2008 Panorama Consulting Group survey [11], 65% of ERP implementations go over budget with nearly one out of five (16%) going 50% or more over budget. Also, companies have historically found implementation time longer than expected. In the same study, only 7% of projects finish on time; 93% of respondents indicated that their implementations took longer than expected, with 68% stating that it took "much longer" than expected. In addition, none of the respondents indicated that their implementations took *less* time than planned. Even more risky is the emotional and physical trauma an implementation will have on a company's personnel, processes, and customers. Often senior executives feel that they are engaging in a "bet-your-company" gamble driving the organization through what can only be described as a virtual skeletal transplant as the old culture and operations are pulled out and the new EBS configuration becomes the standard.

EBS Choices

Considering the opportunities and the risks of implementing an EBS solution, executives need to have a well-defined strategy that includes both system selection as well as objectives to be attained. In a 2009 report [12], the Aberdeen Group identified the main drivers (in order of rank) of ERP software acquisition as cost reduction, need to manage growth, need to improve customer experience, need to improve customer response time, and interoperability issues. When contrasted with the 2007 report, which showed managing growth as the most important and reducing cost second to last, the 2009 report reflects how changes in the business climate, such as the recession of 2008–2009, can impact decisions regarding EBS implementation objectives.

Executives today have five alternatives before them when choosing an EBS solution. Deciding on an option is governed by several factors including length of implementation time, cost, flexibility and complexity of software, quality of the solution, fit to the enterprise, and intensity of staff training.

■ *Build.* In this option a company decides to build its own EBS. In the past this option was distinctly unfavorable, but with today's new composite-application tools or by engaging an inexpensive, high-quality third party developer, such as Infosys, this option has become viable especially among many non-manufacturing organizations.

■ *Buy.* Today this approach is the most common. According to a recent report, there are available about 52 ERP/SCM software products from 32 vendors, and there are many more out there, with new ERP software packages, ERP add-ons, and utilities appearing regularly. These packages can be divided into Tier 1 (very expensive, complex software for very large companies) and Tier 2 (less complex, or software with special industry functionality) for small- to medium-sized companies [13].

■ *Best-of-Breed.* This option consists in acquiring highly specialized point solution, say transportation management (TMS), and then interfacing it with existing systems. The advantage is that a point solution contains significant depth and agility to handle a special business problem than what is available in a standard EBS. The disadvantages center on implementation and integration costs, upgrade maintenance, and risk of vendor viability [14].

■ *Rent.* In this approach a company does not own the software but rather rents it via hosting or an alternative provider. Today, the emergence of new players and the growth of Software-as-a-Service (SaaS) have made this option more attractive. This option enables companies to minimize risk, cut costs associated with IT staffing, leverage deployment flexibility, and enjoy speedy use of the application functionality without long implementation time frames.

■ *Outsource.* Some companies have decided to let service suppliers run key processes for them via business process outsourcing or strategic outsourcing. The goal is to eliminate the cost of IT equipment and overhead. This option is often pursued by very small companies with limited capitalization.

Regardless of the option selected, Chaffey [15] has identified nine major criteria that should be referenced when selecting a software system. They are these:

■ *Functionality:* the array of resident applications and they level of fit to company business processes.

■ *Ease of Use:* the speed by which users can overcome the functional learning curve and the transparency of the applications relative to company processes.

■ *Performance:* the processing speeds by which data can be entered, calculated, maintained, and retrieved.

- *Scalability:* the ability of the user to enable or disable software modules during implementation and as the company's demands on the software expand.
- *Interoperability:* the ease with which the software can be interfaced, integrated, or networked to other software systems.
- *Extensibility:* the ease with which modifications or software enhancements can be made to the base software product.
- *Stability/Reliability:* the software's basic data integrity and freedom from bugs or invalid calculations.
- *Security:* capability of restricting access to users, groups of users, or outside-the-organization partners and customers.
- *Support*: the quality of implementation assistance, ongoing consulting, documentation, and training.
- *Vendor Viability*: financial strength of the vendor organization to provide periodic system enhancements and ongoing support and services.

To ensure effective due diligence, executives need to ensure a match between the needs of the business and the technical solutions available. Best-in-class business system project teams should follow the steps below to help contain their project scope, duration, cost, and overall business risk:

1. Ensure executives outside of IT are involved in the vendor evaluation and planning process. This ensures that the entire executive team will be on board to help identify all of the real benefits and hidden costs of the project.
2. Allocate sufficient time for the evaluation process. Companies should spend at least three to four months on the selection and planning process. Large organizations should plan to spend even more time on this step.
3. Develop an actionable, realistic business case identifying operational business benefits and key performance indicators during and after the implementation.
4. Develop a realistic project plan and implementation timeframe and the mechanisms to keep it accurate and timely. This is a primary cause of project cost overruns.
5. The organization should be prepared for an EBS. An EBS often obsoletes long-standing manual processes and outdated technologies. Such a shock will necessitate a significant change-management effort comprehensive enough to ensure management and the workforce fully understands the risks and possible impact on the existing corporate culture.

Advent of Internet Business Technologies

By the end of the twentieth century, companies became aware that to survive and flourish they needed to expand their businesses beyond the parochial boarders of their own individual enterprises and connect with their trading partners out in the supply chain. Emphasis gradually shifted from a long-time concern with managing fairly static internal organization structures, cost and performance measurements, product design and marketing functions, and a localized view of the customer to a realization that new marketplace and technological forces and the accelerating speed of everything, from communications to product life cycles, were presenting radically new avenues for competitive advantage. By the opening of the twenty-first century, it had become evident that only those companies that could effectively leverage integrative information technologies to open new opportunities for commerce and effect the collaboration of supply chain partner would be able to successfully take advantage of the new global marketplace. The management philosophy that emerged in this new economic environment was SCM and it quickly was adopted as the core strategic model for most enterprises.

At the same time that the concept of SCM optimization was emerging, radical breakthroughs in Internet technologies were providing the communications and networking mediums that enabled the *connectivity* of SCM to move from a decoupled, serialized flow to a real-time integration and synchronization of the competencies of channel network business partners. In the 1990s companies used business process reengineering, TQM (total quality management), ERP, JIT, and intranet and extranet models to create more competitive, channel-oriented strategies. Today, the Internet has propelled SCM to an entirely different dimension by enabling a global capability to pass information and transact business friction-free anywhere, anytime, to customers, suppliers, and trading partners. The application of Internet technologies requires a transformation of tradition business focusing on:

- The *end-to-end integration* of all supply chain functions from product design, through order management, to cash. This integration encompasses the activities of customer management on the delivery end of the supply channel, technologies connecting individual firms with trading partner competencies on the supply end of the channel, and internal transaction management associated with orders, manufacturing, and accounting on the inside of the business.
- The development of *end-to-end infrastructures* that facilitate real-time interaction and fuse together the synchronized passage and convergence of network information. Internet knowledge requires the transformation of infrastructures and technologies from a focus on internal performance measurements and objectives to organizations and information tools that engender unique interconnected networks of value-creating relationships and possess the agility and scalability to meet the constantly changing objectives of today's economic environment.

The bottom line is that integrative information technologies are here to stay. Gone are the days when companies debated the use of Internet technologies as essential to business success: today, to be a viable business companies *must* embrace the power for competitive advantage found in integrative technologies. Surviving in the world of the Internet economy means that companies must not only thoroughly understand the optimal application of integrative technology tools but also the basis of the Internet business landscape.

Defining Internet Business

The application of the Internet to business can best be understood by dividing it into two separate, yet connected concepts and set of practices. The first is termed *e-business* and is defined as a collection of business models and practices enabled by Internet technologies focused on the networking of customers, suppliers, and productive capabilities with the goal of continually improving supply chain performance. The second concept is termed *e-commerce* and is defined as the ability of businesses to buy and sell goods and services over the Internet. Over the past decade these terms have been used as if they were interchangeable. In reality, e-business is a more powerful concept that seeks to utilize the Internet to build integrative, collaborative relationships among supply chain members, while e-commerce is a subset of e-business concerned with the performance of commerce transactions electronically.

All e-business activities occur between consumers and between businesses. e-Business models can be separated into four major categories [16].

- *Business-to-Consumer (B2C).* This model is the most known and applies to any business that utilizes the Internet to sell products or services directly to the consumer. The basic form of B2C model is termed *e-stores* or *e-tailers*. The goal of this model is to imitate an actual store shopping experience where consumers can browse through catalogs or search mechanism to locate, price-compare, and order products and services to be shipped directly to their homes. Some of these B2C sites, such as Amazon.com, are *pure-play* in that they sell only through the Internet. Other types, such as Barnesandnoble.com, sell online as well as from an actual store outlet and are termed a "bricks-and-clicks" Web site.
- *Business-to-Business (B2B).* This model applies to any company that utilizes the Internet to sell products and services to other companies. Also termed *e-procurement,* this model is used by the buy-side organization for the purchase of manufacturing inventories, finished goods, and MRO goods and services based on a preexisting contract (*systematic sourcing*) or random purchasing or (*spot sourcing*) from various types of MRO hubs, yield managers, catalog hubs, and exchanges.
- *Consumer-to-Consumer (C2C).* This model applies to Internet sites that allow customers to buy from each other. C2Cs are consumer-driven and consist of online communities that interact via e-mail groups, Web-based discussion forums, or chat rooms. Currently this area is undergoing dramatic

change as the concept of *social network* gain traction. An example would be ebay.com.

■ *Consumer-to-Business (C2B).* This model applies to any consumer that utilizes the Internet to sell products or services directly to a business. This area is also expected to be dramatically impacted by the growth of online tools like Facebook, Twitter, and YouTube that allow consumers to directly communicate with businesses. An example would be priceline.com that sells products and services to individuals or to companies.

With these basic contours of the Internet marketplace defined, the next section will describe the evolution and content of the above e-business models in much greater detail.

Evolution of Internet Business

Up until the late 1980s, limitations in computer architecture and communications devices forced even the most technically savvy companies to remain fairly concentrated on the management of internal business functions. Although tools fostering business-to-business connectivity, such as EDI, were slowly growing, the enterprise-centric and proprietary nature of computing in this period great inhibited the ability of systems to interconnect with vital information occurring out in the business network. Even simple data components, like inventory balances and forecasts, were communicated with great difficulty to sister warehouses or divisions, let alone to trading partners whose databases resided beyond the barriers of their own information systems. Data transmission was purely a manual affair, and because of the nature of the business philosophies of the time, shared only with the greatest reluctance.

By the late 1990s, however, the marketplace had become aware of a definite acceleration in the speed of change shaping the business environment driven by new management concepts and the emergence of Internet technologies. Management buzzwords such as agile enterprises, virtual organizations, total quality management, business process engineering, and lean manufacturing all focused on a common theme: how to eliminate time and costs from processes while providing businesses with the ability to be more flexible and connected with their supply channel partners. In tandem with these new ideas occurred the rise of technology infrastructures, especially the World Wide Web and the personal computer (PC), permitting companies to escape from the prison of mainframe computers and proprietary databases. Today, the challenge of technology is *integration* and the creation and deployment of Internet-based products and services to network buyers and sellers ever closer together.

The first major technology breakthrough enabling companies to link with other companies was EDI. Despite the recent rise in Internet-enabled data transmission capabilities, EDI constitutes today's most widely used method of supply chain connectivity. EDI provides for the computer-to-computer exchange of business transactions such as customer orders, invoices, and shipping notices. EDI is an *extranet* system and consists of a set of transactions driven by a mutually agreed upon and implemented set of data transfer standards usually transmitted via private value-added networks (VANs). The critical importance of EDI is that the transacting companies can be using business systems that run on different software systems and hardware. The EDI standards act as a "translator" that utilizes the agreed upon transmission protocols to take the data residing in the computer format of the sending company and convert it into the data format used by the business system of the receiving company.

There are various benefits to the use of EDI. To begin with, EDI increases *communications and networking* by enabling channel partners to transmit and receive up-to-date information regarding network business processes electronically, thereby assisting the entire supply chain to leverage the productivities to be found in information networking. Second, EDI *streamlines business transactions* by eliminating paperwork and maintenance redundancies, thereby significantly shrinking cycle times in a wide spectrum of transaction processing activities. Third, EDI provides *increased accuracy*. Because transactions are transferred directly from computer-to-computer, the errors that normally occur as data is manually transferred from business to business are virtually eliminated. Fourth, EDI result in a *reduction in channel information processing* by removing duplication of effort and accelerating information flows that can significantly reduce time and cost between supply channel partners. Fifth, EDI *increases response* by enabling channel members to shrink processing times for customer and supplier orders and to provide for timely information that can be used to update planning schedules throughout the channel. And finally, EDI results in *increased competitive advantage* by enabling the entire supply network to shrink pipeline inventories, reduce capital expenditure, improve ROI, and actualize continuous improvements in customer service.

Although providing an effective method for data exchange between businesses, there are a number of drawbacks to EDI. To begin with, EDI is expensive and time consuming to implement. In addition, the basic data elements of the EDI transaction are centered on transmitting whole packets of information that must be sent, translated, and then received through trading partner systems. The time it takes, often several hours or even days for processing, militates against the real-time flow of information and decision-making necessary for today's business environments. Furthermore, the proprietary nature and cost of EDI renders it a poor supply chain enabler. What was always needed was a way to make information available to the entire supply network at a very low threshold of cost and effort, but which at the same time enabled channel partners to use it to execute strategic decisions.

Sometime in the mid-1990s technologically savvy companies began to explore the use of a new computerized tool that held out the potential to drastically alter forever the way firms fundamentally communicated and conducted business with their customers and suppliers. That tool was the Internet. Fueled by the explosion

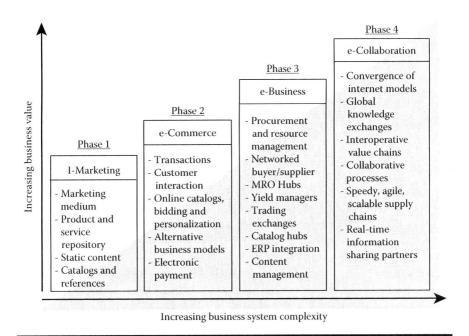

Figure 3.7 **Four phases of Web-enabled e-Business.**

in PC ownership, advancements in communications capabilities, and the shrinking cost of computer hardware and software, companies became aware that a new medium for exchange was dawning, a new medium that would sweep away the traditional mechanisms governing the flow of products and information through the supply chain. The Internet provided businesses with a radically new way of interacting based on the personalized, one-to-one marketing, buying, and selling between individual suppliers and consumers. Web-enabled business can be said to consist of four phases as portrayed in Figure 3.7 and discussed below.

I-Marketing

The first phase of e-business has been termed Internet Marketing (I-Marketing) because it is almost exclusively limited to the presentation of information about companies and their products and services utilizing relatively simple Web-based multimedia functions.

The goal of I-Marketing is simple. Throughout history businesses have been faced with the fundamental problem of how potential customers, separated by space and time, could find out about companies and their range of goods and services. Despite the sophistication of the traditional marketing techniques, there were a number of problems with the approaches. To begin with, by its very nature space and time fragment markets. The use of methods like mass media advertising and direct marketing resulted in silo customer segmentation and hit-or-miss approaches that attempted to provoke a wide band of prospect interest or inform the existing client base. Second, the traditional marketing approaches represented a basically passive approach on the part of customers to search for and learn about new companies and their product/service mixes. Buyers tended to avoid the difficult task of sourcing and comparison in favor of purchasing based on branding and proven personal relationships.

The advent of the Internet-enabled companies to finally escape from the limitations of traditional marketing by providing a revolutionary medium to communicate to customers anywhere, at anytime around the world. This first stage of Internet business was quite simple. Use of I-Marketing browsing enabled customers to search, view graphical presentations, and read static text about companies. I-Marketing Web sites actually are little more than online repositories of information and are often termed "brochureware" because of their similarity to traditional catalogs and other printed product/service publications. Due to their limited functional architectures and business purpose, I-Marketing Web sites do not provide for the entry of transactions or the ability of companies to interact with existing customers or prospects using the site.

Despite the deficiencies, the use of I-Marketing signaled an order of magnitude departure from traditional marketing techniques. To begin with, the ubiquitous use of the Web meant that companies were no longer circumscribed by time and space. A firm's mix of goods and services could be accessed by anyone, anytime, anywhere on earth. This meant that all enterprises across the globe, both large and small, now had a level playing field when it came to advertising their businesses, their products, and their services. I-Marketing also changed dramatically the role of the customer who moved from a passive recipient of marketing information to an active participant in the search for suppliers that best matched a potential matrix of product, pricing, promotional, and collaborative criteria. Finally, I-Marketing-enabled companies to aggregate marketing data from multiple vendors into a common catalog, thereby creating an early version of electronically linked communities of e-marketplaces.

e-Commerce Storefront

While I-Marketing did provide companies with the capability to open exciting new channels of communication with the marketplace, technologically savvy executives soon realized that what was really needed was a way to perform transactions and permit interactions between themselves and the consumer over the Internet. During the second half of the 1990s, advancing Internet capability enabled a new kind of business model—business-to-consumer (B2C). This model was designed specifically to sell and service consumers online. Soon companies like Amazon.com, Best Buy, and Barnes and Noble were offering Web-based storefronts that combined online catalogs and advertising techniques with new technology tools such as Web site personalization, self-service, interactive shopping carts, bid boards, credit card payment, and online communities that permitted actual online shopping.

According to Hoque [17], the new e-business storefronts spawned a whole new set of e-application categories and included the following:

- *e-Tailing and Consumer Portals.* These are the sites today's Internet shopper normally associates with Web-based storefront commerce. The overall object of enterprises in this category is to enable Web-driven fixed price transactions centered on products and services aggregated into catalogs and sold to consumers.
- *Bidding and Auctioning.* Sites in this category perform two possible functions. Some sell products and services through auction-type bidding using bid boards, catalog integration, and chat rooms. Others, like eBay, on SALE, and uBid perform the role of third party cybermediaries who, for a service price, match buyers and sellers.
- *Consumer Care/Customer Management.* These applications provide a wide range of customer support processes and functions focused on enabling a close relationship-building experience with the consumer. These applications include customer profile management, custom content delivery, account management, information gathering, and interactive community building.
- *Electronic Bill Payment (EBP).* These applications assist customers to maintain accounts, pay bills electronically, and assist in personal finance management. A good example of an Internet facilitator in this category is PayPal. Typical EBP features include Internet banking, bill consolidation, payment processing, analysis and reporting, and integration with biller accounting systems.

For many in both the marketplace and the investment community, the B2C enterprise seemed to offer a path to a whole new way of selling and servicing to a global marketplace. The advantages of the e-commerce storefront over traditional "bricks-and-mortar" firms were obvious. A single seller could now reach a global audience that was open for business every day of the year, at any hour. By aggregating goods and deploying Web-based tools, this new brand of marketer could offer customers a dramatically new shopping experience that combined the ease of shopping via personal PC with an immediacy, capability for self-service, access to a potentially enormous repository of goods and services, and information far beyond the capacities of traditional business models.

e-Commerce went through some very difficult times in the early years of the new millennium as the infamous "tech bubble" broke. In the ensuing dotcom carnage, thousands of e-commerce sites shut their virtual doors and some experts predicted years of struggle for online retail ventures. However, over the remainder of the decade B2C commerce has flourished despite concerns over security, taxation, and consumer protection. Shoppers have continued to flock to the Web in increasing numbers. By 2010, consumers are expected to spend $329 billion each year online, according to Forrester Research. What's more, the percentage of US households shopping online is expected to grow from 39% in 2007 to 48% in 2010. Globally, it is estimated that Internet sales will reach $1 trillion by 2012.

The goal of storefront e-commerce is nothing less than the reengineering of the traditional transaction process by gathering and deploying all necessary resources to ensure that the customer receives a complete solution to their needs and an unparalleled buying experience that not only reduces the time and waste involved in the transaction process, but also generates communities-of-interest and full-service consumer processes. Take for instance Amazon.com whose goal is not just to sell products, but to create a shopping "brand" where customers can log on to shop for literally *anything*. In such a culture the real value of the business is found in owning the biggest customer base that contains not only their names and addresses but also their buying behaviors, opinions, and desires to participate in communities of like consumers.

e-Business Marketplaces

e-Business marketplaces (B2B) differ from e-commerce storefronts (B2C) in several ways. The most obvious is that the former is concerned with the transaction of products and services between businesses while the latter is between consumers and various types of e-storefront. Also, the focus of the business relationship is quite different. By definition, e-commerce is concerned with consumer-type buying where the shopper searches electronically from storefront to storefront, often ignoring previous allegiances to store branding. In contrast, B2B marketplaces resemble traditional business purchasing: it is often a long-term, symbiotic, and relationship-based activity where collaboration between stakeholders directed at gain-sharing is critical.

There are several alternative categories of B2B [18]. An *MRO hub* is a public Internet-based product and service aggregator (e.g., W. W. Grainger and McMaster-Carr) that provides nonproduction products and services for a specific vertical or general industry marketplace. All catalogs are hosted on a common hub that businesses connect into. Given their horizontal nature, MRO hubs tend to use "horizontal" third party logistics for delivery. A second type of B2B is *yield managers*. This type of B2B focuses on the spot procurement of manufacturing inventories and tends to be more vertical in nature than MRO hubs. The purpose of a yield manager is to insulate buyers and sellers from production variations by allowing them to scale their operating resources upwards or downwards at short notice by participating in the spot market.

A third type of B2B is a *trading exchange*. This type of B2B focuses on creating spot markets for commodity-type products and services within specific industry verticals. While an exchange maintains buyer–seller relationships, buyers and sellers rarely seek direct relationships. In fact, in many exchanges, buyers and sellers may not even know each others' identities. Exchanges serve a yield-management role, because they allow purchasing managers to smooth out the peaks and valley in demand and supply by "playing the spot market." Exchanges can be divided into three separate types. The selection of a type depends on the overall strategy of the business and how it wants to compete in the marketplace. Exchanges can be described as follows:

- *Independent Trading Exchanges (ITXs).* ITXs can be defined as a many-to-many marketplace composed of buyers and sellers networked through an independent intermediary (Figure 3.8). ITXs are primarily used for managing spot buys, disposing of excess and obsolete inventory, and procuring noncritical goods and services. ITXs operate in industry marketplaces that are highly fragmented and have special user needs, time-sensitive products, geographical limitations, volatile market conditions, and nonstandardized manufacturing or channel delivery processes. ITXs can essentially be divided into two types. The first, *independent vertical exchanges,* addresses industry specific issues and provides industry-specific applications, services, and expertise without significant investment from existing industry players. Examples include CheMatch and ChemConnect in chemicals and plastics, e2open and Partminer in high-tech electronics, RetailExchange and Redtagbiz in retail, and Enermetrix and Altra Energy Technologies in utilities. The second type of ITX is termed *independent horizontal exchanges* and focuses on facilitating procurement economies, products, and services to support business processes that are common across multiple industries.

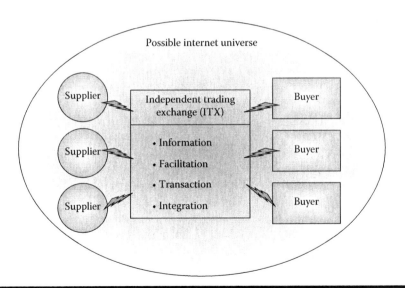

Figure 3.8   Independent trading exchange model.

ITXs are operated by a neutral third party that utilize strong industry and domain expertise to manage relationships and vertical-specific processes. Their business plan is remarkably simple. ITXs offer a neutral site where purchasers and suppliers can buy and sell goods and services. In turn, the ITX collects user fees or transaction commissions for their Web development, promotion, and maintenance efforts. ITXs provide four major levels of functionality: *information* in the form of specialized industry directories, product databases and catalogs, discussion forms and billboards, and professional development; *Facilitation* in their ability to match the specific needs of buyers with the capabilities of suppliers, typically through an auction or chat room; *Transaction* in the form of taking title to the goods and corresponding responsibility for accounts payable and receivable, pricing, terms management, and shipping and order status information; and, *Integration* functionality permitting trading exchange services to fit into a larger supply chain and application integration strategy.

■ *Private Trading Exchanges (PTXs).* A serious problem with an ITX is that it provided only simple buy-and-sell capabilities. Often what many companies really wanted from their B2B exchange was not only ease of doing business but also one-to-one collaborative capabilities with network partners, total visibility throughout the supply chain, seamless integration of applications, and tight security. The answer was the creation of a private exchange. In this model an enterprise and its preferred suppliers would be linked into a closed e-marketplace community with a single point of contact, coordination, and control. Often this type of e-marketplace is driven by a large market dominant company that seeks to facilitate transactions and cut costs while also cementing the loyalties of their own customers and suppliers. Figure 3.9 provides an illustration of the PTX concept.

The decision to construct a PTX is based on several criteria. To begin with, firms with proprietary product/service offerings often feel that a PTX would allow them to use e-business tools while avoiding ITX comparison shoppers as well as protect the product's unique value and brand. Second, companies possessing special process capabilities in areas such as customization or flexible manufacturing capabilities are excellent candidates for a PTX. Third, a PTX is a logical choice for companies with dominant market position or have little to benefit from the aggregation capabilities of an ITX or Consortium. Fourth, a PTX provides enhanced privacy and security regarding pricing and volumes. Fifth, a PTX facilitates linkages across business systems, such as ERP. Sixth, a PTX offers firms the ability to move beyond mere transaction management and build network collaborative capabilities in such areas as inventory management and planning, product design, production planning and scheduling, and logistics. Finally, because PTXs help build among channel partners a sense of collaboration and trust, they can extend a greater level of competitive advantage than can participants in an open exchange.

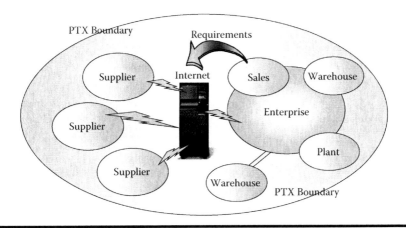

Figure 3.9   Private trading exchange model.

■ *Consortia Trading Exchanges (CTX).* A CTX can be defined as a *some-to-many* network consisting of a few powerful companies and their trading partners organized into a consortium. Historically, CTXs have been formed by very large corporations in highly competitive industries such as automotive, utilities, airlines, high-tech, and chemicals. The goal of a CTX is simple: to combine purchasing power and supply chains in an effort to facilitate the exchange of a wide range of common products and services through the use of Web-based tools, such as aggregation and auction, between vertically organized suppliers and a few large companies. The CTX model can be seen in Figure 3.10.

Functionally, CTXs utilize elements of both private and independent exchanges. A CTX offers control over membership, security, and, most importantly, the ability to build and maintain collaborative capabilities. On the other had, CTXs also enable opportunities for individual members to access each other's trading partners. In addition, consortia e-markets may allow other companies to join the exchange, providing the CTX with an expanding number of members. CTXs, however, can have drawbacks. To begin with, the ability to weld a number of competing superenterprises and their supply chains into a single consortium is fraught with difficulties. Competitive pressures, technology selection, software selection and integration issues, and possible antitrust interference from the government all militate against successful CTX formation. Finally, the establishment of a CTX is an expensive affair and, regardless of the software vendors selected, must often be heavily customized.

The final type of B2B model is a *catalog hub.* The goal of this B2B model is to streamline the systematic sourcing of production inventories within specific vertical industries by putting industry-specific catalogs online, and creating a large

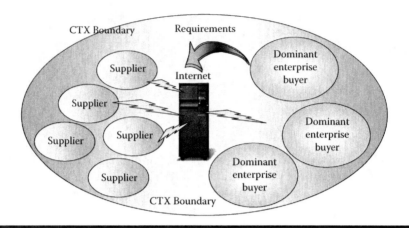

Figure 3.10   Consortium trading exchange model.

universe of supplier catalogs within the vertical. A catalog hub seeks to automate the systematic sourcing process, and create value for buyers by lowering transaction costs. These catalog hubs can be buyer-focused or seller-focused, depending upon who they create more value for. Examples include PlasticsNet.com, Chemdex, and SciQuest. Catalog hubs need to work closely with distributors, especially on specialized fulfillment and logistics requirements for each vertical.

e-Collaboration Marketplaces

The power of Internet marketplaces to increase demand visibility, operational efficiencies, and customer segmentation while simultaneously decreasing costs, replenishment time, and geographical barriers has dramatically changed the nature of SCM. But, while all types of businesses have been able to utilize integrative information technologies to gain significant benefits, it is evident that Internet-enabled collaborative marketplaces are actually in their infancy rather than mature tools for the kind of value chain management they will be capable of in the near future. To date, the attempts companies have made to leverage e-marketplace technologies have been largely focused on improving business with their immediate suppliers. As illustrated in Figure 3.11, these one-to-one relationships consist of linear hand-offs of goods and information from a company to the immediate supply partner, such as supplier-manufacturer or distributor-retailer. The chained trading dyads relationships model, while facilitated by the power of the Internet, nevertheless, is incapable of responding to the value chain needs of groups or tiers of business relationships because their interfaces limit interoperability and their rigid architectures inhibit the synchronization of data from multiple sources.

Activating *Phase 4* e-business requires leveraging the real-time collaboration between supply chain partners possible made possible by the capabilities of the

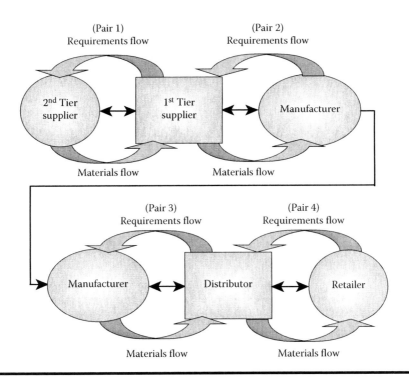

Figure 3.11 Supply chain trading dyads.

Internet. Termed *collaborative commerce* (e-collaboration), the concept seeks to extend the enabling power of business relationships beyond transaction management to true collaboration, visibility, and connectivity across a network of channel allies. As illustrated in Figure 3.12, these tools can be divided into three regions. The first, *Basic B2B Commerce*, consists of application tools that provide marketing information and transaction functions via the Web. While enabling companies to tap into the power of Internet-based commerce, these applications tools provide firms with tactical competitive advantage.

The second region, SCM, seeks to develop the collaborative aspects of integrative information technologies to better manage networked customers and inventories. Briefly, applications at this level can be described as follows:

■ *Collaborative Channel Management.* These tools focus on two e-market objectives: ability to move beyond a marketing strategy that focuses purely on customer segmentation to one that can provide appropriate levels of services to customers with different value and to construct Internet empowered logistics systems that link supplier selection and transportation visibility with company customer service functions.
■ *Collaborative Inventory Management.* The ability to provide inventory visibility beyond chained network dyads is critical to removing the impediments

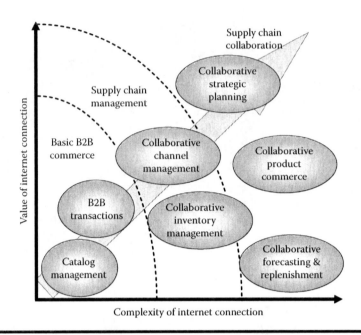

Figure 3.12   Regions of the B2B Collaborative Marketplace.

blocking effective channel inventory management. As changes occur either upstream or downstream in the supply chain, dynamic collaboration will enable network partners to react immediately to ensure customer service while guarding against excess inventories.

The final and most sophisticated of the e-commerce regions is *Supply Chain Collaboration*. Applications in this area seek to leverage the full value of real-time collaborative solutions. These tools can be described as:

- *Collaborative Forecasting and Replenishment.* Broadcasting demand requirements in real-time across the channel network is one of the fundamental objectives of B2B e-marketplaces. Collaborative demand planning enables companies to escape from the restrictions of chains of paired channel relationships that conceal real network requirements and capability-to-promise.
- *Collaborative Product Commerce (CPC).* The needs of product outsourcing, shrinking time-to-market product development cycles, and requirements for increasingly agile manufacturing functions have necessitated that product design and engineering utilize a collaborative approach. CPC is defined as the deployment of cross-channel teams of developers and engineers who are responsible for parts of the overall design. Utilizing Internet tools to provide for information sharing and transactions, the goal is to collectively manage product content, sourcing, and

communications between product OEMs, suppliers, and customers to eliminate redundancy, costs, and time from the product development process.

▪ *Collaborative Strategic Planning.* Perhaps the prime objective of c-commerce is the establishment of collaborative e-marketplaces not only to redesign business and support processes that cross-company boundaries, but also provide for a new vision of the strategic role of the supply value chain. Companies engaged in a c-commerce marketplace will be able to harness Internet technologies to create virtual corporations brought about by any-to-any connections of value-added processes from anywhere in the supply chain system capable of creating immense repositories of competitive advantage.

The *value web* produced by B2B collaborative marketplaces represents a dramatic departure from current views of the supply chain. The goal is to deconstruct the chains of network pairs and reassemble the disconnected amalgam of network business partners into webs of customer-focused suppliers, manufacturers, and distributors. e-Collaborative marketplaces provide any-to-any connections that can drive procurement webs, manufacturing webs, and even linked business strategies. Systems must be robust enough to service a single trading partner and agile enough to evoke wormholes in the fabric of the possible supply universe in the search of any-to-any virtual supply sources capable of linking and unlinking resources in support of critical business processes.

The capacity to access possible resource wormholes and manage and optimize business process webs requires increased technical capabilities. The ability to process transactions and information are only the basics of what the Internet can offer to supply chains. Ultimately, the objective is for companies to share their planning systems and core competencies directly wherever they are on the globe. Among the new technology enablers can be found the following

▪ New XML-based solutions that will allow transaction documents to be quickly generated while enabling sophisticated business rules to be built and modified faster and operated in real-time.

▪ Business intelligence tools capable of supporting, extracting, and validating data in and out of a multiple, heterogeneous systems.

▪ Electronic catalogs with multimedia elements capable of synchronized product update and dynamic market-based pricing.

▪ Operational and supply chain applications capable of enabling coming real-time collaborative planning, forecasting, and replenishment systems.

▪ Technology application providers offering integration/collaboration application services that enable trading partners to limit investment in developing, maintaining, and supporting complex environments required to work with numerous companies across various networks.

Today's e-Business Marketplaces—Summary

As the first decade of the twenty-first century draws to a close, almost all of the hype and drama that characterized the e-business revolution at the turn of the millennium has dissolved away into history. The supposed head-on collision between what had been called the "old economy" of the twentieth century and the "new economy" of the Internet Age never really occurred, as businesses learned to adapt and then exploit the Web-based technologies. The dark specter of being "Amazoned" or "Enroned," of experiencing market share being pirated by an Internet business located somewhere on the other side of the earth, no longer keeps executives up at night. Instead of being a disruptive force, the tremendous advancements in inexpensive Internet technologies has today rendered e-business just business. In fact, the e-business debate now seems old hat as avant-garde companies seek to make a profit by marketing and selling on social networks such as Facebook.

Still, there can be little doubt that e-business is not only here to stay, but that it now underlies the very foundations of business in today's global economy. The benefits are undeniable:

- *Increased Market Supply and Demand Visibility.* Such visibility enables more customer choice, potentially better fit of products to buyers, and a larger market for sellers.
- *Price Benefits from Increased Competition.* Auctions and e-markets can be used to increase price competition and lead to dramatically lower procurement costs for buyers.
- *Increased Operational Efficiencies.* These can be achieved through improved procurement, order processing, and selling processes. Efficiencies can also include faster order cycle times.
- *Improved Partner and Customer Segmentation.* e-Market platforms can be used to transform customer segmentation and provide appropriate levels of services to customers with different value.
- *Improved Supply Chain Collaboration.* e-Market platforms will enable buyers and sellers to work together collaboratively for product design, planning, introduction, marketing campaigns, and life cycle management programs.
- *Synchronized Supply Chains.* Visibility into operating information across the value chain allows companies to drive efficiencies across the entire value chain. These include increased inventory turnover, fast new product introductions, lower WIP inventories, and others.
- *Efficient Payment Transfer.* e-Business greatly facilitates the collection of payment. Often, especially in retail sales, the payment for the goods/services occurs at the moment of purchase either through credit card, P-card, or the use of a third party such as PayPal. Immediate payment can greatly increase the financial "float" of companies like Dell, who can use the funds generated

from customers for investment way before the thirty- to ninety-day window before they have to pay their suppliers.

■ *Impact on Cost.* e-Business can have an enormous impact on an enterprise's infrastructure, inventory, facilities, and transportation. e-Business models that utilize extensive backward and forward integration will find their technology infrastructures and call-center services dramatically increased, while experiencing decreases in inventory through improved supply channel cooperation, reduction in facilities costs by centralizing or outsourcing operations, and increases in direct or partner-based transportation. However, transportation can be almost reduced to nothing for digital businesses, such as NetFlix, Best Buy, and Microsoft, who can offer product delivery directly through Internet downloads.

To continuously leverage the power resident in today's integrative information technologies, however, companies must be careful to close the gaps that prevent full collaborative commerce from being activated. According to Lummus, Melnyk, Vokurka, Burns, and Sandor [19], the following six gaps need to be bridged.

■ *Gap 1: Lack of Alignment and Strategic Visibility.* This gaps includes tighter alignment of operations, logistics, and supply management, better awareness on the part of senior management of the role of SCM, and closer linkages between supply chain design and the overall business plan.
■ *Gap 2: Lack of Supply Chain Models Including Risk Management, Optimization, and Cost Impacts.* This gap includes technology toolsets providing greater insight into supply chain risk and its management.
■ *Gap 3: Inadequate Process Orientation, Including Measurements, Information, and Integration.* This gap requires development of the supply chain as the interrelationship of various processes and how touch points and linkages between functions can be broadened and shared metrics developed.
■ *Gap 4: Insufficient Trust and Relationship-Building Skills.* This gap addresses the need for strong collaboration, trust, and having the "right" relationships in place.
■ *Gap 5: Lack of ongoing frameworks for supply chain architecture and structure.* This gap includes both the dynamic nature of how supply chains are structures and the level of integration and coordination between supply chain partners.
■ *Gap 6: Insufficient management talent and leadership.* This gap requires the development of competency models to better identify and prepare individuals for key supply chain roles.

Impact of e-Business on the Supply Chain

While there are different e-business models, collectively they have had a dramatic effect on the operations, designs, and service factors of conventional SCM. The ability to sell to a global marketplace through the ubiquitous presence of the Internet has placed several unique demands on the supply chain. According to Chopra and Meindl [20], the impact of e-business on SCM can be divided into two spheres: the first associated with several elements oriented around cost/operations management and the second around sales/ service performance. Providing detail answers to these two groupings of factors is essential to the effective pursuit of a profitable e-business initiative.

Customer Service–Driven Elements

The ability to enter and follow up on orders through the Internet is probably the most radical facet of e-business. Among the critical elements of e-sales impacting the supply chain can be found:

1. *Product variety.* An e-commerce site enables customers to choose from a much wider array of products than is possible with a 'bricks and mortar' store. Pursuing a "virtual inventory" model means that the supply chains that actually service the customer will be complex, closely federated by information technologies and contractual relationships, and capable of seamless delivery to the customer.
2. *Product planning.* The ability of e-businesses to utilize networking technologies means that customer demand can be broadcast continuously through the supply chain, providing visibility to requirements all the way back to the manufacturer. This capability will enable supply chains to plan using the customer demand-pull or, at a minimum, more accurate forecasts leading to a much closer match of channel supply and demand.
3. *Shortened Time to Market.* Because an e-business company can introduce new products to the marketplace much faster than a conventional business, the pressure on the supply chain to acquire and distribute them is much greater. Supply chain entities will have to cultivate close relationships with manufacturers and supplying intermediaries so that promotion, pricing, advertising, documentation, forecasting, and other components can be swiftly executed and relayed down the supply chain.
4. *Flexible Pricing, Promotions, and Product Offerings.* Changes to price, promotions, and the product portfolio by the e-business firm must be matched by a mechanism that rapidly communicates these changes to the supply chain. Such changes will quickly have a bullwhip effect on channel inventories if poorly communicated to channel distributors and manufacturers.

Supply Chain Operations–Driven Elements

A decision to move to an e-business channel format will have a significant impact on the structure, objectives, and capabilities of supply chain operations. Among the critical elements resulting from an e-sales format can be found the following:

■ *Customer Response Time.* The ability of customers to quickly browse online catalogs and generate orders through shopping cart technologies places a significant burden of expectation on the supply chain. Unless the product can be directly downloaded (such as computer software), customers might have to wait days to physically receive their purchase. Mechanisms for delivery will need to be devised that are commensurate with the shipping lead times demanded by the customer as well as offered by the competition.

- *Inventory.* One of the critical elements of e-business mentioned above is the ability of firms to offer an extensive portfolio of products. For companies that select to inventory products, an e-business strategy enables them to aggregate inventories at warehouses strategically positioned in key geographical areas. Because demand, regardless of its source, is channeled to these aggregation warehouses, e-businesses can stock significantly less inventory than traditional businesses that must disperse duplicate lot sizes of inventories over multilevel channels of physical locations. For "virtual inventory" e-businesses, those organizations must heavily depend on the strength and robustness of their supply chain stocking and delivery partners. Since product variety is one of an e-business's strategic attributes, stockout of even low-demand, high-variability products risks loss of integrity of site branding.

- *Facilities.* For those e-businesses that choose to stock all or a portion of their product offerings, selling through the Internet imposes several requirements. To begin with, the cost of the facilities (including operating expenses) must be carefully calculated to guarantee the promises of product availability and delivery lead time claimed in the Web site prospectus and experienced over time. While it is true that the business can eliminate the cost of operating an existing network of warehouses in favor of a centralized aggregation warehouse, the complexity of serving an entire customer base, often with very small order lot sizes, will impose a significant burden on existing operational procedures, staff, and information technologies to ensure effective execution. In addition, an e-business strategy may force the organization to deepen forward/backward integration of facilities functions currently performed by other channel companies.

- *Transportation.* For nondownloadable product offerings, e-businesses will have to dramatically improve the quality, capacity, and delivery reliability of transportation both to stocking points and to the marketplace. The failure of e-businesses to actually delivery products to their Web customers in a timely manner was one of the most serious problems that plagued businesses in the early years of the dot.com revolution. Without delivery consistency even the most sophisticated Web site and business proposition will be abandoned for e-businesses that literally can "deliver."

- *Information Technology.* Effective e-business requires the close networking of all members of a supply chain. The Internet can serve as a conduit where critical information such as forecasts, the customer demand-pull, and visibility to disruptions in product and/or transportation movement can be used to foster collaboration and communication. While the cost of SCM systems and networking integration efforts can be steep, businesses must weigh the initial costs versus the ability to execute their business propositions on a continuous basis.

■ *Returns.* Since many e-businesses see themselves as virtual storefronts, they will have to engineer convenient methods for customers to return unwanted products or to engage repair or warrantee service from some node in the supply chain.

Summary and Transition

In their vision of the supply chain of the future, IBM has identified the following three core characteristics [21]:

1. To being with, today's integrative technologies have enabled the supply chain to be *instrumented*. This means that supply chain information will increasingly be generated by sensors, RFID tags, meters, actuators, GPS, and other devices and systems. In terms of visibility, supply chains will be able to "see" more events as they actually occur. Dashboards on devices perhaps not yet invented will display the real-time status of plans, commitments, sources of supply, pipeline inventories, and consumer requirements currently locked away in today's EBS.

2. Integrative technologies will enable supply chains to be more *interconnected*. This means that tomorrow's smarter supply chains will take advantage of unprecedented levels of interaction not only with customers, suppliers, and business systems in general, but also with physical objects that flowing through the supply chain. Besides creating a more holistic view of the supply chain, this extensive interconnectivity will also facilitate collaboration on a massive scale. Worldwide networks of supply chains will be able to plan and make decisions based on real-time data dashboards without clumsy batch simulation or reporting.

3. Integrative technologies will enable supply chains to be more *intelligent*. This means that business systems will possess the increased capability to assist executives in evaluating trade-offs, enabling them to examine myriad constraints and alternatives, and then to simulate in real-time various courses of action before decisions are made. These technology-enabled "smart" supply chains will also be capable of learning and making some decisions by themselves, without human involvement. For example, a system might reconfigure supply chain networks automatically based on user-defined rules and thresholds when disruptions occur anywhere in the supply channel. This intelligence will enable supply chains to move past sense-and-respond to predict-and-act modes of network planning and execution.

Realizing such a vision of highly integrated, information-generating supply chains requires the existence of enterprise business software suites composed of modular architectures that can be configured to meet any particular industry or specific business process requirement. The goal of the system architecture is to facilitate, either through existing application suites or through software partners, the creation of configurable, scalable, highly flexible information business systems that provide enterprises with the capability to respond effectively with value solutions and collaborative relationships at all points in the supply channel network. Supply channel connectivity and networking requires a single version of data encompassing demand, logistics, demand-capability alignment, production and processes, delivery, and supplier intelligence across those firms comprising the entire end-to-end supply network. Generating a single view of the supply chain requires information technologies that enable collection, processing, access, and manipulation of complex views of data necessary for determining optimal supply chain design and execution configurations.

The technology tools capable of providing integrated, networked suites of supply chain applications can be separated into three spheres. The first sphere is composed of the core business functions necessary to run the business. Today's ERP and SCM enterprise system enable companies to access highly componentized, modifiable business models easily configured to meet the challenges of supply chain operational efficiency, channel visibility, and marketplace change. In the second sphere can be found applications like CRM, APS, and Internet B2B that provide connectivity, information sharing, event tracking, exception management, and dynamic optimization to reduce lead times, wastes, and increase supply chain agility. Included are technologies that assist in the execution of regulatory compliance, security, and oversight. This area includes shipment tracking, channel disruption management, supply chain improvement, and regulatory functions available in such toolsets as supplier collaboration portals, 3PL, and transportation systems. The final sphere, automated data collection, consists of toolsets such as EDI and RFID that will not only allow transsupply chain tracking of a product's positioning and other related physical attributes, but more importantly, it will enable a company to define future products, services, markets, and competitive advantage. In this environment the information, and not the product/service, will constitute the key value for tomorrow's enterprise.

Finally, it would be difficult to think of today's information technologies without the enabling power of the Internet. By utilizing Web-based applications available to anyone, anywhere at a very low threshold of cost and effort, the connectivity necessary to drive true SCM integration became possible for the first time. The Web-enabled business could utilize four distinct, yet fully compatible regions of e-business to manage their supply chains. Through I-Marketing, companies could leverage the ubiquitous use of the Web to transcend the limitations of tradition marketing to reach out to any customers, anytime, anywhere. Through e-commerce, companies could perform transactions and permit interactions between themselves and the customer over the Internet. Through e-business, companies could apply the tools of B2C commerce to the B2B environment by leveraging catalog hubs and exchanges. Finally, through e-collaboration, entire supply chains could now be able to move beyond the linear supply chain to a model fostering the integration, synchronization, and collaboration of supply network partner strategies and competencies.

Integrative information technologies have enabled the basic concept of SCM to evolve into three possible approaches coalescing around cost leadership, operations performance, and customer-driven organizations. In the next chapter, these three approaches are discussed in detail.

Notes

1. These five principles can be found in Ross, David Frederick, *Distribution Planning and Control: Managing in the Era of Supply Chain Management.* (Norwell MA: Kluwer Academic Press, 2004), pp. 752–754.
2. There are many descriptions of what constitutes an effective system. One of the most important is Wight, Oliver W., *Manufacturing Resource Planning: MRPII ñ Unlocking Americaís Productivity Potential.* Essex Junction, VT: Oliver Wight Limited Publications, Inc., 1982, pp. 100–106. See also Bowersox, Donald J. and Closs, David J., *Logistical Management: The Integrated Supply Chain Process.* New York: The McGraw-Hill Companies, Inc, 1996, pp. 190–193.
3. For an interesting summary of early software systems see Davenport, Thomas H., *Mission Critical: Realizing the Promise of Enterprise Systems.* (Boston, Massachusetts: Harvard Business School Press, 2000, pp.1–54.
4. Davenport, p. 23.
5. Forrester Research, "The Value of a Comprehensive Integration Solution", (Cambridge, MA: Forrester Research, March 11, 2009).
6. McWhirter, Douglas, "Middleware: Directing Enterprise Traffic," *Customer Relationship Management,* 5, 8, 2001, 32.
7. For a detailed review of BPM see "BPM Technology Taxonomy: A Guided Tour to the Application of BPM," SAP/Accenture White Paper (March 2009).
8. These four points can be found in Haag, Stephen, Paige Baltzan, and Amy Philips, *Business Driven Technology.* (New York: McGraw-Hill-Irwin, 2006), pp. 120-121.
9. "Top 10 Supply Chain Management Vendors ñ 2009: Profiles of the Leading SCM Vendors," *Business Software,* Accessed July 5, 2009 at http://www.business-software. com/erp-reports/supply-chain-management.php.
10. "2008 ERP Report, Part 4," *Panorama Consulting Group.* (Denver CO: 2008), p. 3.
11. "2008 ERP Report, Part 1," *Panorama Consulting Group.* (Denver CO: 2008), p. 2.
12. Jutras, Cindy, "ERP in Manufacturing 2009: Extending Beyond Traditional Boundaries," (Abredeen Group. June 2009).
13. "ERP Software for Manufacturers: Product Directory," *SearchManufacuringERP .com* (2009), accessed June 8, 2009 http://viewer.bitpipe.com/viewer/viewDocument. do?accessId= 10643984.

14. An excellent analysis of the state of the best-of-breed software market can be found in Murphy, Jean V., "ERP or Best-of-Breed? The Question is the Same, But the Answer May Not Be," *Global Logistics and Supply Chain Management Strategies*, 12, 12, 2008, pp. 34–37.
15. Chaffey, Dave, *E-Business and E-Commerce Management.* (New York: Prentice Hall, 2007), p. 573.
16. This section references the e-business breakdown found in Haag, *et al.*, pp. 310–319.
17. For much greater detail see Hoque, Faisal, *e-Enterprise: Business Models, Architecture, and Components."* (New York: Cambridge University Press, 2000), pp. 58–87.
18. For a full treatment of B2B models see Kaplan, Steven and Mohanbir Sawhney, "B2B E-Commerce Hubs: Towards a Taxonomy of Business Models," accessed July 11, 2090 at http://faculty.chicagobooth.edu/steven.kaplan/research/taxonomy.pdf
19. Lummus, Rhonda, Steven A. Melnyk, Robert J. Vokurka, Laird Burns, and Joe Sander, "Getting Ready for Tomorrow's Supply Chain," *Supply Chain Management Review*, Vol. 11, No. 6, (September, 2007), 48–55.
20. Chopra, Sunil and Meindl, Peter, *Supply Chain Management: Strategy, Planning, and Operation*, 4th ed. (Boston, MA: Prentice Hall, 2010), pp. 86–99.
21. Moffat, Robert W., "The Smarter Supply Chain of the Future: Global Chief Supply Chain Officer Study," *IBM White Paper* (Rochester, NY: IBM, 2009), pp. 34–54.

Chapter 4

Technology-Driven Supply Chain Evolution: Building Lean, Adaptive, Demand-Driven Supply Networks

Technology is the driver of supply chain management (SCM). Technology tools enable companies to automate supply chain functions to remove redundancies and costs, generate information that assists supply chain managers plan, execute, and evaluate, and network the capabilities, skills, and experience of people thereby empowering business teams to cut across functional and enterprise barriers and interweave common and specialized knowledge to solve supply chain problems. Technology also provides supply chains with a common application business system framework. Today's enterprise business systems (EBS) permit companies to not only effectively run internal processes, but also to synchronize demand and supply up and down the channel network and link their systems with the core and secondary EBS functions of supply chain partners. Such an approach enables entire supply chains to compete as if they were a single, seamless entity focused on channel cost reduction and superior customer service.

Since its inception, the linkage of SCM and integrative information technologies has enabled SCM to evolve to meet the changing needs of the marketplace.

Originally conceived as an approach to more effectively execute logistics functions, SCM today describes both a business philosophy and a practical operations toolkit. Despite the codification it has undergone in what has become an enormous body of literature, SCM continues to morph in response to changes in business environment and the constant breakthroughs occurring in information technologies.

Activating technology-enabled SCM requires the adoption of practical approaches that utilize the basic tenets of SCM to fashion new strategies and best practices that will provide fresh insights into today's supply chain issues and enable a pathway revealing tomorrow's opportunities. Leveraging the evolving nature of SCM involves finding solutions to the following six challenges [1]:

1. Creating new avenues for supply chains to be more intimate with the customer by understanding what constitutes *individual* customer value and providing an unsurpassed buying experience
2. Ensuring the long-term profitability of the supply chain and each network partner by removing barriers to higher levels of supply chain efficiency and effectiveness
3. Increasing the level of collaboration and access to channel partner competencies by sharing data, intelligence, and more closely aligning organizations
4. Architecting supply chains consisting of productive assets that are agile and flexible enough to respond to any marketplace challenge
5. Increasing the velocity of inventory and delivery network processes while continually reducing costs
6. Enabling greater levels of connectivity to more closely interweave the knowledge and skills of cross-channel teams

While there are many variants, supply chain strategists have narrowed the response to these challenges to three distinct approaches. The first has coalesced around *cost management* and encompasses a set of principles and tools designed to increase supply chain productivity and profitability by ruthlessly reducing wastes found anywhere in the channel network. This model is known as *lean SCM*. The second approach, *operations performance,* is focused on ensuring supply chain execution functions are as agile as possible in the face of demand variability. This model is concerned with supply flexibility and is known as *adaptive SCM*. The final approach, *customer-centered*, is concerned with the continuous development of supply chain capabilities and resources to provide total value to the customer. This model is known as *Demand-Driven Supply Network Management.*

This chapter explores the content of what it means to be lean, adaptive, and demand-driven. The key focus is on exploring how integrative technologies can provide organizations and their supply chains with access to demand and supply signals that assist them to reduce the latency that grows in a supply channel from the point a business, environmental, transactional, or out-of-bounds metric occurs to the point that it reaches the final downstream node in the channel network. The speed of receipt, processing, simulation, decision-making, and communication of event-resolution and then the ability of businesses and their supply channels to react to optimize resources and reduce the threat of disruption to channel flows is at the core of today's SCM.

The Lean Supply Chain

While SCM has been defined as a powerful technology-enabled business strategy capable of engendering radically new sources of competitive value based on the real-time convergence of networks of suppliers and customers into collaborative supply chain systems, perhaps the most practical approach to SCM is to consider it as a mechanism for the continuous reduction of cost and process improvement across the supply chain. Regardless of the technical elegance and strategic capabilities of SCM, what goes on in a supply chain must pass the simple litmus test applied to any process: *Do the functions being performed add value to the customers, the companies involved, and the supply chain in general?* In other words, do the inventories, policies and procedures, supply and delivery processes, and operating cultures of a supply channel provide rich sources of continuous value or do they produce too little value and too much cost. The concept of lean SCM directly centers on this concern with the continuous reduction of waste and the need for continuous process improvement not only inside the organization, but also with customers and suppliers found anywhere in the supply chain.

Lean SCM provides the know-how enabling companies to tackle the difficult task of creating the strategies, developing the cross-channel plans, and optimizing the supply network's collective capabilities that will enable them to reach superior levels of performance. In fact, the need to leverage lean SCM to shrink costs and drive improvement has never been greater. In the recessive economic climate of 2008–2009 companies have identified the pursuit of lean principles as their foremost objective. According to an Aberdeen Group survey (Table 4.1), over 79% of the respondents felt that operational costs were their main concern while pursuing lean to grow revenues gathered only 26%. Suggested solutions were identified as extending lean principles across both the organization and the entire supply chain, emphasizing business capabilities that facilitated the continuous flow of information and product throughput the supply chain, and deploying lean technology enablers to aggregate, update, and distribute optimized plans while promoting visibility and marketplace intelligence.

Anatomy of Lean Principles

The story of the foundation of the concept and process management toolsets constituting Lean is well known to all professionals in the field of production and

Table 4.1   2009 Pressures for Lean Implementation

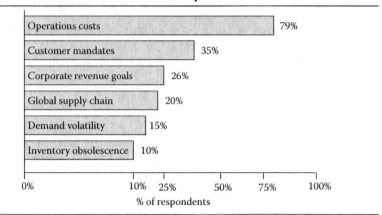

Source: From Nari Viswanathan and Matthew Littlefield, *Aberdeen Group Whitepaper*, p. 4, April 2009.

distribution management. Springing out of the teachings on quality management of W. Edwards Deming and the assembly-line practices of the Ford Motor Company, the concept was adopted by Japanese manufacturers and nurtured to maturity in the 1980s into what has been called the *Toyota Production System* (TPS). Over the decades, lean has been enhanced by the addition of the concepts and techniques to be found in total quality management (TQM), Six Sigma, and theory of constraints (TOC). Today, lean has emerged as a *philosophy* preaching the total elimination of all wastes and the optimization of all resources, a *toolbox* of techniques for process improvement, and a *system* through which companies and their business partners can deliver continuous improvement and customer satisfaction.

Perhaps the best summary of lean can be found in the following five principles detailed in Womack and Jones's ground-breaking book, *Lean Thinking* [2]:

1. *Produce Value.* The value a company offers must be determined from the perspective of the customer—whether that value is lowest cost, best delivery performance, highest quality, an engaging buying experience, or a unique solution to a product/service requirement.
2. *Optimize the Value Stream.* Production and fulfillment processes must be closely mapped to expose any barriers to optimizing value to the customer from concept to product launch, from raw materials to finished product, from order to delivery.
3. *Convert the Process to Flow.* Once waste is eliminated, the goal is to replace "batch and queue" thinking and related performance measurements with a mindset dedicated to enabling the continuous flow of goods and services through the channel pipeline.

4. *Activate the Demand-Pull.* Espousing the process flow mindset means producing and delivering exactly what the customer wants just when the customer wants it; this means migrating from demand driven by forecasts to demand driven by the customer "pulling" products to meet their individual wants and needs.

5. *Perfection of All Products, Processes, and Services.* With the first four principles in place, supply chains can focus their attention on continuously improving efficiency, cost, cycle times, and quality.

These principles provide companies with a practical road map to apply lean concepts and tools to generate truly "customer-focused organizations" capable of delivering increased levels of customer value. Some of the benefits delivered by lean include the following:

- Reduced lead times
- Improved delivery performance
- Increased sales revenue
- Lower operating costs and increased profits
- Improved customer satisfaction and supplier relations
- Increased inventory turns and a drastic reduction in inventory
- Better employee morale and increased employee retention
- Improved quality
- Creation of additional working capital for new products reduced physical space requirements

The process of realizing the potential of lean can be a daunting task for many organizations as it requires companies to rethink fundamental assumptions about processes and performance. Although lean thinking is absolutely critical for the survival of every company, it has become evident that true realization of lean lies in extending its principles and tools to the supply chains to which they belong.

Defining Lean SCM

While many companies can point to the benefits of a successful internal lean initiative, it can be argued that for lean thinking to drive the level of increased value and continuous improvement it promises, companies *must* look beyond the narrow boundaries of their own enterprises and encompass the supply chains in which they are intertwined. In fact, regardless of the impact of the initial benefits on a local lean project, the advantages will rapidly diminish in value as they encounter the entropy and friction caused by interaction with the wasteful processes of channel partners.

Table 4.2   Comparing Lean and Lean SCM

Lean Production	Lean SCM
Reduction of processing wastes and removal of non-value-added activities	Reduction of processing wastes, lead times, and costs everywhere in the supply chain
Optimization of shop floor resources, setup reduction and removal of unnecessary movement	Optimization of product movement and cost reduction across all supply channel processes
Utilization of structured toolsets such as the Poka-yoke, Six Sigma, and the 5 "Ss" to improve enterprise processes	Utilization of the kit of structured lean toolsets to improve processes at each node in the supply chain as a collective correlative
Reduction of component and finished goods inventories through Kanban, continuous flow, and demand-pull	Reduction of total channel inventories through demand-pull driven by end-to-end channel visibility

Source: Karl B. Manrodt, Jeff Abott, and Kate Vitasek, "Understanding the Lean Supply Chain: Beginning the Journey," *APICS White Paper* (November 2005), p. 6.

What makes a supply chain lean and how does it differ from "Lean production?" To begin with, as illustrated in Table 4.2, a lean supply chain seeks to reduce wastes found anywhere in the supply network, standardize processes across traditional, vertical organizations, and optimize core resources. Lean supply chains seek to create customer-winning value at the lowest cost through the real-time synchronization of product/service demand with the optimum channel supplier. Achieving such an objective requires supply chains to be both *responsive* (capable of meeting changes in customer needs for requirements such as alternative delivery quantities and transport modes) as well as *flexible* (adapting assets, pursuing outsourcing, and deploying dynamic pricing and promotions). Finally, lean supply chains are dedicated to the continuous improvement of people and processes throughout the extended supply chain.

All of these elements can be combined to construct a useful definition of lean SCM. For example, Manrodt, Abott, and Vitasek define it as

a set of organizations directly linked by upstream and downstream flows of products, services, finances, and information that collaboratively work to reduce cost and waste by efficiently and effectively pulling what is needed to meet the needs of the individual customer. [3]

In a similar vein, Poirier, Bauer, and Houser consider the lean supply chain

> a collaborating set of businesses linked by the end-to-end flows of products and service, information and knowledge, and finances, resulting in total enterprise optimization gained through total elimination of waste and increased revenues gained through greater customer satisfaction. [4]

This text offers a simple definition—*a supply and delivery network of firms capable of supplying the right product at the right cost at the right time to the customer with as little waste as possible.*

Six Components of Lean Supply Chains

While the above definitions provide a concise foundation, lean SCM can be better understood when it is deconstructed into six competencies as illustrated in Figure 4.1.

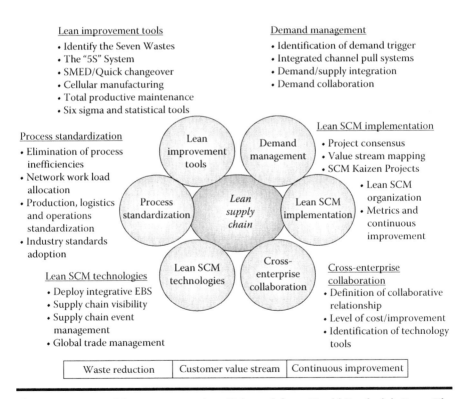

Lean improvement tools
- Identify the Seven Wastes
- The "5S" System
- SMED/Quick changeover
- Cellular manufacturing
- Total productive maintenance
- Six sigma and statistical tools

Demand management
- Identification of demand trigger
- Integrated channel pull systems
- Demand/supply integration
- Demand collaboration

Process standardization
- Elimination of process inefficiencies
- Network work load allocation
- Production, logistics and operations standardization
- Industry standards adoption

Lean SCM implementation
- Project consensus
- Value stream mapping
- SCM Kaizen Projects
- Lean SCM organization
- Metrics and continuous improvement

Lean SCM technologies
- Deploy integrative EBS
- Supply chain visibility
- Supply chain event management
- Global trade management

Cross-enterprise collaboration
- Definition of collaborative relationship
- Level of cost/improvement
- Identification of technology tools

Lean improvement tools	Demand management	Lean SCM implementation
Process standardization	Lean supply chain	Lean SCM implementation
Lean SCM technologies	Cross-enterprise collaboration	

| Waste reduction | Customer value stream | Continuous improvement |

Figure 4.1 Lean SCM competencies. (Adapted from David Frederick Ross, *The Intimate Supply Chain*, CRC Press, Boca Raton, FL, 2008, p. 116.)

Lean Improvement Tools

Lean SCM is about the reduction of waste found anywhere in the supply chain and the continuous pursuit of process improvement. Waste can occur not only in physical elements such as excess inventory and inefficient processes, but also in time, motion, and even digital waste. According to the TPS, there are potentially "seven wastes" found in any process: *overproduction, transportation, unnecessary inventory, inappropriate processing, waiting, excess motion*, and *defects*. Lean advocates have argued that an eighth waste exists: *underutilization of employees' minds and ideas* [5]. An understanding of what contributes waste to a supply chain is the starting point for the development of improvement initiates, or what is called a process *Kaizen* (or "event").

Lean advocates have identified a toolbox of methods designed to attack wastes anywhere in the company or the supply chain. Among the methods can be found the following:

- "*5S" System of Improvement.* The 5S's are lean concepts derived from the Japanese words: *seiri* (sort), *seiton* (set in order), *seiso* (shine or purity), *seiketsu* (standardize), and *shitsuke* (sustain) [6]. The 5S philosophy focuses on establishing effective work place organization and standardization of work procedures centered on simplifying the work environment, reducing waste and nonvalue activity while improving quality efficiency and safety. The 5S system must be pursued rigorously by each supply chain partner if a lean SCM initiative is to succeed.
- *SMED/Quick Changeover.* While primarily concerned with the reduction of machine set-up times, the concept of quick changeover of processes to reduce operational lead times can be applied to increasing the performance of any process anywhere in the supply chain [7].
- *Process Flow Analysis.* This tool utilizes value-stream mapping to detail all steps in a process and illuminate barriers to process flow. This tool is critical to lean SCM to identify and remove barriers between channel nodes and delivery points to accelerate the flow of goods to the customer.
- *Total Productive Maintenance (TPM).* Lean tool is used to ensure the proper maintenance and availability of all equipment everywhere in the supply chain.
- *Six Sigma and Statistical Methods.* The reduction of process errors to six decimal points and the availability of mechanisms to ensure total quality across the supply chain constitute the scorecard as to the continuing success of a lean SCM effort.

Process Standardization

A key objective of lean SCM is the ability to pursue the continuous elimination of waste by eliminating error through the rationalization and standardization of all internal and external business processes. Standardization enables companies to effectively apply Kaizen methods to any process and track, measure, and demonstrate the effects of the Kaizen initiative. Standardization also enables identification of all inhibitors of flow, such as batch and queue processing, unnecessary transportation, and product storage. Industry standards should be used whenever possible, and supply channel partners should participate in confirming the standards to be used by the entire supply chain. Standards should not be limited solely to products and processes, but should be expanded to determine how information is shared across the supply chain.

As the supply chain becomes more complex and outsourcing expands, lean efficiencies increasingly are coming to depend on collaborative efforts aimed at cross-channel value enhancement and waste reduction. Reducing supply chain wastes by eliminating batch-and-queue thinking and activating "pull-systems" enables the entire supply chain to reduce costs to the customer. By uncovering channel redundancies, supply chain partners can collaborate to standardize critical processes and shift work to the most efficient node in the supply network. By improving the extended value stream, all channel partners will be strengthened and their individual contribution to the supply network and the customer will be dramatically enhanced.

Lean SCM Technologies

Once a lean SCM initiative begins, the internal and extended value stream mapping processes should precede the acquisition of an EBS solution. An effective mapping exercise will ensure the internal and supply chain practices and business process flows are well understood and can be matched with the capabilities of either an MRP or SCM core solution and the necessary secondary Web-based applications such as customer relationship management (CRM), advanced planning systems (APS), and global trade management (GTM) (Figure 4.2). The configured solution should provide a linked platform that would facilitate the capability of supply chain partners to easily integrate their business systems to form a channel network. The key is the creation of real-time, channelwide visibility to demand and supply, unplanned channel events, and common performance measurements.

EBS functionality could assist in rolling out a lean program to the supply chain by doing the following:

- *Operations.* Resources can be freed up by streamlining processes while reducing costs through grater automation, standardization, and consolidation.
- *Networking Sourcing.* Business-to-business (B2B) applications can network supplier management, sourcing, contract management, and operational procurement processes, thereby eliminating the serial hand-off involved in request for quotation (RFQ) and open PO management.
- *Optimizing Sales and Marketing.* SaaS, CRM, and business-to-consumer (B2C) functionality can assist in aligning and integrating account planning, forecasting, price and margin management, available-to-promise (ATP) and

SAP lean processes are supported by a variety of solutions including these:

- **SAP BusinessObjects™ enterprise performance management solutions** help companies capitalize on the value of their corporate data by providing organizational alignment, visibility, and greater confidence.
- The **SAP ERP application** offers comprehensive financial management, talent management, and support for end-to-end operational process control for companies across a broad range of industries, including the most complex multinationals.
- The **SAP Supplier Relationship Management application** automates, simplifies, and accelerates procure-to-pay processes for goods and services.
- The **SAP Customer Relationship Management application** helps companies use their customer information for optimal customer interactions in sales, marketing, and service.
- The **SAP Supply Chain Management application** helps transform a linear, sequential supply chain into a responsive supply network.
- The **SAP Manufacturing solution** provides a powerful set of software that helps manufacturers integrate their operations with the rest of the enterprise.
- The **SAP Product Lifecycle Management application** provides 360-degree support for all product-related processes – from the first product idea through manufacturing and service.

Figure 4.2 SAP's lean application suite. (This list is featured in Chakib Bouhdary, Jeff Flammer, and David Strothmann, "Enabling the Lean Enterprise: A Three-Tiered Approach to Improving Your Operations," *SAP White Paper*, p.8, 2009.)

capable-to-promise (CTP) visibility, quotes, order management, and financial settlement. The goal is to remove redundancies and open visibility to the sales and customer management process.

■ *Connect Collaborative Demand and Supply Planning.* Internet tools provide the ability to link channel partner plans with network events through integrated and end-to-end demand and supply planning.

■ *Optimize Inventory Assets.* Integrative information tools enable channel members to continuously share order, inventory-to-sales, and shipment data channel partners. Network nodes would use real-time analyses of these data to evaluate total orders in the supply chain, balance inventory-to-sales levels in the channel, manage service fill rates, and measure the performance of individual businesses and the supply chain in general.

Cross-Enterprise Collaboration

The breadth and complexity of a lean SCM undertaking normally mirrors the intensity of supply chain collaboration existing between channel partners. According to Treacy and Dobrin [8], the best way to determine the extent and depth of channel collaboration is to divide it into two spheres of ascending intensity. The first sphere is concerned with *technical* collaboration. Relationships here can range from manual connectivity all the way through the use of electronic data interchange (EDI) and Internet tools providing visibility to data across supply networks, server-to-server links, and, finally, process management applications that enable true real-time channel information and transaction synchronization. The second sphere is concerned with *business* collaboration. On the low end, forms of collaborative practices are minimal and traditional competitive values dominate. From this level the degree of collaboration intensifies beginning with increasing communication to facilitate joint operations, to coordination where companies in the supply chain use the competencies of network partners, to cooperation where channel partners work together as if they were a single company.

In a similar vein, Prahalad and Ramaswamy [9] feel that the content of collaborative relationships exists on several levels. As illustrated in Figure 4.3, it is comparatively easy to adapt this model to possible lean SCM projects. In the first level, *internal focus*, can be found companies who intermittently engage in collaborative relationships to reduce local costs and time or enhance customer satisfaction but normally will not pursue a lean/Kaizen outside the four walls of their company. In addition in such firms technology is not perceived as an enhancer to competitive advantage. Companies at level two, *transactional/informational collaboration,* understand that channel technology enablers can dramatically shrink costs, improve cycle times, and enhance customer satisfaction by linking interchannel processes. Businesses at this level can launch lean/Kaizen projects that seek to reduce wastes in channel areas such as transportation, warehousing, and inventory.

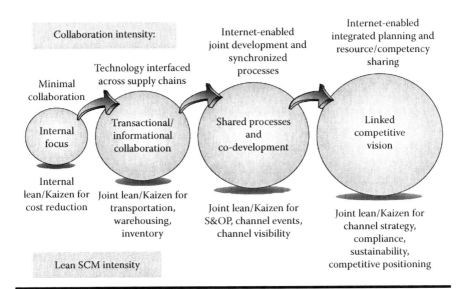

Figure 4.3 Collaborative levels. (Adapted from C. K. Prahalad and Venkatram Ramaswamy, "The Collaboration Continuum," *Optimize Magazine* **(November 2001), pp. 31–39.)**

In level three, *shared processes and codevelopment*, channel network partners seek to engage technology to more deeply integrate individual resources and competencies in order to synchronize competitive capabilities. Participating firms can initiate lean/Kaizen activities that will improve competitiveness by improving demand management, sales and operations plans (S&OP), and channel visibility. Finally, in the fourth level, *linked competitive vision,* collaborating partners will utilize lean/Kaizen projects to move into new dimensions of joint strategy and marketplace development, compliance and transparency, performance and risk management, sustainability, and shared resources. On this level, companies will leverage highly integrated, complex technologies that provide for common applications and infrastructures, unified information access, and rapid knowledge creation and deployment.

Lean SCM Implementation

Companies frequently launch a lean initiative thinking they are now operating in a lean environment. Unfortunately, while a lean/Kaizen can be crisply focused and enforced *within* an organization, it will only produce short-term, localized results if it is not expanded to encompass the entire supply chain. But while results will be significant, a cross-channel lean project is extremely more difficult to design and implement than an internal one. The dynamics present complex challenges and involve consensus on a plan whose participants reside on multiple levels consisting of independent companies, each pursuing individual and often contradictory objectives. Launching a truly effective lean project requires comprehensive cross-enterprise collaboration if it is to succeed. Achieving this objective requires an approach that can be broken down into three layers as seen in Figure 4.4.

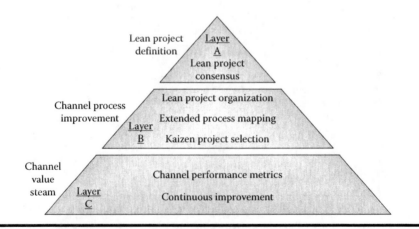

Figure 4.4   Lean cross-enterprise implementation layers.

In the first layer consensus needs to be reached among internal and external channel constituents concerning objectives, goals, and the nature of the lean SCM project organization. The overall objective can be defined as ensuring the entire supply chain achieves optimum business benefit from the lean project on a global scale through direct and indirect management of all facets of the project life cycle. Among the supporting objectives are defining the strategy standards, interface processes/protocols, cost control, prioritized benefits, and ground rules and guidelines for cross-enterprise project integration. Critical to this stage is *extended value-stream mapping.* This step will illuminate the actual process flows within the supply chain and pinpointing and quantifying areas for improvement, communication and collaboration between channel members will actually determine the level of ongoing success. Once this process is completed, agreement on a set of Kaizen tools needs to be reached. This step should detail what lean toolsets are to be used, conducting a Kaizen event, application of new technologies, lean accounting techniques, problem-solving, statistical methods charting performance, and team building.

Effective management of the lean initiative requires an effective cross-enterprise project organization. A lean SCM project organization can germinate from several sources. Poirier, Bauer, and Hauser [10] discuss the establishment of a *partnering diagnostic laboratory* (PDL) to assist supply chain participants to effectively establish consensus on improvements to existing and potential cross-channel processes. A PDL consists of the channel parties involved and a knowledgeable facilitator who can lead a lean/Kaizen encompassing a single or mixture of product development, technical, transactional, procurement, or logistics lean efforts. In a similar fashion Dolcemascolo [11] advocates the formation of *supplier associations*. A supplier association is a network of supply chain partners organized to share the information and expertise necessary to sustain a lean system across the entire supply channel. The key functions of the association are the establishment of a top management team, utilization of committees targeted at specific supply chain Kaizen objectives, performance of quality audits, and consulting and advisory assistance.

The third and final layer of a lean SCM project is establishment of the metrics and performance standards to guide ongoing Kaizen efforts and a mechanism for the identification of continuous improvement. While absolutely critical, consensus about performance are very difficult o achieve. Performance targets pursued by one channel business seldom directly match the objectives occurring elsewhere in the channel. However, the standard metrics employed (such as channel total cost, customer service, inventory turns, purchase days-to-cash) should gauge the performance of supply chain activities necessary to promote profitable growth while seeking to optimize channel functions to promote total supply chain performance. In the end the metrics should enable executives to determine the overall performance and health of their own organizations and that of the supply chain in general and illuminate new areas for continuous improvement.

Demand Management

Of the five principles of a lean thinking determined by Womack and Jones, the two most important revolve around the customer and the management of demand. The first principle states that the only value added to a supply chain is value that is understood and valued by the customer. When designing a lean SCM initiative it is essential to ensure it directly provides what customers want and need: otherwise it is waste and should be eliminated. The resulting customer-side benefits can be found improved on-time delivery performance, reduced order-to-shipment turnaround time, improved product quality, reduction in returns, potential lower pricing due to lower operational costs, and ability of customers to increase order volumes.

Principle four of lean thinking states that *push* supply system should be replaced by the *demand-pull*. The mechanism of the demand-pull is set in motion at the point of sale, and then the requirement is pulled from upstream delivery channel partners, node-by-node, all the way back to the producer. Enabling the demand-pull requires lean differentiating processes that are

1. *Visible and Transparent.* Lean supply chains need to be immediately alerted when out-of-bounds events occur and will cause costs and wastes to appear anywhere along the supply channel continuum.
2. *Demand-Driven.* Demand in the supply channel is no longer driven by forecasts but only occurs when interchannel or customer demand is present.
3. *Instrumented.* Information about marketplace conditions needs to be driven by information technology applications that can alert not just the next node but the entire channel simultaneously of impending demand changes.
4. *Integrated.* Lean supply chains are organized around demand management process teams whose activities are aligned around value streams focused on creating optimum value for the customer.

According to Manrodt, Abott, and Vitasek [12], there are four technology tools to assist companies seeking to enable the flow of the demand signal from the end customer all the way back to the original product supplier.

The first, point-of-sale (POS) systems, enables real-time communication of the customer demand signal at the moment the sales occurs. In turn, the companies can instantaneously respond to the demand-pull by then starting the replenishment process through the supply network based on actual usage. While only a small fraction of today's supply chains have operating POS environments, many are looking to the second technology, *demand collaboration.* This tool is focused on utilizing forms of electronic communication, such as the Internet, to pass real-time sales data to supply chain partners. In addition, supply chains can also share S&OP that provide a window into expected demand and supply planning. Finally, companies can share the mechanics of their *inventory management practices* located in their business system with channel partners. For example, quoted customer delivery lead times and manufacturing process *takt* times will provide a detailed knowledge of the pulse of demand and supply cycles from each channel member. The communication of other information, such as costing, engineering change management, new product offerings, and individual company target inventories, will assist supply chains to realize individual and SCM lean objectives.

The goal of demand-pull management is to provide advanced warning of demand occurring anywhere in the supply network and enable supply channel nodes to assess their collective ability to respond effectively. This lean SCM competency directly attacks waste created by the bullwhip effect caused when inventory is added to the original demand signal as it moves upstream in the supply channel. Effective demand management removes wastes by reducing channel uncertainty, variability in fulfillment processes, supply lead time, and linking channel partners together in networks that enable them to respond as a team to the demands emerging in the marketplace.

The Impact of Lean SCM

By relentlessly pursuing the six lean competencies portrayed in Figure 4.1 companies can realize the three success factors shown at the bottom of the figure. To begin with, application of the suite of lean tools enables cross-channel businesses to effectively pursue *waste reduction* at all supply chain levels. The design of lean SCM initiatives focused around customer wants and needs and deployment of technology tools enabling demand to be pulled through the supply chain will keep all supply network nodes focused squarely on how to continuously build and sustain a *stream of value* to the customer. And finally, well-designed lean supply chain implementation projects and the capability to broaden and enrich cross-channel communications concerning quality, change management, collaboration opportunities, and joint metrics will enable supply chains to maintain a focus on *continuous improvement* as they drive toward network competitiveness and profitability.

Enterprises that embed lean techniques into core business processes and institutionalized them into their own and their supply chain cultures can expect significant benefits. According to an Aberdeen survey on lean SCM practices [13], all organizations undertaking a lean SCM initiative exceeded expectations relating to reductions in inventory, assets, and product development costs, while increasing product quality, channel flexibility, and customer service. Achieving such results requires companies to grow a culture of lean/Kaizen both inside and outside organizational boundaries, devise methods to quantify existing and new improvement projects, gain deeper commitment and collaboration of channel partners, and apply information technologies that enable synchronization and visibility of channel network demand planning and demand-pull mechanisms, optimization of channel inventory, and the most appropriate usage of outsourcing for warehousing, transportation and logistics.

Achieving such results requires communication and resolve among all members of the supply chain. Best-in-class supply chains not only embed lean into their business processes, but also enhance these efforts with the deployment of supporting technology solutions, integrate Kaizen programs for continuous improvement into their cultures, and design key metrics that drive the operational excellence. Lean supply chains continuously seek to expand on the following steps to success.

1. *Champion a Lean Culture Everywhere in the Supply Chain.* This step is concerned with the elimination of barriers to lean by growing and empowering cross-functional teams with the task of both promoting and applying the lean SCM philosophy. These teams not only live the philosophy but are also imbued with the mission to spread lean throughout the supply chain network.

2. *Deepen Supply Chain Collaboration.* Collaboration ensures not only that suppliers, intermediaries, and customers are included in lean SCM initiatives but that they are expected to be leaders and active participants. Deep collaboration ensures that process improvements are selected that continuously shrink the costs and cycle times of suppliers while increasing customer value.

3. *Implement Supportive Information Technology Solutions.* Technology tools enable channel partners to work collaboratively to ensure synchronization of the highest levels of customer value at the lowest cost with collective channel resources and competencies. Enterprise resource planning (ERP) and SCM systems ensure a cohesive technology infrastructure. Business intelligence tools provide for the centralization of and access to knowledge repositories for the entire supply chain. Networking supply chain planning, execution, and event management applications provide mechanisms for the extended channel demand-pull, access to production and delivery schedules, and visibility to disruptive events.

4. *Lean SCM Performance Measurement.* The performance and value-generating power of lean resides in the ability of organizations to pursue continuous improvement. Although gaining consensus and tracking cross-channel metrics is challenging, without agreed-upon performance measurements the impact of lean project is unknown. Customer service, postponement, lead times, warehouse efficiencies, transportation costs, and average channel inventory, are all candidates for lean supply chain metrics.

5. *Engage in Long-Term Lean Strategies.* While the activation of a lean initiative often produces immediate benefits on the company level, the true value of lean can only be realized by a comprehensive long-term commitment by the entire supply chain. In the end, a lean supply chain is better positioned to redirect resources, undertake innovation, and respond rapidly to capitalize on marketplace opportunities brought about by business environment changes, channel restructuring due to events like mergers and acquisition, and the rise of new competitors.

Adaptive Supply Chains

The objective of lean SCM can be summarized as consisting of two fundamental drivers: the *elimination of waste* found anywhere in the channel network and the *pursuit of continuous improvement* targeted at optimizing the value stream. However, even as companies strive for greater efficiencies, today's marketplace turbulence is also requiring them to strive for greater *flexibility* when responding to a growing number of threats to the viability of their supply chains. Effectively meeting this challenge is driving businesses to move beyond lean to the next stage of SCM: they need to embrace the concept of *adaptive SCM*.

This stage involves replacing less responsive linear supply chains with highly agile networks capable of rapidly adapting to marketplace changes. Adaptive networks seek to be *responsive* (i.e., able to meet changes in customer needs for alternate delivery quantities, transport modes, returns, etc.) and *flexible* (i.e., able to manipulate productive assets, outsource, deploy dynamic pricing, promotions, etc.). In addition they also need to be able to proactively reconfiguring themselves in response to market events, such as introduction of a disruptive product or service, regulatory and environmental policies, financial uncertainty, and massive market restructuring, without compromising on operational efficiencies.

Understanding Supply Chain Risk

In today's business environment the management of risk has been dramatically growing as perhaps the number one reason companies have become more focused on their ability to more flexible and adaptive. Much of it has to do with the acceleration of complexity in all facets of business. As the impact of such sources of supply chain risk as natural disasters, global recession and the tightening of credit markets, strikes and bankruptcies, the price and supply of oil, and port congestion and infrastructure inadequacy dominate the headlines, the heightened exposure of supply networks brought about by a decade-long mandate to decrease inventories and cost everywhere in the supply chain has become painfully visible. Simply put, unplanned events occurring anywhere in an organization's extended supply chain, from process failures to faulty sourcing strategies, can result in catastrophic supply chain disruption.

As illustrated in Figure 4.5, risk in today's supply chain can be located in four areas.

■ *Business Environment.* There can be little doubt that the prime drivers of risk today can be found in globalization and increasing business complexity. Globalization has required businesses to cope with unprecedented levels of coordination among geographic operations incorporating everything from cultural nuances to languages, currencies, and regulatory (tax, import/export) structures to expanding lead times and corresponding increase in costs,

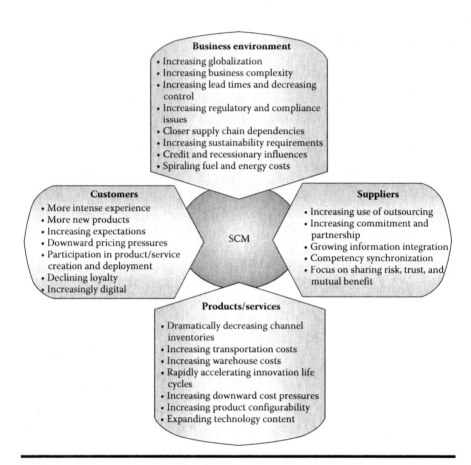

Figure 4.5 Risks to SCM.

uncertainty, and erosion of control. As the second decade of the twenty-first century opens, companies also have to come to grips with ballooning fuel and energy costs, tightening of global credit markets and financial recession, and eenvironment-related tax credits and regulation of carbon emissions.

■ *Customers.* Several factors coalescing around the growing power of the customer have made supply management problematical. To begin with the hypercompetitiveness of today's global markets have increased the volatility of demand, escalated requirements for product customization and variety, shortened product life cycles, and intensified expectations of a personalized buying experience. In addition, the ubiquitous power of the Internet has resulted in declining customer loyalty and continuous downward spiraling of prices. Such marketplace realities have resulted in a swelling of the supplier base, increased dependence of channel partners to maintain branding and promotional marketing, and product management with all of the attendant complications of integrating components into the final product, and managing larger and larger volumes of purchase orders, invoices, and so on.

■ *Suppliers.* While one of the prime targets of lean SCM in the search to continuously reduce costs, improve quality, and shrink cycle times, supplier management has also been a critical source of risk. As companies expand their use of outsourcing strategies, intensify their search for supply partners willing to broaden and deepen commitment and partnership, engineer joint process improvement, and more closely synchronize special competencies enabling whole network ecosystems to better manage risk while sharing benefits, companies' growing dependence on lean supplier management practices pose severe risks to supply chain effectiveness. One of the key mantras of lean—*single sourcing*—while enabling reductions in price and administrative costs, can also increase the vulnerability of supply chains if the supplier can not perform or even goes out of business. Beyond supply continuity disruption and price volatility, supplier consolidation can inadvertently create bottlenecks in the supply chain that can stress resources and suppliers to the breaking point. Even corporate moves, such as an acquisition, can introduce risk with the addition of relatively unknown suppliers and new assets that can traverse the entire supply chain.

Then again, while outsourcing can reduce costs and activate new sources of productive competencies, it also can dramatically expand the possibility of supply chain disruption. Persuading even close partners to engage in investments to change information systems and performance metrics, as well as to cooperate in sharing risks, building trust, and positing mutual benefits, are difficult initiatives to achieve. The result can be an escalation in supply chain complexity and decline in process visibility resulting in possible disruptions due to inefficiencies stemming from elongated transportation distances, governmental restrictions, and other factors driven by the vagaries at the heart of time and space.

■ *Products and Services.* Of all the risks before a supply chain, the twin problems of excess inventory and inventory shortage are perhaps the most damaging. Production lines can be halted, transportation schedules disrupted, and customers left unsatisfied. On the other hand, elongated global supply chains threaten to increase warehousing and transportation costs at a time when the marketplace is pressuring for lower prices. Even the burgeoning density of information content in our products and services, which requires an ever-growing information storage capacity and knowledge management capability, directly contributes to supply channel risk [14].

While there is little doubt that the application of lean to the supply chain has reduced overall costs and diminished the sting of the bullwhip effect, it has also left companies with little margin for error and susceptible to serious disruption arising from even minor disruption. Furthermore, a focus on sole sourcing and outsourcing has increased dependencies on a single supplier. Such dependency reduces the ability of supply chains to respond effectively to change and makes them insensitive and rigid in the face of a global business environment increasingly dominated by volatility in tastes and fashion.

As the probability of risk accelerates, the need for supply chain contingency planning has moved from an optional to a required strategy and a wealth of literature (as well as college courses!) has arisen to address the issue. Possible solutions run the gamut of pragmatic approaches that deploy modeling and systems analysis to provide outcomes that are measurable and analytical rather than subjective [15] to other approaches that divide supply chain risk into three areas (strategic, operational, and event-driven) so that the impact of risk can be assessed from overall enterprise, medium-range business operations, and short-term daily transactional levels. Regardless of the method, the goal is to implement initiatives to enable supply chains to be more resilient and efficient by identifying and profiling risk variables, quantifying risk for business decision-making, and enabling technology solutions so they can adjust their supply chains intelligently to today's changing economic and market conditions.

Advent of the Adaptive Supply Chain

Constructively managing risk requires companies to move beyond managing supply chains based solely on the criteria of throughput and low cost to a strategy that places agility and nimbleness as the central operating attributes. Despite their espousal of SCM principles, most supply chains remain islands of optimization, are linear and sequential, have intelligence only as to the status of their immediate customers and suppliers, and are unable to sense or respond to changes happening in other sectors of the supply chain until it is too late to meaningfully respond. Countering these supply chain deficiencies is driving businesses to move a perspective where success is measured by a supply chain's ability to be strategically and operationally nimble while pursuing efficiency and total process improvement. This mandate involves replacing less responsive linear supply chains with highly agile networks capable of rapidly adapting operations and resources to be more *responsive* and *flexible* as well as *resilient* to withstand the threats posed by marketplace disruptions.

According to a study by SAP [16], adaptive supply chains possess three critical attributes: management of visibility, management of velocity, and management of variability. These attributes are in turn mapped to three key information enablers: quality of information, timeliness of information, and depth of information. *Visibility* management enables businesses to extract relevant demand and supply intelligence from multiple systems across the network and broadcast it in real-time to all connected participants to guide channel actions. Once data is visible, the next step is increasing the *velocity* by which intelligence and assets can be moved through the supply network based on timely information. And finally, as supply chains become more dynamic, they will be better able to manage the *variability* occurring in customer demand and supply chain capabilities as robust information about orders, inventory, plans, delivery, profit margins and other data is available to all channel members.

With these attributes in mind, *adaptive supply chain* can be defined as

> a virtual networked supply chain community possessing the capacity to sense marketplace changes as they occur anywhere in the channel and then to communicate through interactive information sharing and synchronized functions the critical intelligence necessary to enable rapid planning, decision-making, and alternative action execution to leverage risk to secure competitive advantage.

Agile supply chains are inherently dynamic, aggressively change-focused, and growth-oriented. In an agile channel network all participants can sense and transmit intelligence of local shifts in demand and supply concurrently to all channel points as they occur so that the entire ecosystem can respond intelligently to emerging risks. Adaptive supply chains cannot be implemented by just one company: they succeed by networking channel partners through real-time information and collaborative technologies to both reduce cost and drive collective value to the customer.

Adaptive supply chains possess the following operational characteristics.

- *Demand Flexibility.* Agile supply chains are capable of sensing and responding to demand as it actually occurs. Conventional supply chains utilize sophisticated forecasting techniques to predict demand and then drive the results through their supply channels. In contrast, agile supply chains utilize demand-gathering, planning, and execution technologies to capture real-time information that enables them to sense both planned and unplanned demand events as they occur. Such intelligence in turn enables them to rapidly adapt and synchronize *marketing factors*, such as inventory substitution, promotions, pricing, and auctions and exchanges, and *operations factors*, such as network substitution, outsourcing, and logistics, to met new demand patterns, and activate visibility, collaborative, and analytical toolsets that enable every node in the network to keep the right products flowing to the right customers. Finally, demand driven means tapping into CRM and *social networking* technologies to gain an intimate understanding of actual customer wants and needs so that gaps are identified and channel resources effectively allocated or reconfigured.
- *Supply Flexibility.* Adaptive supply chains require companies to link themselves together into networked federations focused on intense collaborative integration between buyers and suppliers. The goal is to erect mechanisms by which suppliers can focus their special competencies on accelerating joint product development, sourcing, order management, and delivery that enables rapid deployment of inventories and transportation capabilities to respond to demand as it actually happens at every point in the network. Suppliers that are "seamlessly"

integrated into the channel system can continuously work on reducing lead times, possess the knowledge to and are capable of quickly responding to product changes, and are committed to transparency and openness.

■ *Delivery Flexibility.* The supply chain delivery process has often been described as a pipeline through which product flows. Unfortunately, most supply chain delivery functions have been designed for cost management rather than agility. The consequence has been a disconnect between customer order management and delivery caused by the inability of many supply chains to effectively respond to disruptive events. Adaptive supply chains can overcome these deficiencies by shrinking cycle times, synchronizing all logistics, transportation, and fulfillment operations, deployment of sense-and-respond technologies enabling rapid change, and utilizing radio frequency identification (RFID)-enabled processes that will pinpoint in real-time information from people and physical objects.

■ *Organizational Flexibility.* At the core of the adaptive supply chain stands organizations agile enough to rapidly alter resources and competencies in response to real-time demand events while concurrently devising alternative, intelligent plans to be implemented by the entire supply network. Perhaps the most important enablers of organizational flexibility are the implementation of adaptive planning and execution software. *Adaptive planning* requires creating logistics plans that ensure lowest total delivered cost and profitability across all channel partners, streamlining operations by coordinating all activities—purchasing, transportation, and inventory—based upon the expected consumption of products, and the coordination of suppliers, logistics services providers, and carriers to optimize financial and operational trade-offs. *Adaptive execution* provides for the implementation of the adaptive plan by monitoring events as they occur, coordinating understanding and assessment of the alternatives, and ensuring rapid joint action for optimal recovery for affected plan elements while minimizing the impact on areas unaffected by the disruption.

Components of Adaptive Supply Chain Management

Adaptive SCM [17] presupposes the activation of close collaborative relationships between channel members, the application of connective technologies that not only provide data concerning demand and supply events, but also merge intelligence, planning, and execution of plans, and a common goal to provide the customer with an exceptional buying experience that welds them to a product and a delivery channel. Regardless of whether they are purely transaction-based with minimal dependency on other channel players or organized into closely integrated partner networks, all agile supply channels must be willing to design, continuously improve, and execute channel processes and customer-facing values.

In contrast to these traditional methods of managing supply and delivery channels, which depend on forecasting and inventory buffers, adaptive networks seek to counter demand variability by abandoning static plans in favor of mechanisms that provide immediate feedback regarding the marketplace and the flexibility to quickly devise alternative plans that can be rapidly communicated to channel partners. As illustrated in Figure 4.6, adaptive networks possess information tools that permit them to *sense* changing conditions in the supply and demand channel when they occur. Illuminating potential or actual problems anywhere in the network requires *technologies* that quickly provide visibility to threats concurrently to all channel partners. Data is crunched to reveal potential troubles and then aggregated to disclose systemic patterns of variation and contributing factors. Awareness of variation and understanding of the level of potential risk is then driven by system *alerts* that provide detailed action messages to all network planners. Exception messaging supplies warnings on both tactical and strategic levels. Strategically, alerts may reveal emerging challenges that endanger the fundamental assumptions of the supply chain. Tactically, alerts may indicate that market conditions have altered the supply chain's ability to execute originally planned courses of action.

Exception messages enable channel planners to assess the degree of required *plan* revision. For example, action to correct a tactical problem, such as an inventory shortage at a network supply node, would result in an emergency effort to relocate stocks

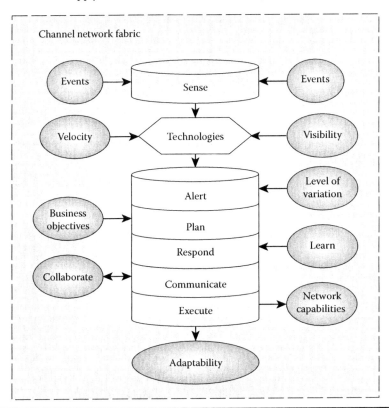

Figure 4.6   Components of adaptive SCM.

from other channel warehouses to the affected location. On the other hand, a major disruption at a critical supplier might require a fundamental revision of the entire supply chain strategy. The goal is to ensure that events occurring anywhere in the supply network are visible to all participants so that effective analysis, ending in changes to the plan, can begin. Actually, the intensity of the resulting *response* is dependent on the information supplied by the live event. The simplest response may be correcting an out-of-balance process. A more complex response would involve replanning a local supply/delivery point in order to put the plan back on track. Finally, a significant disruptive event may require complete reformulation of the entire channel strategy. In all cases, event data should be complete enough so that standard operating procedures and planning optimization tools can rapidly and accurately assess the resulting impact on performance of a fundamental change in plans.

Before a determined response can be executed, it is critical that possible solutions be *communicated* to affected supply network participants. The speed of response by channel partners to plan revision is driven by two factors: the level of existing network *collaboration* and the availability of communications tools. The more integrated and collaborative channel relations the quicker alternative plans can be communicated and consensus reached. The actual message must be delivered via communications enablers that present data and analysis in a format and style that can be easily transposed into local planning and diagnostic dialects. Once solutions have been agreed upon, plan *execution* can follow. Disruptions can be handled by adjusting the standard channel process by applying an alternate solution to the problem area and directing it back on track to the original plan. Incremental, focused event adaptation permits event obstacle resolution without adversely disturbing and destabilizing standard processes occurring at other network points. If the volume or magnitude of adaptive planning becomes too great over time, fundamental change to the overall supple network is probably warranted.

Motorola and adaptive supply chains

"Motorola wireless mobility solutions address some of the most challenging issues for the distribution supply chain: thinning margins, shorter product lifecycles, accurate demand fore-casts, increased customer expectations, and a global economy. With motorola mobility solutions, you can count on real business value by synchronizing the many aspects of your supply chain and providing workers and managers with the tools and real-time information needed to improve business processes and create adaptive supply chain strategies to reduce costs and meet changing consumer demands.

Motorola mobility solutions create a tighter collaboration across the supply chain that results in a more agile, flexible supply chain capable of improving quality, reducing the cost of doing business and improving customer service and satisfaction levels for a real competitive advantage."

Source: "The Adaptive Supply Chain: Increase Supply Chain Visibility and Meet Customer Demands with Mobile Technologies," *Motorola White Paper:* (2008), p. 7. (see http://motorola.com/supplychainmobility).

Advantages of Adaptive Supply Chain Management

Adaptive SCM enables companies to directly address channel risk by providing them with continuous intelligence as to the real-time status of channel demand and supply and the mechanisms to rapidly reconfigure operations anywhere in the supply network. The ability to leverage event management technologies to provide visibility to channel disruptions and activate collaborative decision-making provides businesses with the following critical channel management advantages:

a. *Customer Focus:* enabling rapid respond to changing marketplace conditions so that targeted customer satisfaction and profitability targets are maintained regardless of possible disruptions occurring in the business environment

b. *Event Visibility:* providing visibility to evolving demand and supply changes so that channel networks can rapidly adapt processes without significant channel disruptions

c. *Operations Flexibility:* enabling supply chains to monitor logistics bottlenecks and adjust plans to avoid congestion, treat in-transit inventory as available for inventory planning purposes, and reroute in-transit shipments to avoid congestion or meet changing customer demands

d. *Lean Strategies:* engaging lean concepts and practices to activate flexibility and nimbleness through the reduction of channel wastes, increased efficiencies, and inauguration of continuous improvement initiatives while pursuing supply and delivery optimization

e. *Increased Channel Collaboration:* establishing closer cooperation and information sharing between network partners thereby activating processes for rapid decision-making and problem event resolution as well as the reduction of overall channel costs, acceleration of product design, reduction of time to market, and increase of channel service as a competitive differentiator

f. *Exploit New Revenue Opportunities:* generating flexible supply and delivery networks capable of pursuing emerging marketplace opportunities much more effectively than companies that follow a more rigid channel strategy

Demand-Driven Supply Networks

The implementation of lean, adaptive SCM strategies are squarely focused on optimizing *process value chains* and *value delivery networks* by reducing costs, continuously improving processes, and rendering entire supply chains more flexible and nimble to respond rapidly to changes in the business environment. However, while the lean, adaptive SCM model does indeed enable companies and their supply chains to gain supply efficiencies and flexibility, in today's global environment, they constitute relatively short-term strategies for accelerating the supply pipeline and conserving cash and other assets and can not be considered as constituting a long-term competitive differentiator. Being *supply flexible* is no longer enough: supply chains today need also to be *demand flexible* as well. Being demand flexible means that companies must move beyond a focus purely on operational optimization and restructure their supply chains to be able to sense and proactively respond to actual demand signals arising from the *demand channel* rather than just building to forecast or manipulating productive and inventory assets to counter emerging disruptions in the supply network.

Transitioning from channel management styles oriented around supply optimization to the architecting of networks oriented around customer demand optimization represents a significant expansion of the traditional role, scope, and skill set of the channel manager. In today's world chief supply chain officers (CSO) are not only responsible for supply chain asset utilization, they are also involved in enhancing collaboration between sales and operations everywhere in the channel, closer logistics and transportation coordination to increase customer delivery speed and efficiency, and new forms of technology capable of sensing and communicating demand as it occurs anywhere, anytime in the supply network. Today the head of the supply chain, in most major manufacturers and retailers, is influencing margins, time to market, and customer retention as well as improving and accelerating the flow of operations strategies. Simply, SCM as a science now has expanded to encompass procurement, manufacturing, distribution, and even field service to the point to where it is actually just one big process for customer fulfillment.

As illustrated in Figure 4.7, a DDSN attempts to connect and focus the three fundamental components of SCM—the *demand channel*, the *process value chain*, and the *value delivery network*.

Defining Demand-Driven Supply Networks

The inclusion of the demand channel into the SCM concept has been termed the demand-driven supply network (DDSN). Introduced in 2003 by AMR Research, a DDSN can be defined as "a system of technologies and business processes that sense and respond to real-time demand across a network of customers, suppliers, and employees" [18]. DDSN organizations are more *demand sensing,* capable of more *demand shaping,* and able to execute a more profitable *demand response* than companies that are simply supply-centered. A DDSN possesses the following attributes:

■ *Business Philosophy.* In the past, supply chains were configured to maximize large stocks of inventory to be sold to customers based on forecasted

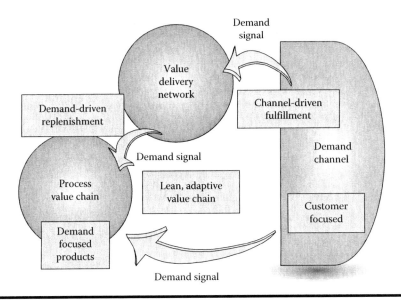

Figure 4.7   Components of a DDSN.

estimates of demand. These supply chains were fairly rigid and monolithic, dependent on linear channel optimization techniques. They were also driven by productivity-centered factory schedules or "push" system allocations, and prone to the bullwhip effect as demand uncertainty cascaded through the supply chain. Today, an expanding global marketplace is requiring organizations to create supply chains that are more flexible, responsive, and synchronized to actual customer demand. A DDSN strategy centers on the configuration of supply chain processes, infrastructure, and information flows that are driven by the demand channel rather than by the constraints of factories and distribution intermediaries located upstream in the supply network.

■ *Technology.* A DDSN requires technologies that enable businesses to closely synchronize demand and supply. DDSN technologies provide real-time demand sensing and visibility to changing marketplace requirements. An effective DDSN should also possess technologies that enable the organization to quickly determine profitable tradeoffs and compromises in response to unplanned events occurring in the demand channel. Finally, technologies should enable planners global access to actionable data, information toolsets permitting rapid and collective decision-making based on "what-if" alternatives, and comprehensive scoring mechanisms that provide for clear alternatives when choosing between competing courses of action.

Visibility to the last mile

The control of inventory through the supply chain has always been exposed at the very last stage – delivery to the retailer. What happens when retailers cut inventory to the bone? They don't have what you want about 8% of the time you step into their store; out of stocks are about double that rate for promotional tems. Most relailers handle that situation in one of three ways: they send someone to scrounge around the stock room just in case; they take your name and promise to call to see if other stores have what you're looking for, or they throw up their hands and shrug.

To solve this problem sterling commerce introduced a new application that replaces the shoulder shrug with a layer of intelligence, visibility, and automation to save the sale. Once sales clerks open the application they gets global visibility into the inventory network as defined by the retailer. That could be visibility into neigh-boring stores, all the stores in a district or region, or inventory in the ware-houses.

At this point retailer decides how to complete the sale, either by having a warehouse deliver thc product directly to the customer's home or place of business or to a nearby store for pickup, or by having a similar store send the product to the customer's home.

"The critical backbone to the solution is the combination of our order management, global inventory visibility and store management operations," says Jim Bengier from Sterling. "We've bundled them together to provide a complete solution."

But in addition to addressing the out-of-stock issue that retailers wrestle with, it completes that last mile, connecting what's going on inside retail stores forward to the customer's home and backwards to the distribution center. "This is the starting point," Bengier says. "But, if a retailer wants to do this, we could extend visibility all the way back to the manufacturer."

Source: "Sterling Commerce Solves Retailers' Out-of-Stock Dilemma," http://www.sterlingcommerce.com/about/news/Storage/20091026Sterling+Commerce+Solves+Retailers+Out-of-Stock+Dilemma.htm

- *Demand.* The fundamental characteristics of a DDSN resides in its ability to rapidly recognize all forms of demand arising from the marketplace. Demand can come from channel forecasts, booked orders, or emerging opportunities. By using monitoring tools like customer analytics and demand-event notification, supply chains can not only understand demand, but even be able to shape it. Knowing what drives the customer purchase and requirements for product features, price points, and buying habits keeps supply chains focused on real-time demand. In the end, being demand-sensitive is more than just filling orders: it is using demand signals to scale processes and resources quickly across the entire supply network.
- *Collaborative Network.* While a DDSN more closely integrates marketing, customer management, and operations inside the organization as to what products and capacities are needed to match the demand-pull, it also acts as a supply chain unifying force. DDSN engage suppliers, delivery

capabilities, and customers in collaborative value-creating relationships. DDSNs require network supply nodes possessed of the ability to rapidly reconfigure internal processes and external channel points of delivery by tapping into the competencies found in outsourced manufacturers, supply partners engaged in sourcing and collaborative product design, and third party logistics providers. The goal of executing "demand-pull systems" is realized by channel supply points capable of rapidly aligning demand with the network's portfolio of resources to service the unique needs and priorities of the customer.

To summarize, AMR Research characterizes a DDSN as "a supply chain driven by the voice of the customer," constructed to serve "the downstream source of demand rather than the upstream supply constraints of factories and distribution systems." A DDSN enables supply and delivery channel nodes to understand and adapt to demand anywhere in the supply chain as it is actually occurring.

Competencies of a DDSN

The demand-driven supply chain has emerged in response to the significant risks associated with not being able to effectively respond to the customer. Companies that expect to thrive in the twenty-first century will be those that understand that it is the customer and not the producer or distributor who determines marketplace direction. It is the customer who has assumed the power to direct the design of product and service content, pricing, transaction management, and information transfer. Rather than simply a marketing category, today's customers expect to be treated as an individual, and requires suppliers to provide them with configurable, solutions-oriented bundles of products, services, and information custom designed to meet their unique wants and needs. Finally, today's customers are simply demanding more control over the buying experience, easy to use order management tools that empower them to design their own solutions, flawless and speedy fulfillment, robust information content, ease of search, ordering, self-service follow-up, and effortless methods for financial settlement.

Effectively responding to today's customer requires businesses and their supply chains to possess the following demand-driven competencies as illustrated in Figure 4.8. A detailed discussion is as follows.

Demand-Driven

In yesterday's "product-centered" business environment, the "voice of the customer" was rarely considered. In that era, the focus was on brand management, mass production, economies of scale and scope, and push replenishment systems. Demand for products was shaped by forecasting, advertising, promotions, and

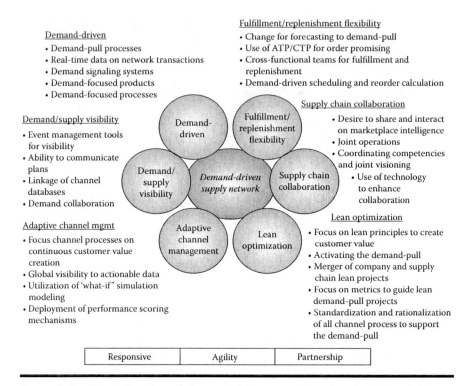

Demand-driven
- Demand-pull processes
- Real-time data on network transactions
- Demand signaling systems
- Demand-focused products
- Demand-focused processes

Fulfillment/replenishment flexibility
- Change for forecasting to demand-pull
- Use of ATP/CTP for order promising
- Cross-functional teams for fulfillment and replenishment
- Demand-driven scheduling and reorder calculation

Demand/supply visibility
- Event management tools for visibility
- Ability to communicate plans
- Linkage of channel databases
- Demand collaboration

Supply chain collaboration
- Desire to share and interact on marketplace intelligence
- Joint operations
- Coordinating competencies and joint visioning
- Use of technology to enhance collaboration

Adaptive channel mgmt
- Focus channel processes on continuous customer value creation
- Global visibility to actionable data
- Utilization of 'what-if" simulation modeling
- Deployment of performance scoring mechanisms

Lean optimization
- Focus on lean principles to create customer value
- Activating the demand-pull
- Merger of company and supply chain lean projects
- Focus on metrics to guide lean demand-pull projects
- Standardization and rationalization of all channel process to support the demand-pull

| Responsive | Agility | Partnership |

Figure 4.8 Competencies of demand-driven supply networks (Adapted from David Frederick Ross, *The Intimate Supply Chain*, CRC Press, Boca Raton, FL, 2008, p. 154.)

pricing—customers could any color automobile they wanted as long as it was black! Requirements for customization, last-minute changes, a unique, individualized experience, and flexibility in ordering, payment, and delivery were subservient to a "one-size-fits-all" mentality.

In contrast, today's demand-driven networks have replaced the factory-driven model of the twentieth century with a customer-centric pull model. Being demand-driven means that organizations are dead-focused on knowing what influences customer purchasing and their requirements for product features, price points, delivery, and buying habits. But putting the customer at the center of the supply chain can only occur when businesses have in place lean, adaptive supply chains and information technologies that enable them to receive timely and accurate demand signals from any point in the channel network. Knowledge of actual demand enables companies to quickly reconfigure their supply chains to facilitate the fast-flow alignment of goods and information that matches customer service need for distinct product/service solutions. The speed and accuracy of response to each channel customer touch point presents channel players with the opportunity to expand the customer experience and deepen value-creating relationships.

Demand-driven networks succeed by creating *value* for their customers, their network partners, and their own companies by enabling the continuous realignment of all channel processes, infrastructure, and information flows to serve the downstream source of demand rather than the upstream constraints of factories and distribution systems. Such an objective is achieved by viewing the supply and delivery channel not simply as a pipeline for the transfer of standardized goods and services, but as an integrated network capable of quickly locating and adapting resources that treat each customer requirement as unique. Supply chains that can quickly react to customer needs, present unique buying experiences, and continuously provide the innovative product/solution mixes will be those who will be able to lock in brand awareness, create an exceptional customer service benchmark, and are recognized as the supplier of choice among peers.

Demand/Supply Visibility

Perhaps the number one priority of a demand-driven network is improving *visibility* to demand as it occurs anywhere in the supply channel. Agile supply networks excel in deploying demand-gathering, planning, and execution technologies that reveal events as they actually occur. This information, in turn, enables them to rapidly reconfigure and synchronize supply and delivery resources through automated exception handling, directed workflows, directed resolution, and overall network management [19]. Performance-wise, companies with highly visible supply chains have customer service levels of at least 96%, reduced inventory levels of 20–30%, and are twice as likely as their competitors to have an on-time delivery rate of 95% or higher [20].

At the heart of supply chain visibility is found technology tools that *connect* supply network partners and focus them on customer value delivery. Connectivity here can refer to a wide range of data transfer processes. In its ideal form, connectivity would consist of a single electronic communications hub linking individual ERP, CRM, S&OP, warehouse management, and transportation systems. These technologies would provide for the merger, harvesting, and analysis of supply chain data intelligence associated with critical data such as demand forecasts, customer and supplier order statuses, shipment status, production schedules, and finished goods levels essential to effective business decision-making. Regardless of the technology, effective connectivity requires the application of the following enablers [21]:

■ *Demand Forecasting.* Although actual demand signals provide critical data driving the demand-pull, effective forecasting is still needed for long-term planning and new product introduction. Today's forecasting tools can draw upon simple models or complex algorithms that attempt to project future demand driven by data from sales, invoice, POS, lost sales, and promotion histories.

■ *Demand Shaping.* This enabler permits supply chain nodes to simulate demand intensity through the use of sales techniques such as channel promotions, bonus and incentives, pricing, and advertising and marketing strategies.

■ *Alert Signal Management.* Once notification of demand or supply abnormalities occurs, the close networking of trading partners provides for the efficient rebalancing of channel resources. Visibility tools include alert-driven signals revealing unplanned product shortages (or excesses), emergency plant shutdowns, process failures, unexpected outlier demand, and evidence of wide variance of actual demand and supply against the plan.

■ *Predictive Analytics and Simulation.* The ability to capture, analyze, and simulate actual supply chain performance is essential to provide visibility to the health of the customer value channel.

As examples, companies such as P&G, Wal-Mart, and Cisco are taking demand and supply visibility way beyond just receiving forecasts from their customers. Utilizing techniques such as shopper loyalty cards and POS data, these companies are tapping into enormous reservoirs revealing what their customers' retime needs are. Wal-Mart, for example, provides P&G access to its huge database of POS information. Other organizations are working closely with key suppliers and customers to develop linked S&OP tools so they can see the pulse of supply and demand in the supply channel. P&G calls this ability to "sense" demand as it happens "joint value creation," says Roddy Martin, senior vice president and research fellow at AMR Research. "The new challenge for the supply chain is in fact getting and translating real demand data" [22].

Enabling the emergence of effective demand/supply visibility management is a complex affair requiring the linkage of several information technologies. At its foundation each company within the supply chain must have a solid IT infrastructure that provides for system integrity and usability. Next, individual companies must possess a comprehensive EBS that provides for data collection, process standardization, and planning and execution applications. Finally, meaningful supply chain visibility can only emerge when the systems of channel partners can be connected together individual irrespective of hardware and software.

SAP, Technology and DDSN

SAP can help companies lay the foundation for DDSN. SAP solutions and services provide the following support:

- *Harmonization* – The SAP NetWeaver platform provides the foundation for harmonizing processes and data. In particular, Netweaver can help companies effectively manage master data across the entire supply and demand network to generate "one version of the truth."

- *Integration and internal collaboration* – SAP software provides the right collaboration and analytic tools to support user productivity, along with predefined analysis configurations and rapid integration to other applications.

- *External collaboration* – A key capability for DDSN is extension of the internal corporate network to suppliers and customers. The network then seamlessly communicates shifts in demand to suppliers who in turn respond with the materials necessary for manufacturing to meet the demand.

The SAP Advanced Planning & Optimization component ties demand planning and supply network optimization to production planning and detailed scheduling, as well as transportation planning. The SAP Inventory Collaboration Hub component helps companies develop responsive replenishment and supplier-managed inventory.

- *Adaptability* – Adaptability is the key to the long-term viability of DDSN. When DDSN functions most effectively, it immediately adapts to changes in demand – all along the supply chain. SAP solutions that foster demand forecasting also help develop an advanced forecasting capability. That leads to load building and eventually to replenishment with order generation, subdaily planning, and dynamic sourcing.

Source: Panley, Mark and Boerner, Stefan, "Demand-Driven Supply Networks: Advancing Supply Chain Management," SAP White Paper (2006), http://www. sap.com/asia/industries/pdf/ met_exec_demand_driven_SCM.pdf

Adaptive Channel Management

Supply chains can not hope to effectively leverage the demand-pull signal without having the agile infrastructures, scalable resources, and speed of information transfer necessary to continuously align material suppliers, contract manufacturers, and logistics providers spread across a global landscape. The key to adaptability in today's complex, multienterprise supply networks is the ability to quickly make tradeoffs and compromises to resolve the many issues ranging from new product introduction to emergency orders that can clog the supply chain. Optimizing supply chain agility requires three critical capabilities:

1. *Global Visibility to Actionable Data.* This competency requires connectivity between all internal and external supply chain systems combined with proactive alerting to enable channel action teams to constantly and continuously capture data as to all forms of changes, from inventory to purchase agreements, occurring in the supply chain. Such intelligence empowers strategists across the channel network to make intelligent, reality-based judgments affecting everything from major channel realignment to disaster and recovery.

2. *Ability to Rapidly and Collaboratively Assess an Array of Possible "what-if" Alternatives.* Today's information technologies infrastructures enable close client interface so that action teams both inside and outside the enterprise can communicate simulation details and potential results so that affected channel partners can participate in identifying optimal courses of action and weighing consequences.

3. *Capability to Deploy a Comprehensive Performance-Scoring Mechanism.* This competency leverages shared metrics and performance scorecards that accurately predicts the impact of possible responses and weighs alternatives against company goals and demand requirements so the best course of action can be implemented that meet company and supply chain goals and profit targets. Analytics assist in helping supply chains know their performance and where bottlenecks are emerging. Metrics enable companies to be agile and flexible by providing a detailed window on how the supply chain is performing real-time and illuminates areas for reconfiguration.

Demand-driven agility enables supply chains to effect timely, meaningful changes to the channel model so they can rapidly respond to changes in demand, supply, and product. Some changes may be long-term, such as moving distribution closer to a customer to optimize shipping time and lower transportation costs. Some may be driven by planning, such as when supply chains can couple their productive and delivery functions with forecasted and daily actual demand triggers. Then again, immediate marketplace changes might enable agile delivery channels to offer substitute products or execute postponement strategies; manufacturing capacities might be shifted to products with rising demand to drive higher volumes for products customers are currently demanding.

Lean Optimization

Being "demand-driven" means that companies are engaged with providing customers with the best products and the optimal delivery system. The demand-driven concept directly applies to the essential principles regarding lean customer responsiveness: providing the *value* actually desired by customers, optimizing the *value stream* for the delivery of each product/service, engineering continuous *flow* to speed response times, enabling the customer to *pull* value from the supply chain, and actualizing an endless search for perfection [23]. Such demand-driven objectives can only be achieved when entire supply ecosystems continuously clear away constraints, collapse processing times, remove redundant operations steps, even eliminate entire supply chain levels when necessary in order to optimize process, product, and delivery capabilities.

By applying lean principles, companies can effectively pursue *waste reduction* at all supply chain levels, leverage supply chain partnerships and technology tools to continuously build and sustain a high-velocity *stream of value* to the customer, and finally, deploy cross-channel metrics for effective quality, change management, and collaboration to maintain a focus on network *continuous improvement*. The companion to lean is the pursuit of quality. This process begins with an assessment of existing conformance to quality targets, delivery performance, and total cost, then proceeds to quality approaches that link lean, Six Sigma, and other toolsets, and concludes with the application of quality targets to the entire supply chain. Lean optimization enables the delivery of benefits to customers in the form of better quality, shorter lead-times, expanding breadth of product lines, simplified pricing and payment, higher inventory turns, lower product costs, and lower transaction costs.

Supply Chain Collaboration

The adoption of the lean, adaptive, demand-driven supply chain model requires companies to transform their supply chains from linear, sequential processes into collaborative networks in which communities of customer-centric companies share knowledge, intelligently adapt to changing market conditions, and proactively respond to more unpredictable business environments. Demand-driven collaboration takes place between a company and its suppliers, contract partners, and customers as illustrated in Figure 4.9. Suppliers and contract manufacturer/service provider collaboration streamlines the flow of replenishment information, materials, components, transactions, visibility to order status, and financial settlement. These functions in turn support customer collaboration centered on the ability to sense demand signals and automatically replenish the customer's requirements on the basis of actual demand.

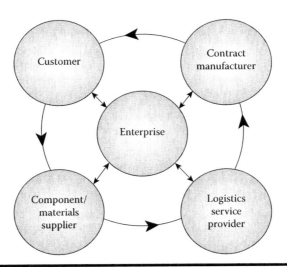

Figure 4.9 Supply network collaboration participants.

Effectively supporting collaborative demand networks are a variety of technology tools. For example, with a customer collaboration strategy, replenishment processes become more responsive and are triggered primarily by actual customer demand information from such sources as POS and electronic product code (EPC). These tools enhance visibility across the entire supply chain and enable channel suppliers to manage and execute marketing events, such as promotions and deal, as well as disruptions and last minute changes. Visibility also enables channel players to assess automatically the impact of out-of-stock information, apply the resulting information for sales forecasting and drive replenishment scheduling and communicate the results to business partners. Information regarding changing patterns in customer consumption can also assist to more efficiently control vendor-managed inventory (VMI) at the customer's site. Finally, supply chains can also leverage collaborative planning, forecasting, and replenishment (CPFR) and S&OP technologies to directly export forecast/demand requirements, production performance, and available inventories directly into suppliers' systems.

Fulfillment/Replenishment Flexibility

The management of fulfillment and replenishment functions has traditionally been marked by extreme variability regardless of the precision of today's computerized tools. The scenario is all too familiar: forecasts are created and loaded into the planning system, materials and production are scheduled, and, based on the outcome, warehousing and shipping resources are calculated to provide optimal customer demand fulfillment at the moment of truth when the customer picks the product from the shelf. In reality, regardless of system sophistication, this ideal rarely materializes. Part of the problem is timeliness and accuracy of the data. Part is depending on forecasts that reflect the *past* buying habits of customers. Then again, part is result of customers simply changing tastes or reacting to the latest "new thing."

Demand-driven networks require companies to abandon statistical/forecast-driven forms of planning for inventories and fulfillment. The ability to quickly modify plans and alter supply execution is a hallmark of world-class supply chain companies. Through the connectivity enabled by POS systems, RFID, event management tools, dynamic S&OP, and collaborate forecasting and planning, DDSN provides the entire supply chain with visibility to demand and product flows and enables supply channel nodes to sense possible problem situations caused by changes in demand or supply capacities. As is illustrated in Figure 4.10, effective DDSN management rests on an information infrastructure that provides a network for the entire supply chain and enables visibility to channel events, exception messaging to alert planners of out-of-bounds situations, and automated decision-making utilizing prebuilt scenarios and alternative courses of action that will enable businesses to dynamically respond to plan (or replan) in real-time.

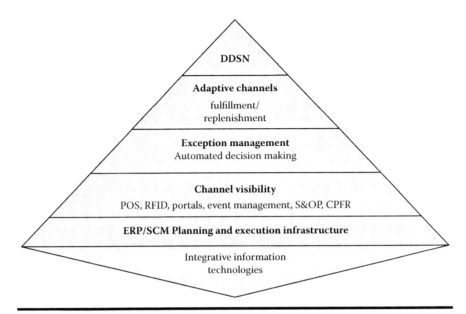

Figure 4.10   Optimized fulfillment/replenishment planning.

A DDSN enables channel fulfillment processes to become driven by the actual pull of the customer order as it its impact cascades down through the supply channel. This capability means that supply partners can now manage total channel demand activity and not just a single order being placed on some channel sales point. The overall goal is not only visibility of the customer order to the supply chain, but also the ability to synchronize demand as it matures with existing channel strategies, service-level agreements, order automation techniques available (EDI, Web, VMI, phone), and product availability propositions. Among the steps to building effective DDSN fulfillment can be found the following:

- Visibility at the point of purchase for in-stock availability and exception messaging for impending stock-out. This should include clear policies regarding product substitutions and shortages before they occur.
- Reconfiguration of customer order processing systems to pull-based systems based on sense and respond.
- Use of POS/scanning tools to identify promotions fulfillment opportunities and drive new orders based on actual demand to improve promotional effectiveness.
- Building cross-functional awareness of sales, marketing, and fulfillment groups so they have a clear understanding of how the demand-pull will drive fulfillment.
- Ensuring total accuracy of order promising and tracking mechanisms such as ATP and CTP.

- Channel order status is visible to customer management, operations, and fulfillment teams through synchronization of orders, advanced shipping notification, and delivery status messages.
- Delivery functions must be capable of supporting demand-pull logistics functions such as cross-docking, drop ship/direct shipment, order switching, and order shipment accuracy.

DDSN replenishment can be defined as the synchronization of supply channel sourcing, manufacturing, and distribution with the demand-pull of the customer. The movement from planning fulfillment based on forecasts to pull-based demand requires the sensing and cascading of the demand signal down through the supply channel and its translation into a replenishment signal to guide the priorities of manufacturing and distribution channel nodes. Among the steps to building effective DDSN fulfillment can be found the following:

- Integration of the demand-pull with channel replenishment (ERP/SCM) business systems so that demand is driven into planning on a timely and accurate manner.
- Replenishment processes capable of sensing impending stock-out conditions and then feeding the data automatically into ERP and APS to make demand-driven replenishment a reality.
- Manufacturing and distribution teams have access to scorecards providing visibility to fulfillment targets.
- Replenishment resources (assets and labor) are designed for flexibility and practice lean improvement principles.
- The status of production and distribution is visible across operations, fulfillment, and customer management teams thereby promoting joint ownership of fulfillment decisions.

Demand-driven supply networks succeed by relentlessly pursuing the above six critical competencies. Effective execution of these competencies enables supply chains to effectively pursue the three critical drivers shown at the bottom of Figure 4.8. To begin with, customer-focused, demand/supply technologies provide the depth of visibility necessary for companies to sense real-time information about demand and supply events that, in turn, enables them to be *responsive* to demand as it occurs at each node along the channel network continuum. Operations excellence and fulfillment/replenishment flexibility promote *agility* by endowing supply chains with nimbleness, simplicity, and speed to rapidly execute adjustments to demand and supply capabilities as demand shifts over time. Finally, collaboration and optimization intensify visibility and agility and help foster effective supply chain *partnership* that can enhance and optimize the channel network's competitiveness and profitability.

Advantages of Demand-Driven Supply Networks

A DDSN seeks to reduce risk and respond directly to the threat of marketplace variability by combining lean toolsets and adaptive processes to synchronize supply and delivery flexibility with the demand-pull cascading through the supply chain. Implementing demand-driven concepts plays a significant role in the success or failure of a form. For example, based on SAP benchmark data [24], supply chains guided by a DDSN philosophy have increased fill rates and reduced stock outs by 3–10%, reduced inventories by 17–15%, improved asset utilization by 10–15%, decreased cash-to-cash cycles by 10–13%, and reduced waste and obsolescence by 35–50%. In addition DDSN leaders have a higher percentage of perfect order fills, more accurate and timely marketing information, faster response to changes in demand, and execute faster, more effective product introductions and phase-outs.

An effective DDSN helps everyone in the supply chain to create greater value for the customer, forge closer relationships with network partners, and reduce costs. For distributors, demand-driven strategies enable companies to escape from the tyranny of the bullwhip effect. Besides reductions in nonprofitable inventories, channel distributors and retailers can also fine-tune stocks so that they can concentrate on stocking exactly what the marketplace wants to buy. Increased agility ensures that delivery nodes can create deeper customer relationships while increasing inventories of full-margins stocks that maximizing revenues and profits. Finally, the demand-driven delivery network is able to drastically reduce transportation costs by minimizing channel transfers caused by stocking imbalances and overstock returns.

For materials suppliers and manufacturers, DDSN strategies assist channel product producers become more agile so they can operate with shorter production runs, develop more flexible response times and planning cycles, and better manage capacities. By converting production functions from a dependence on long-term plans based on forecasts to pull-based models, manufacturers can focus on making what customers actual want instead of just pushing products into the delivery channel that so often results in obsolescence, excess carrying costs, and price markdowns. Finally, demand-driven strategies enable producers to focus on product innovation, create close linkages to customers and suppliers in the development life cycle, and synchronize product introductions with channel marketing efforts that highlight brand differentiation and enhance customer loyalty.

Summary and Transition

Driven by new marketplace challenges and the growing networking capabilities of information technologies, companies in the mid-1990s began to dramatically rethink the concept and practice of managing their supply chains. The driving force of this new concept, SCM, sought to merge operations and organizations across company boundaries and link critical supply chain partner competencies in the search for new avenues of competitive advantage. However, despite the innovative power of the original definition, SCM has proven to be too general of a concept to counter the pressures of globalization, financial markets demanding more effective use of capital, accelerating innovation cycles, and customers demanding to be treated as unique individuals characterizing today's marketplace. Competitive SCM now requires companies to move beyond the standard definitions and embrace new ideas and technologies if it is to provide new sources of market leadership.

The object of this chapter is to detail the evolution of SCM from its logistics foundations into three new channel process models capable of providing supply chains with fresh ideas and management practices to respond successfully to today's increasingly complex global environment. The first model can be said to be concerned with *cost management* and encompasses a set of principles and tools designed to increase supply chain productivity and profitability by ruthlessly reducing wastes found anywhere in the channel network and the establishment of a culture of continuous improvement. This process model is known as *lean SCM.* The second approach, *operations performance,* is focused on ensuring supply chain execution functions are as agile as possible in the face of demand variability. This method is concerned with supply flexibility and is known as *Adaptive Supply Chain Management.* The final approach, *customer-centered*, is concerned with the continuous development of supply chain capabilities and resources to provide total value to the customer. This method is known as *DDSN management.*

As is emphasized throughout the chapter, at the heart of the three SCM process models stands the enabling power of today's integrative information technologies. What makes these advanced forms of SCM so dynamic is that they utilize technology to enable a basket of supply chain competencies including connectivity, visibility, networking collaboration, fast-flow operations execution, and optimization. Common to all of these competencies is the ability to manage demand and supply events as they occur anywhere in the organization or in the supply chain. Events can arise from such sources as global positioning systems (GPS) or RFID signals, POS data, photoelectric cells or other monitoring devices, online orders, or other sources. The key is to shorten the span of latency that begins to build from the moment an event occurs until it the data is received at the furthest downstream node in the channel ecosystem. The continuous shrinking of supply chain information latency is the objective of lean, adaptive, demand-driven networks and the utilization of technologies to automate and informate event information the benchmark separating those supply chains capable of rapid change and restructuring from channel laggers.

In the end, lean, adaptive, demand-driven supply chain create value through five critical capabilities [25]. The first is centered on the ability to "sense" demand in the supply network through joint value creation/demand visibility into markets, segments, and customers. The second capability resides in the availability of agile, flexible supply network components to support demand-driven networks. The third is the ability to leverage innovation drivers such as quality-by-design (QbD), design for manufacture, distribution, and market. The fourth capability strives to leverage S&OP and network design as core processes to deliver profitable and balanced trade-offs across the value network. And finally, the last capability centers on the continuous pursuit of agility to profitably shape demand across the supply network.

As we see in the next chapter, competitive advantage today will go to those supply chains that are not only lean, adaptive, and demand-driven, but that also cultivate intimacy with the customer. Customer intimacy opens up an entirely new region of SCM by seeking to manage the customer's experience with a company and its products. Customer intimacy seeks to build rich relations with customers at every network touch point by delivering information, service, innovative products, and interactions that result in compelling experiences that build loyalty and add value to the network community.

Notes

1. These challenges were identified in David Frederick Ross, *The Intimate Supply Chain: Leveraging the Supply Chain to Manage the Customer Experience.* (Boca Raton, FL: CRC Press, 2008), p. 112.
2. James P. Womack and Jones, Daniel T., *Lean Thinking: Banish Waste and Create Wealth in Your Corporation.* (New York: Simon & Schuster, 1996), pp. 15–26.
3. Karl B. Manrodt, Abott, Jeff, and Vitasek, Kate, "Understanding the Lean Supply Chain: Beginning the Journey," *APICS White Paper,* (November, 2005), p. 7.
4. Poirier, Charles C., Bauer, Michael J., and Houser, William F., *The Wall Street Diet: Making Your Business Lean and Healthy.* (San Francisco, CA: Berrett-Koehler Publishers, Inc., 2006), p. 63.
5. See the excellent summaries of the meaning of Lean waste in Taiichi Ohno, *The Toyota Production System,* (New York: Taylor & Francis, 1988), pp. 1–143; Darren Dolcemascolo, *Improving the Extended Value Stream: Lean for the Entire Supply Chain.* (New York: Productivity Press, 2006), p. 134; and Joel Sutherland, and Bob Bennett, "The Seven Deadly Supply Chain Wastes," *Supply Chain Management Review,* 12, 5, July/August 2008, 38–44.
6. Hiroyuki Hirano, *5S for Operators: 5 Pillars of the Visual Workplace.* (Portland, OR: Productivity Press, 1996).
7. The concept of SMED or *single-minute exchange of dies* was introduced by Shigeo Shingo in his book *A Revolution in Manufacturing: The SMED System.* (Portland, OR: Productivity Press, 1985).
8. Michael Treacy and David Dobrin, "Make Progress in Small Steps." *Optimize Magazine,* December 2001, 53–60.
9. C.K. Prahalad, and Venkatram Ramaswamy, "The Collaboration Continuum," *Optimize Magazine,* November 2001, 31–39.
10. Poirer, *et al.,* pp. 179–184.
11. Dolcemascolo, *Improving the Extended Value Stream,* pp. 169–180.
12. Manrodt, *et al.,* "Understanding the Lean Supply Chain," pp. 10–12.
13. According to the research, best-in-class companies felt that their Lean SCM projects increase product quality by 35%, reduced inventories by 26%, improved customer

service by 25%, and reduced order cycle times by 22%. Maura Buxton, and Cindy Jutras, "The Lean Supply Chain Report," *Aberdeen White Paper,* (September, 2006), p. 12.

14. One group of researchers found that inventory shortages had a deleterious affect on company profitability and financial well-being. It was discovered that stock prices for companies with disruptive parts shortages under-performed their benchmarks by an average of 25%. In addition they experienced a median decrease in operating income of 31%, a decrease in sales of 1.2 percent, and an increase in costs of 1.7 %. Vinod R. Singhal and Kevin Hendricks, "The Effect of Supply Chain Glitches on Shareholder Wealth," *Journal of Operations Management.* (Vol. 21), 501–522.

15. An excellent example of a pragmatic approach called "SMART" can be found in Dirk De Waart, "Getting Smart About Risk Management," *Supply Chain Management Review,* Vol. 10, No. 8. (November 2006), 27–33.

16. "Adaptive Supply Chain Networks," *SAP AG White Paper,* (2002), pp. 8–9. This paper can be found at accessed August 10, 2009 http://www.sap.com/usa/solutions/business-suite/scm/ pdf/50056466.pdf.

17. This section has been summarized from Ross, *Intimate Supply Chain,* pp. 143–146.

18. Lora Cecere, Debra Hofman, Roddy Martin, and Laura Preslan, "The Handbook for Becoming Demand Driven," *AMR Research White Paper,* (July 2005), p. 1.

19. See the comments in Tony J. Ross, Mary C. Holcomb, and Brian Fugate, "Connectivity: Enabling Visibility in the Adaptive Supply Chain." *Capgemini White Paper,* (2005), pp. 1–6, accessed August 20, 2009 http://www.capgemini.com/resources/thought_leadership/ connectivity_enabling_visibility_in_the_adaptive_supply_chain/.

20. This analysis can be found in "The Supply Chain Visibility Roadmap," *Aberdeen Group White Paper* (Aberdeen Group, November 2006), p.2.

21. These points have been summarized from Ross, *Intimate,* p. 156.

22. Tam Harbert, "Why the Leaders Love Value Chain Management." *Supply Chain Management Review,* Vol. 13, No. 8 (November, 2009), pp. 12–17.

23. These five principles can be found in James P. Womack, and Daniel T. Jones, *Lean Solutions: How Companies and Customers Can Create Wealth Together.* (New York: Free Press, 2005), p. 2.

24. Mark Panley, and Stefan Boerner, "Demand-Driven Supply Networks: Advancing Supply Chain Management," SAP White Paper (2006), p. 1, Accessed August 25, 2009, http://www.sap.com/asia/industries/pdf/met_exec_demand_driven_SCM.pdf.

25. Referenced from AMR Research "Value Chain Transformation" (June 29, 2009), reproduced in Harbert, p. 17.

Chapter 5

Customer and Service Management: Utilizing CRM to Drive Value to the Customer

The effective management of the customer has become the dominant objective for firms seeking to sustain leadership in their markets and industries. With their expectations set by world class companies and interactive technologies, today's customers are demanding to be treated as unique individuals and requiring their supply chains to consistently provide high-quality, configurable combinations of products, services, and information that are capable of evolving as their needs change. Companies know that unless they can structure agile infrastructures and supply chains that can guarantee personalization, quick-response delivery, and the ability to provide unique sources of marketplace value their customers will quickly migrate to alternative suppliers.

At the start of the second decade of the twenty-first century, the power of the customer has the power of the customer has grown dramatically and has morphed into new dimensions amplified by the Internet and social networking revolution. The ubiquitous presence of the Web implies that whole supply chains are expected to provide all around 7/24/365 business coverage. Customers now assume they can view marketing materials, catalogs, and price lists, and place orders as well as comparison shop, execute aggregate buys, participate in online auctions, receive a variety of information from correspondence to training, review delivery status, and check on financial information through the Internet. Social networking tools now enable them to bypass Internet, not to mention traditional marketing methods, of responding to companies, their products, and their policies in an open forum through blogs, Facebook, Twitter, YouTube, and other social networking media.

Responding to such a diverse array of requirements has forced most companies to explore radically new ways to reach and understand their customers. This movement has spawned a new science of customer management—*customer relationship management* (CRM)—and has simultaneously transformed and posed radically new challenges to how companies should be structured to execute the functions of marketing, sales, and service.

Defining the concepts and computerized toolsets available to manage effectively today's customer are explored in this chapter. The chapter begins with an attempt to define CRM, detail its prominent characteristics, and outline its primary mission. Next, the discussion shifts to an attempt to paint a portrait of today's customer. The profile that emerges shows that customers are *value driven*, that they are looking for strong partnerships with their suppliers, and that they want to be treated as unique individuals. Retaining loyal customers and effectively searching for new ones are best achieved by a *customer-centric* organization. The steps for creating and nurturing such an organization are then outlined. Following, the chapter is then focused on the array of CRM technology applications available to today's businesses in their pursuit of marketing, sales, and service initiatives. Among the technologies covered are Internet sales, sales force automation (SFA), service, partnership relationship management, electronic billing and payment, and CRM analytics. The chapter ends with an introduction to two of the most important innovations in customer management: customer experience management (CEM) and social networking.

Creating the Customer-Centric Supply Chain

While technology tools, such as the Internet and Google, have enabled customers to become more sophisticated by providing them with a variety of choices and unprecedented access to information, they have also enabled customers to become more capricious in their buying habits and less inclined to remain faithful to past supplier relationships. To counter these marketplace realities, many companies are in a life and death struggle to continuously develop business models that bring not only their organizations but also the entire supply chain closer to the customer in the search for the right mechanisms to attract and build sustainable customer loyalty. Achieving this goal requires that enterprises and their business networks focus on how they can become more "customer-centric." It requires them to reengineer their strategic plans and measurements to ensure customer focus. Finally, it requires them to search for mechanisms to converge marketing, sales, and service functions to architect real-time, synchronized product and service fulfillment systems that provide increased buyer–seller interaction and add expanding value to the customer.

The Advent of Customer Relationship Management (CRM)

All businesses, whether product or services oriented, have a single, all-encompassing goal: *retaining loyal customers and utilizing whatever means possible to acquire new customers.* Realizing this goal in today's fiercely competitive marketplace is easier said than done. Until very recently, most companies focused their energies on selling products and services regardless of who was doing the buying. Nowadays, the tables have dramatically turned. The growing power of customers, facilitated by the Internet, to chose (and to change) who they buy from has required companies to shift their strategic focus from "what" they are selling to "who" they are selling to.

This dramatic transformation in the goals of marketing, sales, and service from a product to a customer-centric focus has coalesced around the CRM concept. Far more than simply a methodology for improving sales and service effectiveness, the objective of CRM is to enable the continuous architecting of the value-generating productivities of enterprises and the supply chain networks in which they are participants in the search to build profitable, sustainable relationships with customers. Such a statement about CRM is very broad indeed and begs for a more detailed definition. However, while there is much debate about CRM, defining it in clear, universally accepted terms has yet to occur. While industry analysts, consultants, and practitioners alike are agreed that it is simply not just a technology, there is a wide divergence among those same professionals as to a precise definition of the full meaning of CRM. Some believe it is a computerized method of segmenting customers to improve selling efficiency. Others feel it is a method of database marketing to locate which customers would be a match for a specific product/service offering [1].

In addition, CRM is hardly a stationary topic. Like many business models, such as lean or supply chain management, CRM today has been enhanced by the inclusion of new technologies that have altered former structures as well as the evolution of new concepts such as CEM and Social CRM. One CRM expert has gone so far as to divide CRM into "old" or traditional CRM—CRM 1.0—and the growth of Social CRM, or CRM 2.0, built around the search and mobility tools associated with Web 2.0 functionality.

In understanding the meaning of CRM, it would perhaps be most fruitful to view the leading definitions. According to Greenberg [2],

> CRM is a complete system that (1) provides a means and method to enhance the experience of the individual customers so that they will remain customers for life, (2) provides both technological and functional means of identifying, capturing, and retaining customers, and (3) provides a unified view of the customer across an enterprise.

Dyche feels that CRM can be defined as "the infrastructure that enables delineation of and increase in customer value, and the correct means by which to motivate valuable customers to remain loyal—indeed to buy again" [3]. The final definition comes from Renner, Accenture's global CRM practice managing partner, who sees CRM as encompassing "all of the activities that go into identifying, attracting, and retaining customers, and focuses on aligning the whole organization to building profitable, lasting relationships with customers" [4].

CRM as a business philosophy and a practical customer service management (CSM) set of applications can be characterized as follows:

1. *CRM Is a Strategic Tool.* Historically, CRM has been perceived as a strategic technology focused on increasing profitability, enhancing the marketing plan, and expanding competitiveness. While the software applications provide marketers with critical tools to gather, segment, and query customer sales data for effective decision-making, its real value resides in the strategic advantage it provides the organization. According to Michael Boyd, director of CRM at Eddie Bauer,

   > Our experience tells us that CRM is in no way, shape, or form a software application. Fundamentally, it is a business strategy to try to optimize profitability, revenue, and satisfaction at an individual level. Everything in an organization—every single process, every single application—is a tool that can be used to serve the CRM goal [5].

   CRM is a comprehensive toolkit encompassing marketing, sales, service, and supporting technologies focused on forging customer relationships that provide mutual value, revenue, efficiency, and unique solutions to business problems.

2. *CRM Is Focused on Facilitating the Customer Service Process.* Being more responsive to the customer requires that sales and service functions be able to make effective customer management decisions and design superior responsiveness based on their capability to identify what brings value to the customer. Often success requires the availability of metrics and analytical tools that provide a comprehensive, cohesive, and centralized portrait of the customer.

3. *CRM Is Focused on Optimizing the Customer's Experience.* Perhaps the fundamental objective of CRM is the goal of "owning the customer experience." CRM initiatives that continually win customers can mean anything from providing a level of personalized service and customized products to utilizing advertising, ease in ordering a product, or ensuring a service call-back that will positively influence a customer's perception of the buying experience. The end result is to make customers feel they are dealing with a winner and are personally connected to their supplier.

4. *CRM Provides a Window into the Customer.* An effective CRM system ensures that all service nodes along the supply chain that can influence the customer

experience are provided with critical information about the customer, what that customer value the most, and how they can ensure the customer has a positive buying experience each and every time. Intelligence as to customer-winning attributes, such as buying habits, pricing and promotions, channel preferences, and historical contact information, must be all-pervasive, integrated, and insightful.

5. *CRM Assists Suppliers to Measure Customer Profitability.* Effective customer management requires that companies be able to determine which customer segments, if not each individual customer, are profitable and which are not, what product/service values drive profitability for each customer, and how marketers can architect processes that consistently deliver to each customer the values they desire the most.

6. *CRM Is about Partnership Management.* Effective customer management is about knowing the needs, values, and visions of each customer. CRM is about nurturing mutually beneficial, long-term relationships intimate enough to provide improvement opportunities and tailored solutions to meet mutual needs beyond physical product and service delivery. The end-point is to build unbreakable customer loyalty regardless of what actions are pursued by the competition.

7. *CRM Is a Major Facilitator of Supply Chain Collaboration.* No customer transaction can be executed in a vacuum but is actually an instance in what is often a long chain of events as products and information progress from one supply chain entity to the next. Firms that can create integrated, synchronized processes that satisfy the customer seamlessly across the supply channel network will be the ones that will have the most loyal customers, be the most attractive to new customers, have the most effective collaborative relationships, generate the highest revenues, and have sustainable competitive advantage.

Mapping the Cluster of CRM Components

To assist in better understanding the mission of CRM, it might be useful to sketch out a map of the functions associated with CRM found in the typical organization. As will be discussed, CRM is not concerned with a particular aspect of customer management, but actually encapsulates several related business processes and technologies and directs them to search for ways to optimize the customer experience. As detailed in Figure 5.1, the *customer* is the fulcrum of all CRM processes and acts as the centrifugal force attracting the customer value-producing functions of the firm. Clustered around the customer are seven critical technology-driven processes.

■ *EBS.* The *enterprise business system* (EBS) provides the "backbone" for all aspects of customer management. A firm's EBS consists of five critical applications. The pivotal customer-facing application provided by an EBS is the *customer database*. The database contains a complete profile of the customer,

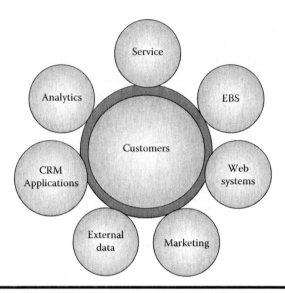

Figure 5.1 CRM management sphere.

from contact information and accounts receivable data to order management and shipping preferences. The second application, *transaction maintenance,* enables the entry and maintenance of sales orders and today is often driven by Internet-driven shopping functions. The results of sales transactions are then kept in the sales history file for use by standard EBS reporting functions and CRM analytical tools. The third application is the availability of easily configurable yet very powerful *displays* of the status of open and closed sales orders. In the fourth component, *information,* the EBS provides sales management personnel with visibility to such data files as pricing, promotions, and inventory balances. Finally, the EBS contains *financial detail* used for accounts receivable balance information, payment aging, interest charges, collections, insurance, and financial analysis.

■ *Web Systems.* More and more of today's forward-looking companies provide their customers with easy to use Web sites. Effectively constructed Web sites enable customers to visit catalogs, enter orders, review pricing, configure orders, participate in auctions, and perform a host of self-service functions from order status review to online learning.

■ *Marketing.* The ability to communicate product, brand, service, and company information is at the heart of customer management. Marketing's role is to identify the wants and needs of the customer, determine which target markets the business can best serve, decide on the appropriate mix of products, services, and programs to offer these markets, and generally motivate the organization to continuously focus on optimal customer service. In addition, marketing is concerned with identifying what value each customer expects to receive from the company's products and services, what are the firm's selling, campaign, and pricing strategies, and how to generate profits by ensuring customer satisfaction.

- *External Data.* The ability to sustain competitive leadership requires the continuous unfolding of collaborative relationships both within the organization and across resellers, suppliers, and channel support partners. Information from these internal and network nodes are critical in devising everything from promotional/product bundling, financing, and packaging design, to fulfillment, merchandising, and transportation.

- *CRM Applications.* CRM technology can be separated into three segments. The first, *operations CRM*, consists of the traditional functions of customer service, ordering, invoicing/billing, and sales statistics found in the EBS backbone. This also includes *e-CRM* Internet-driven applications like portals and exchanges, e-mail, EDI, SFA, and wireless customer management. The second, *collaborative CRM*, focuses on channel spanning functions such as forecasting and process design. The third segment, *analytical CRM*, consists in the capture, storage, extraction, reporting, and analysis of historical customer data.

- *Analytics.* Effective management of the customer requires a way to access accurate and timely business intelligence. Information can come from sales activities and can include databases containing customer prospecting, product lists, and payment data. It can also come from marketing and can include information such as sales revenues, customer segmentation, campaign responses, and promotions history. Finally, information used for analysis and reporting can come from service and can consist of customer contacts, support request incidents, and survey responses. *Business intelligence*, or as some say, *data warehousing*, however, is not the same thing as CRM. The difference, according to Dyche [6], is that CRM "integrates information with business action." The goal of CRM analytics is to deploy the ability to *act* on the data and analysis mined from customer and marketing repositories to improve business processes so that they are more customer-centric.

- *Service.* The last component in the cluster of CRM functions is CSM. Being able to efficiently and effectively respond to the customer *after* the sale is critical in keeping current customers and acquiring new ones. Whether they are termed contact centers, CIC, or customer care centers, companies have increasingly come to realize that the strength of their support functions is instrumental in enriching their customer relationships. Recently, Internet and other communications technologies have been applied to the service function in the form of automated contact centers, computer telephony integration, CTI Web-based self-service, cyberagents, and electronic service surveys.

The goal of CRM is to provide a 360-degree view of the customer. The cluster of CRM components attempts to provide companies with an understanding of who their customers really are, what they value in a business relationship, what solutions they wish to buy, and how they want to interact in the sales and service process. An accurate and intimate knowledge of each customer's behavior, preferences, and sales history will significantly assist businesses move their customers from being simply buyers of goods and services to loyal partners who keep coming back for more to value chain collaborators who see their suppliers as the primary contact node in an integrated, seamless channel focused on total customer satisfaction.

Understanding Today's Customer Dynamics

CRM is about providing companies with the ability to explore new ways of responding to the realities of the expanding power of today's customer. While, on the surface of things, much emphasis has been placed on customers' immediate concerns, such as personalization, superservice, convenient solutions, and product and service customization, today's customer is in actuality being driven by two very powerful needs. To begin with, today's customer is *value driven*. While it is true that customers will continue to search for the best value when they make a purchasing decision, it does not mean that price will be the sole determining factor. In fact the contents of what provides value to the customer can be decomposed into three regions [7].

- *Economic Value.* Customers receive economic value when they can leverage a product or service to generate additional value beyond the initial cost. This is particularly the case involving business-to-business commerce. Negatively, value can also be acquired when the customer applies the product/service to reduce internally associated costs, the savings of which exceed the original cost of the purchase.
- *Solution Value.* The acquisition of a product or service can provide benefit to the customer by providing access to certain desired functions or features. Value in this region is found in the capability of the function/feature to provide a desired level of performance or capability. While such value may in some cases be difficult to quantify, in other cases it may be possible to calculate the cost of certain features against alternatives to determine those with the optimum trade-off.
- *Psychological Value.* There can be no doubt that intangible factors can have a significant impact on customers' perception of product/service value. Psychological preferences are often subliminal (e.g., brand loyalty or image), and lead customers to believe they are receiving value beyond direct economic or solution-driven benefits. Normally focused around the concept of *brand,* customers feel that they receive increased benefits and reduced risk when they buy from a known supplier. Brand becomes increasingly more important as the differentiation between product/service feature, function, and cost erodes through market and technological maturity.

For the supplier, it is critical to move beyond just knowing past raw customer buying behavior data to a position of understanding what are the personal needs, wants, and preferences that constitute value for each customer. Once such a "value profile" has been populated, a meaningful *value proposition* can then be drafted that will detail how each individual customer's perception of value can be consistently realized.

Second, despite all the hype about the fickleness of the consumer, today's customer is more than ever looking to build strong *relationships* with their suppliers. In the past, customers often focused their purchasing habits around product and service brands. While today's consumer will still bypass cheaper generic alternatives to buy Tide or Coca-Cola, increasingly brand loyalty is being transferred from products to the provider. Whether it is Amazon.com, Nordstrom's, or Federal Express, consumers are now looking to their suppliers as "brands" that they can consistently count on to provide expected value regardless of the actual products or services purchased. This means that businesses must continuously reinvent their "brand-image" in often radically new ways that makes their customers feel as if they are always getting a superlative level of value, that they are in control of their purchasing experience, that they have confidence in what they buy and who they are buying from. Customers want their suppliers to customize their offerings to fit their individual needs and to feel that their participation in the actual purchase process provides a sense of empowerment as well as partnership.

The risks of ignoring these critical customer expectations can be catastrophic. In the past, businesses atomized each customer transaction and considered them independent of those executed in the past, the ones to follow, or any other related transaction. Today, customers are very aware of the service, products, and personal experience they receive. A single failure may drive them to an easily accessible competitor and today's social networking enablers make it easy for them to tell their friends and associates about their decision. Unsatisfied customers can be a significant negative force in today's Internet-empowered business environment that can destroy trust and abruptly end what once had been a long-lived relationship. The opposite is also true. Companies that let their customers know that by attending to their individualized needs they not only want their customers' business, but want to establish a trustworthy relationship and will be able to weather the storms of an uncertain economy and the encroachment of global competitors.

While the mission of CRM is to look for new ways to retain existing and woo new customers, from the outset it is critical to note that CRM has an accompanying object, and that, in the words of a recent Gartner Group Research Note Tutorial, "is the optimization of profitability. It begins with the premise that not all customers are created equal." In the past, companies treated their customers as if they were all the same. Each received the same level of service, each were charged the same for the products offered regardless of whether they afforded the company a profit. In reality a small percentage of a firm's customer base provides most of the profits. This means that anywhere from 70 to 80% of the customer base either provides little or no profit; as much as 40% actually will cost more to service than the profits received. Today's best companies know who their profitable customers are and will focus their businesses on retaining them. At Dell Computer, for example,

> not all customers at Dell are created equal, nor are they treated equally. Dell's data enable it to know the ultimate fact about its largest customers—exactly how profitable they are. The more money a customer brings in, the better treatment it gets; for instance, someone who buys servers and storage from Dell is more likely to get a special package that includes PCs and portables. Other industries … have also begun to offer better service and process to large accounts, but they don't like to say so. Dell is willing not only to admit it but also to say that some accounts may not be worth its time [8].

Knowing which customers are profitable and which are not is as important as the ability of an enterprise to tailor its resources and capabilities to respond to the individual needs of each customer. As a yardstick, customer profitability enables companies to differentiate the levels of relationship and accompanying service to be rendered to each customer so that their needs and the corresponding value to the business can both be realized. An interesting way to understand the relation between customer cost and profitability is the use of a metric called the lifetime customer value (LCV). The formula for the metric is as follows: the total sales revenue of a customer over the lifetime of their relationship, discounted by interest and inflation rates as appropriate [9]. Companies pursuing intimate customer strategies view LCV as detailing not only the value *of* a customer to the firm but also the value of the firm *to* the customer. Instead of focusing on metrics that reveal traditional concerns with sales/share, product profitability, and satisfaction, customer-oriented metrics require marketing teams to concentrate on customer profitability, rate and cost of customer acquisition, profit and growth in customer margins, and cost and success in customer retention.

The bottom line is to identify good customers who will have high LCVs, segregate them from mediocre to bad customers, and targeted for much greater attention. For example, AMR Research, Boston, says that credit card companies and retailers have already begun to utilize CRM technology tools to retain high- and mid-value customers while pushing away the bottom low-value ones. Higher-value companies receive what AMR calls "personalized fulfillment." This model strives to customize the entire experience of the customer ranging from customer profiles and order generation to fulfillment, logistics, and returns. Conversely, bottom-value customers will increasingly be charged for every service [10]. Cokins [11] feels that the application of activity-based costing profiles, which act like business electrocardiograms, can help identify not only which products, but also which customers are the most profitable. By generating metrics that provide a measure of the "cost-to-serve," a customer "profit and loss" statement can be constructed. Once this data has been attained, marketers then can construct strategies for each customer that result in (1) managing their cost-to-serve to a lower level, (2) reducing their services, or (3) raising prices or shifting the customers' buying to higher-margin products and service lines.

Extending utilization of CRM activities to the management of the customer out into the supply chain is potentially one of CRM's strongest functions. In its simplest form, sharing information about customer transactions can assist each channel business node manage the product and information flow along the demand and fulfillment network. For example, call-center or sales force order promising and management is much more effective when accurate information about product/service availability and delivery can be accurately made based on visibility into trading partner capabilities. By combining the metrics and relationship-building capabilities of CRM and the electronic linkages provided by today's Internet technologies, the supply chain can become not only more responsive but facilitate continuous product and fulfillment planning. In a wider sense, the information gathered via sales force and marketing surveys, and electronically through billing, customer information systems, and call-center data can provide channel partners with insights into the customer that can assist in product development and process design.

Creating the Customer-Centric Organization

In 1999, e-tailer Buy.com had been gripped in a downward spiral of unprofitability. The company's original strategy—sell products under cost and generate revenues through advertising—had proven, like it had for many a dot-com e-tailer, to be untenable. While Buy.com's low prices were attractive, the company was notorious for its almost nonexistent service and support, which resulted in a poor track record of customer loyalty. In an effort to reverse this perception company management embarked on a multifront strategy that included implementing new technology tools, building a dedicated call-center, increasing information availability at customer contact points, and initiating various cost-cutting programs. The results were dramatic. By 2001 order processing cost had dramatically declined, service had skyrocketed (both Forrester Research and Gomez.com ranked Buy.com No. 1 for quality of support), its margins had risen to the point that it was ranked the second largest multicategory Internet retailer in the industry, and its CRM technologies had dramatically increased the quality of the experience customers enjoyed when they visited Buy.com's Web site. "The key to our success," recalled Tom Silvell, vice president of customer support,

> is that we built our programs and technologies around what customers wanted and needed instead of letting our programs and technologies drive their behavior. This tactic helped transition us from a price-sensitive shop to one focused on the customer experience, on offering value to clients, and on providing quality merchandise at reasonable prices [12].

At the beginning of 2010, Buy.com could boast a strong international presence, "The Lowest Prices on Earth," over 12 million customers, and awarded one of the top 100 best retail Web sites of 2009 by Internet Retailer [13].

The dramatic change in Buy.com's fortunes has been the direct result of the implementation of an effective CRM strategy dedicated to the creation of a customer-centric organization. Achieving such a level of customer focus is not easy. According to Michael Maoz from Gartner Group, only 5% of companies today can say that they have implemented similar levels of customer service. Creating a customer-centric company capable of consistently delivering customer value while building customer loyalty is a multiphased process that involves reshaping the infrastructures of both individual organizations and accompanying supply chains as much as it does implementing computerized CRM functions. The following steps should be considered in architecting such individual companies and supply chains systems.

1. *Establish a Customer-Centric Organization.* Migrating the enterprise from a product- to a customer-centric focus will require changes in the way companies have managed everything from customer service to product design. Literally every customer touch point needs to be oriented around how each business function can continuously foster customer service. An emerging management position organizations have been establishing to achieve customer-centric organizations is the chief customer officer (CCO). Basically, this position acts as a liaison between customers and the firm. Requiring strong operational, marketing, and financial skills, the CCO, according to Manring [14], will identify customer touch points, define and enforce service standards, assist customers to navigate the organization, and search for methods to enrich the customer experience. Strategically, the CCO will be responsible for "integrating and leveraging customer information across the organization or owning and managing customer segments as units of optimization."

2. *Determine Existing Customer Positioning.* Understanding the customer is fundamental to a customer-centric focus. The goal is to unearth what each customer values and from these metrics to design the products, services, and communication infrastructure that will drive increasing customer loyalty. The process should begin by measuring the customer landscape. This can be accomplished by identifying the best customers with the greatest LTV. Their buying values should be detailed and contrasted with low LTV or lost customers. Next, qualitative research through surveys, face-to-face interviews, and other techniques should be conducted with each customer segment identified. The goal is to learn first hand how customers view their relationship with your company and with the competition. Finally, quantitative research tools should be applied to reveal concrete metrics associated with the notions of needs, behaviors, motivations, and attitudes identified in the qualitative review. A critical problem is knowing what to do with this data. In this area

a CCO could spearhead the analysis and devise action plans to turn the feedback into results or utilize it for strategic planning or resource allocation.

3. *Devise a Map of Customer Segments.* The qualitative and analytical data arising from the above step should provide a clear geography of the customer base and illuminate key drivers, such as convenience, price, reliability, and so on, of loyalty, value, and satisfaction. The goal is to focus company service efforts around processes that support and encourage the buying behavior of the firm's top customers and how they can be applied to less profitable marketplace segments. By pinpointing what provides true customer differentiation, competitive advantage factors can be leveraged to consistently enhance customer value at every touch point across the internal and supply chain organization. In addition the map should also reveal the effectiveness of current company product and service strategies and the core competencies of the organization. Businesses should be diligent in assuring there is not a mismatch between their offerings and what the customer base truly values.

4. *Develop and Implement the Solution.* An effective CRM program should be tireless in searching for opportunities for enhancing customer experiences. Transforming these programs into meaningful marketplace initiatives, promotions, and points of customer contact that improve company visibility, confirm customer value expectations, and cement loyalties is the next step in the process of generating a customer-centric organization. Unfortunately, there is no boilerplate methodology that companies can easily snap into place. Each company must painstakingly investigate its own customer-centricity strengths and weaknesses. In structuring these programs companies must be careful to ensure financial profitability by matching sound business scenarios with measurements such as ROI and net present value (NPV). Once these CRM initiatives have been verified, marketers can then begin the process of utilizing the CRM customer data warehouse to begin mining the data in an effort to locate the company's best customers and how retention and new customer acquisition programs can be best applied.

5. *Monitor, Measure, and Refine.* The elements of the previous step need to be performed iteratively. The driver of CRM project review is obviously the metrics arising out of the record of customer contact and exchange. Without such processes in place the quest will ultimately fail. Marketers must be careful to continuously research and document what is working and what is not by utilizing the analytical tools available within most CRM applications. These tools should provide ongoing quantitative tracking of buying patterns, customer attitudes, and degrees of satisfaction for all market segments and points of contact. One analyst recommends conducting focus groups at least once a year with the best and the worst customers, as well as the internal service staff [15]. Such a procedure will enable effective monitoring of the qualitative input to assist in massaging the quantitative results of performance metrics.

The successful implementation of a customer-centric organization requires that everyone in the business be aware and prepared to execute the enterprise's CRM strategy. Driving the CRM strategy requires, in turn, the firm support and active participation of senior management who needs to provide the vision and to focus the energies of the organization on communicating the CRM initiative to customers and partners as well as to the internal staff. Without such direction and sponsorship, most CRM programs will quickly decay and revert to previous "silo" operating methods.

Applying Technology to CRM

CRM can be divided into three major functions: *marketing*, the activities associated with creating company branding, identifying the customer, selecting product/service offerings, and designing promotions, advertising, and pricing; *sales*, the actual selling and distribution of products and services; and *service*, activities encompassing customer support, call-center management, and customer communication. Together, the mission of these functions is to inform the organization of who its customers are, how to better understand what customers want and need, what is to be the product and service mix to be taken to the market, and how to provide the ongoing services and values that provide profitability and expand relationships. These functions also detail the technologies that will be used to market to the customer base, conduct transactions, respond to customer service issues, collect marketplace metrics, and format customer contact information for review and analysis. These functions also assist in the development of the strategies governing how the supply channel network is to be constructed and the nature of trading relationships. Finally, these functions should provide the entire organization with the information and motivation necessary to continuously reshape the enterprise's perception of customer service, reengineer vestiges of silo management styles, and architect infrastructures that foster customer collaboration.

In the past, the functions of marketing, sales, and service were, at best, loosely connected with each other and utilized varying levels of technology to transact business, collect information, and communicate with the customer. Even the software tools that had evolved, such as ERP, sales force automation (SFA) and call-center applications, were developed in isolation or heavily focused on the transaction engine while leaving the marketing and service component fairly underdeveloped. Marketing in the "Industrial Age" focused on direct contact with the customer and relied heavily on printed matter such as catalogs, direct marketing, and mass media advertising. Until the 1980s, sales had relatively little to do with technology and perceived their function as centered around salesmanship and leveraging personal relationships. Finally, customer service, while always open to adopting the latest technologies to communicate with the customer, was often separated from the product producing and sales functions of the business. Up to just a few years ago, customer service consisted mainly in employing banks of service reps fielding customer inquiries by mail, phone, or fax [16].

While it can be argued that many companies have for years utilized CRM methods to deal with their customers, the rise of Internet technologies has rendered obsolete many of the traditional concepts of customer management through the creation of new technology toolsets that have significantly expanded existing CRM functions and capabilities. Today's Internet-driven applications provide companies with radically new avenues to gain visibility to customer value, retain and attract new customers, enhance transaction and service capabilities, and generate integrated, customer-centric infrastructures that enable businesses to realize opportunities for profitability while providing the customer with a level of seamless end-to-end service impossible less that a decade ago. In fact, over the past decade or so CRM software has been one of the hottest segments in the business solutions marketplace. According to a July 2008 report published by AMR Research, the estimate for CRM software sales alone topped $14 billion, a 12% increase over 2006 revenues (as a benchmark, the 2001 revenues were $5.6 billion). To show the trend, AMR projected software sales of $22 billion by 2012, despite the severe recession of 2008–2009.

The rest of this section is devoted to exploring the geography of what can be termed as *Internet-based* CRM or e-CRM. Figure 5.2 provides an illustration of the major functions of CRM and associated Web-enabled applications.

Figure 5.2   Range of CRM functions.

CRM and Internet Sales

Until fairly recently, the sales process was pretty much an affair that had little to do with technology and everything to do with the ability of the individual sales-person to win deals by leveraging their personal sales savvy combined with their knowledge of products, the marketplace, pricing, and the competition. Today, Web-driven applications have opened radically different opportunities for tech-nology-assisted selling. Chaffey describes this movement of CRM to the Internet as "e-CRM" and defines it as "using digital communications technologies to maxi-mize sales to existing customers and encourage continued usage of online services" [17]. The seemingly unstoppable movement to the Internet and customer-driven sales does not mean that the salesperson has become obsolete. The salesperson role, in fact, has become more important than ever in developing circles of closely defined business and mutually supportive relationships between their companies and their customer base. According to Poirier and Bauer [18], the role of the sales force in an e-business environment takes on added significance by performing the following functions:

- Acting as an advocate for the customer
- Exploration of personalization and mass customization to focus in on cus-tomer buying habits
- Providing information about company products and services
- Coordinating company resources to ensure superlative response to customer needs, and linking channel resources with customer demand requirements
- Acting as an initiator for the conveyance of information regarding process improvement changes from the company to the demand channel targeted at realizing mutual advantage
- Providing a medium by which critical company resources in the form of mar-keting information, training, logistics opportunities, customer and supply channel diagnostics, and collaborative planning initiatives are made available to each customer
- Managing online service quality to ensure buyers always enjoy a winning experience an motivate them to return
- Managing the multichannel customer experience to blend in different media to ensure customer satisfaction each and every time

Besides changing the role of the sales force, technology-assisted selling will broaden both selling and buying opportunities. Web applications enable companies to sell directly to the customer thereby bypassing costly channel intermediaries. Further, through the device of real-time technologies, companies will be able to improve effectiveness and better utilize resources. In addition, according to Sawhney and Zabin [19], "technology-enabled selling will be used increasingly to synchronize and integrate all selling channels used by the enterprise, including telesales, the Net, resellers, and the direct sales force, through the use of a common customer relationship repository, a common applications infrastructure, and a shared business process."

For the customer, technology-enabled selling opens other doors for productivities. To begin with, Web-based search engines have significantly enabled customers to find new suppliers and easily view the range of their product and service offerings. Additionally, Internet applications have dramatically simplified the ordering process and streamlined open-order inquiry. Web-enabled communication tools have made it easy for customers and suppliers to engage in bidirectional communication, a feature that increases one-to-one personalization of the transaction experience. Finally, the Web offers customers options for a buying experience unattainable in the past. Attributes such as 24/7/365 service, real-time information, online customer support, instantaneous availability of documentation, self-service, and Web-page personalization offer customers new ways of realizing the value propositions that meet their individual needs.

One of the primary Web-based tools offered by companies today is the customer *portal*. A portal is basically a Web-based application that aggregates information, third party resources, and reference materials arranged in a specific Web content that can be customized and personalized to sell to and service prospects, a known customer, or customer segments across multiple channels. Beyond portals, customers can also directly access seller services through independent, private, and consortium exchanges. Some of the basic application functions available in online sales can be described as follows [20]:

- *Online catalogs* provide customers with the opportunity to research and compare the array of products, prices, and services offered by a supplier.
- *Online order processing* is the most widely known form of e-CRM. It provides prospects and customers with online access to supplier product information, pricing, and fulfillment capabilities. Web-based shopping provides customers with tools to comparison shop, search for desired quality and service requirements, view product/service aggregations, participate in online auctions, and access-related product/service mixes through on-screen portals. For suppliers, Web-enabled selling permits the development of what Tom Peters calls *micro-brands* or customized Web sites that appeal to very narrow groups of customers. In addition, Internet selling enables company's to receive a detailed picture of their customers' buying habits and experiences that can be used for cross-selling, up-selling, and customer service.
- *Online order configurability* enables customers to design their own products and services through special configuration capabilities.
- *Lead capture and profiling* provides detailed repositories of prospect inquiries, customer sales, and profile information that can be mined to provide information for Web site personalization or marketing follow-up.

- *Online surveys* enable marketers to quickly test the attitudes and possible behavior of prospects and customers critical for Web site customization and market segmentation.
- *Literature fulfillment* provides customers with easy access to company and product/service information that can be downloaded or sent via e-mail to qualifying prospects and customers.
- *e-Mail marketing* enables companies to leverage captured prospect/customer information to establish customized marketing campaigns communicated to the marketplace via e-mail.

The Internet is critical in assisting companies to deliver tailored responses to their marketplaces by effectively sorting good customers (profitable/valuable) from the bad (unprofitable/nonvaluable). Once stratification of the customer base is completed, businesses can then architect an individualized response commensurate with the expected level of customer profitability potential. The goal of the entire process is the development of communities of customers who continue to visit and buy from the Internet site either because they are genuinely interested in what the site offers uniquely to them, or because they have become economically tied to the site. The savviest Internet companies will continue to enhance their capability to *discern* the best customers and *differentiate* their response through the use of Internet capabilities to structure "digital loyal networks." Such networks have the ability to collect, manage, and shares information seamlessly across organizational boundaries with customers and suppliers. All of this is *digitally* enabled by the Internet and new technology platforms for supply chain and CRM.

Sales Force Automation (SFA)

The advent of SFA has been credited as being the foundation for today's e-CRM business model. Beginning in the early 1990s, SFA was conceived as an electronic method to collect and analyze customer information from marketing and contact center organizations that in turn could be used to advance opportunities for customer retention and acquisition as well as enhance marketplace relationships and revenues. In addition, the sales force needed automation tools that could assist them to more effectively manage their existing accounts, prospect for new customers, track the impact of pricing, promotions, campaigns, forecasts, and other sales efforts on their pipelines, generate meaningful analysis and statistics from their sales database, become more mobile, organize their contact lists, and have real-time customer information in an easily accessed presentation. According to Dyche [21], the mission of SFA "was to put account information directly in the hands of field sales staff, making them responsible for it, and ultimately rendering them (and the rest of the company) more profitable."

Although early SFA applications were plagued by downtime due to cumbersome data downloading, less than timely information, and often the inability to send data back to backbone business system, today's technologies have overcome these limitations and now are capable of driving powerful SFA systems capable of synchronizing data from unconnected sources, such as laptops, mobile devices, and desktops, and utilizing flexible and scalable databases, such as Microsoft SQL or Oracle, and memory-resident PC applications equipped with scoreboards and reporting functionality that can exploit powerful engines, such as HTML and Java, to drive real-time information sharing. While the SFA marketplace contains a number of software vendors and competing products, they all posses to some degree the following functionality:

- *Contact Management.* This application is one of the original components of the SFA product suite. The basic function of the software is to enable the organization and management of prospect and customer data, such as names, addresses, phone numbers, titles, and so forth, the creation and display of organizational charts, the ability to maintain marketing notes, identification of decision makers, and capability to link to supplementary databases. Today's packages also provide sales reps with enhanced contact lists and calendars, and the functionality to merge them with customer contact efforts or automated workflow programs capable of assigning and routing appointments. According to Dyche [22], "The real value of contact management CRM is in its capability to track not only where customers are but also *who* they are in terms of their influence and decision-making clout."

- *Account Management.* Often individual sales reps and managers are responsible for large territories and hundreds of customers. Account management applications are designed to provide detail information regarding account data and sales activity that can be accessed on-demand. In addition, these tools permit managers to effectively develop and assign field sales and marketing teams to match customer characteristics.

- *Sales Process/Activity Management.* Many SFA applications provide imbedded, customizable sales process methodologies designed to serve as a road map guiding sales activity management. Each of the steps comprises an aspect of the sales cycle and details a defined set of activities to be followed by each sales rep. In addition, SFA tools also can ensure that major sales events, such as product demos or proposal deadlines, trigger alarms as they become due and remind sales reps of closing dates. While such tools currently lack sophistication and deep functionality, they do assist in promoting sales process standardization and, ultimately, greater productivity.

- *Opportunity Management.* Also known as *pipeline management,* this aspect of SFA is concerned with applications that assist in converting leads into sales. In general, these toolsets detail the specific opportunity, the company involved, the assigned sales team, the revenue credits, the status of the opportunity, and the proposed closing date. Some applications provide for the automated distribution of leads to sales teams, who the competition is and what their advantages/disadvantages are, product/service/pricing competitive matrices, and even the probability of a successful closing. Still other tools provide performance metrics compiling for each sales person/team opportunities won and lost.

■ *Quotation Management.* When not available from an EBS backbone, SFA systems can assist in the development of quotations for complex orders requiring product configuration and pricing. Some vendors provide applications that use graphical tools to map and calculate the quotation process. Once the order has been completed, the order can be transmitted via e-mail or the Internet for management authorization and inventory and process availability check, and then quickly returned back to the sales rep for final review and signoff by the prospect.

■ *Knowledge Management.* Much of the software composing today's suite of SFA products is oriented around standardizing and automating sales processes. However, effective sales management also requires access to resources that provide sources of information that reside in each company and are difficult to automate. Such information might include documentation such as policy handbooks, sales/marketing presentation materials, standardized forms and templates such as contracts and estimating, historical sales and marketing reporting, and industry and competitor analysis. Often termed *knowledge management* systems, these applications can act as a repository for all forms of information that can be easily added to and referenced through online tools such as Lotus Notes or Web-based browsers.

e-CRM Marketing

Effective marketing is and will always be founded on a simple premise: customers are won by personalizing the communication between the seller and the buyer and customizing the product and service offerings so they directly appeal to the desires and needs of individual customers. In a preindustrial economy, selling is always a one-to-one affair and is characterized by personal contact whereby the buyer examines physically the array of available goods and services and the seller negotiates an individual contract to sell. In the Industrial Age, the concept of *brand* and *mass marketing* replaced personal relationship and direct review of available goods and services. Mass marketing meant standardization of products and services as well as pricing, and assumed uniformity of customer wants and needs. The prospect for marketplace success was focused on the availability and choice of the product and service mixes companies offered. Although by the mid-1990s modifications to the mass marketing approach, such as *direct-marketing, target marketing,* and *relationship marketing,* began to point the way toward a return to one-to-one buyer–seller contact, marketers lacked the mechanism to initiate what could be termed *personal marketing.* This approach can be defined as the capability of companies to present their goods and services customized to fit the distinct personal interests and needs of the customer. A critical feature of this strategy is that the array of offerings is presented with the permission of the customer.

With the advent of the Internet, marketers were finally empowered with a mechanism to activate *personal marketing*. What had always been needed was a medium whereby the interactive, two-way dialog between customer and supplier, so necessary for the establishment of true one-to-one relationships, could be established. According to Fingar, Kumar, and Sharma [23],

> *customization* is the byword of the 21st century marketing revolution. By interacting with customers electronically, their buying behavior can be evaluated and responses to their needs can be tailored. Customization provides value to customers by allowing them to find solutions that better fit their needs and saves them time in searching for their solutions. ... Not only can a solution be pinpointed for a customer, but also as the relationship grows, the more a business knows about individual buying behavior. As a result of the growing relationship, cross-selling opportunities will abound. With the Net, the savvy marketer can sense and respond to customer needs in real-time, one-to-one. ... In the world of electronic consumer markets the success factor mantra is: relationship, relationship, and relationship.

When it is considered that the cost of gaining a new customer is five to eight times greater than marketing to an existing customer, companies who can leverage the power of *personal marketing* are infinite better positioned to keep their customer base intact.

Perhaps the importance of the Internet to marketing can be best seen in the concept of *brand management*. According to Taylor and Terhune [24], brand can be defined as a complex set of elements, "including awareness or recognition, customer loyalty, image or brand traits, name and logo design, personal benefits, positioning in relation to competitors, media presence, pricing relative to value, perceived quality, reported satisfaction via word of mouth, reputation, and perceived popularity." In the past the concept of a brand was linked to the properties to be found in a company's products and services. Today, the Internet provides the power to deliver targeted brand messages, interactive experiences, and lifestyle appeals that make it possible not just to offer unique, personalized products, but also to define narrow groups of customers, or *microbrands*, that seek to create a one-to-one match between the needs/wants of the customer and the capabilities of a set of products and services. In the Internet Age, e-businesses have tried to establish their Web sites as brands. Dot-coms, such as Buy.com, Barnes&Noble.com, and Amazon.com, have become successful because they have been able to provide their customers with a unique experience and have generated an emotional loyalty. Such companies have created interactive experiences that provide customers with a quick and complete solution and, in the effort, established themselves in the psyche of the customer. "The bottom line," state Taylor and Terhune, "is that as we know more about who people are, we can use the power of the Internet to create thousands of combinations of product, service, and packaging characteristics. To craft a message that is as flexible and multifaceted as the human experience itself" [25].

The explosion of Amazon.com on the scene was the harbinger of change to the concept of brand. Many analysts have credited Jeff Bezos as being the first to understand that customer relationships could escape from being physical to being virtual. Bezos had been the first to understand that he could win the customer by providing them with individual attention and a killer marketing strategy. "First of all," states Voth,

> you could get stuff cheap. You could buy your *New York Times* best-seller at or below the price you would pay at Barnes and Nobles, and you could buy them tax-free. You had a seamless customer experience—so you didn't have to wait in line, you could always find what you wanted and you could easily send gifts to your friends and families. Amazon would take care of the wrapping. Christmas and birthday shopping suddenly became much easier. Amazon was able to capitalize on the fact that shopping for books and music are a function of time. This changed the fundamentals [26].

Such a strategy guides companies, such as Dell Computers and Cisco, which have come to understand that individualizing and enhancing customer relationships will cement marketplace loyalties and expand the LTV of their customers.

Automating the marketing function requires the use of software applications that enable companies to compile, search, and utilize customer databases to define who the customer is and then generate targeted marketing campaigns via e-mail, e-fax, the Web, the telephone, or other technology tools to reach the marketplace. The focus of what has come to be known as enterprise marketing automation (EMA) is *campaign management.* In the past, campaign management was an labor intensive affair where customer databases were reviewed and a campaign based on a carpet-bombing strategy launched to pulverize the marketplace. Analyzing the impact of the campaign often took months or even years. Today, EMA provides the capability to automate the entire campaign process. The suite of toolsets available include customer intelligence and data extraction, campaign definition, detailed campaign planning and program launch, scheduling of activities and continuous performance measurement, and response management. While many of the activities appear similar to traditional marketing campaign processes, the major difference is that EMA utilizes the Internet to capture, extract, and analyze campaign inputs. By tracking campaign results over time, marketers are then better equipped to construct future campaigns that can enhance one-to-one marketing relationships.

The major components of an EMA-driven marketing campaign can be described as follows:

- *Promotions.* EMA provides the ability to bring the promotional side of a campaign directly before a customer as never before. Whether it be giveaways, contests, or discounting, *opt in, opt out* capabilities on the Web page give an immediacy to customers' willingness to engage in the promotion impossible with paper-based or telemarketing-type methods. Once data are captured, it can be directly input into the marketing database and used for ongoing review and campaign modification.
- *Cross-Selling and Up-Selling.* Cross-selling is the practice of offering to the customer related products or services during the buying process. Up-selling is the practice of motivating customers to purchase more expensive (and more profitable) products. To be effective, Web site and buying exchanges must be able to analyze the customer and prepare alterative offerings that will truly arouse their interest.
- *Marketing Events.* In the past, trade shows and exhibitions provided customers with opportunities to view new products and services. Today, marketers can broadcast the latest marketing information through online newsletters, Web-based seminars, and special webcasts.
- *Customer Retention.* While companies spend lavishly to attract new customers (it is estimated that over $180 billion is spent each year in the United States on advertising alone), it is with bitter resignation that marketers must accept the fact that statistically as high as 50% of their customers will be lost over a 5-year period. Utilizing EMA toolsets can assist companies not only to isolate and rank customers most likely to leave but also to weigh the possible impact of promotional efforts on this class of customers. The goal is to mine the customer data and devise models that can assist in the prediction of customer behavior.
- *Response Management.* Once data from a marketing campaign begins to stream in, marketers need to be able to utilize the information to perform several crucial tasks. First, they must be able to gather, extract, and analyze the data. Second, they must be able to determine the impact of the campaign by calculating actual customer profitability. A value model, such as a customer's lifetime value (LTV), can dramatically assist in the process of making sense of the deluge of data collected. And, finally, the marketing automation tools must be able to assist in refining and possibly altering the course of the campaign.

Customer Service Management (CSM)

The ongoing management of the customer once the sale has been completed has traditionally been organized around the *customer service* function. The impact of *customer care* on the continuing success of a business has been widely known and is part of service folklore.

- "The average company loses half its customers over a 5 year period."
- "Reducing defections 5% can boost profits from 25 to 85%."
- "Yet companies typically spend five times more on customer acquisition than on retention."
- "Sixty Five Percent to 85% of customers who defect say they were satisfied with their former supplier."
- "Totally satisfied customers are six times more likely to repurchase than satisfied customers."
- "A happy customer will tell five people about their experience, while each dissatisfied customer will tell nine."
- "US on-line businesses lost more than $6.1 billion is potential sales in 1999 due to poor customer service at their Web sites" [27].

Regardless of the accuracy of these metrics, they do reveal an essential reality: total customer care is the cornerstone of the customer-centric organization. According to Fingar, Kumar, and Sharma [28], the mission of customer care functions include the following:

- Improve customer service while reducing costs.
- Put the customer in control by providing self-service and solution-centered support.
- Segment customer behavior one-to-one to individualize goods and services.
- Earn customer loyalty to gain a lifetime of business.

Over the past 25 years, the purpose, scope, and mission of CSM have changed dramatically. In the beginning, customer service consisted in receiving and answering personally correspondence with customers who had questions or problems about products or information. Next came the *help desk,* where, instead of writing, customers could talk directly to a service rep about their issues. By the 1990s the purpose and function of CSM had evolved beyond just an 800 telephone number to encompass a wide field of customer care objectives and activities. Known as contact centers, or customer interaction centers (CIC), service functions sought to deploy a range of multimedia tools to not only relate order and account status, but also to manage every component affecting the customer from product information to maintenance, warranties, and upgrades.

Today, the capabilities of CICs have been pushed to a new dimension with the advent of exciting new toolsets, such as the Internet, wireless communications, speech recognition, and video, to join older technologies such as phone, caller ID, fax, e-mail, and EDI (Figure 5.3). Such applications provide customers with even more opportunities for control of service dimensions while enabling companies

Figure 5.3   Dell Computers Internet service site.

to integrate all avenues of customer interaction on a central platform. Self-service opens a new dimension of customer service at less cost while service databases improve knowledge of customer behavior that enable the delivery of customized sales and service one customer at a time.

Over the past few years, CIC has been transformed from a bank of service reps connected to customers by phone and fax to a highly automated communications center. The mission of these applications is to enable companies to activate open, productive dialog with the customer that are *personalized* in that they reflect each individual customer's needs, *self-activating* in that they permit the customer to successfully self-service their questions, *immediacy* in that critical information can be conveyed in real-time, and *intimate* in that the customer feels the supplier is sincerely concerned about their issues and that the outcome will provide a basis for future sales and service interaction. The following seeks to detail the technologies currently in use.

- *Automatic Call Distribution (ACD).* This technology provides for routing incoming customer calls to the proper service resources based on defining characteristics. This toolset seeks to minimize customer waiting by monitoring the call queue, automatically switching calls to available resources, matching the call requirements to service reps with particular areas of expertise, and even prioritizing calls to favor high-profile customers.
- *Interactive Voice Response (IVR).* These systems provide 24/7/365 routing of service calls based on the customer's response typed on the telephone keypad.

The objective of these applications is to provide information or to qualify and route a call without human interaction. A new and more sophisticated tool, automated speech recognition, provides callers with the ability to communicate their questions verbally without having to use the telephone keypad.

■ *Computer Telephony Integration (CTI).* These applications provide the technologies necessary to integrate data with telephones. For example, it is CTI that enables a service phone call to be routed to a particular service rep or other resource.

■ *Internet Call Management.* The use of Web-based self-service has enabled companies to escape from the frustrations associated with IVR-driven keypadding. The advantage of Web-activated service is that the customer is able to enjoy a significant level of self-driven interactivity. Information ranging from proactive notification of new products to troubleshooting tools, support guides, and online forums has changed the scope of service management. CICs can also overcome customer frustrations with the service Web site by including a "Call me" button that provides for in-person contact. Lands' End, for example, enables customers to chat online with a service rep.

■ *Service Cyberagents, Bots, and Avatars.* While in its infancy, the use of automated intelligent agents is expected to expand dramatically in the forthcoming years. The goal is to equip bots with specific expertise, instructability, simplified reasoning, and the ability to cooperate with other bots to guide cyberagents in solving problems for customers.

■ *Call-Center Analytics.* A critical part of effective customer service is being able to assemble a holistic view of the customer. Realizing such an objective will mean correlating massive amounts of Web data with information in other databases. For example, the CRM system will contain the customer profile that, when combined with behavioral Web activity, will enable service reps to model the customer and architect the service criteria necessary to respond effectively to customer needs. Functions in this area are part of what has come to be called CIC's *Workforce Optimization Suite.*

■ *Performance Measurement.* To be effective, CSM systems must contain tools for service performance monitoring. Applications must possess analytics gathering to record and evaluate customer service interactions as well as metrics to evaluate, measure, and manage service rep quality and productivity.

The requirements of today's customer and advances in technology have transformed the scope and mission of customer service. A few years ago, companies considered their customer service functions as purely a cost center and a drain on profitability. In contrast, today's enterprise views customer service as a necessary investment in cementing customer loyalties and assuring maximum customer value. As evidence, the changing of the function's name from "help desk" to "customer integration center" belies the central position it occupies it architecting the interconnectedness of the customer-centric organization. CICs in the view of some companies have actually become a vital link in the supply chain. According to Dave Csira, vice president and general manager of e-services for USCO Logistics, Naugatuck, Connecticut, "we have made our call center responsible for forward planning so that in essence they are the forefront edge of the supply chain for us." [29]

CRM and the Supply Chain

While CRM functions are primarily turned inward, several applications are essential in facilitating the management of customer and channel network partners. Among these toolsets can be found the following.

Partner Relationship Management (PRM)

During the 1990s, many sectors of the economy had begun to explore ways to disintermediate wholesale/distribution partners in the search to streamline operations, cut costs, and increase revenues. Spurred by advances in information technology, the growth of warehouse clubs and mass merchants, excess capacities, and the emergence of competing channel formats, the wholesale/distribution industry had begun to seriously feel the competition. By the end of the decade the Internet seemed to portend the day when all forms of channel intermediary could be eliminated. In reality, when the business practices and the numbers were examined, it was obvious companies had actually become more and not less dependent on their channel partners. For example, according to AMR Research, by the end of 2001 dealers, agents, resellers, brokers, and other forms of indirect sales channels represented between 40% and 70% of many company's revenues. Even in the high-tech sector, about 60% of sales were estimated to have come through indirect sales partners.

The reason for what has been termed the *reintermediation* of the supply channel is simple. No single company can hope to fill all of the needs of its customers. Channel partners solve this problem through their ability to personalize and customize the customer experience by providing products and services from many producers. Take for example Amazon.com and Yahoo!: they are succeeding because they have the capability to offer unique value to the customer. Instead of bypassing channel intermediaries, companies have become acutely aware of the need to architect more closely structured partnerships with channel partners, dealers, and resellers. This growing movement to search for management methods and software technologies to expand partner relationships has coalesced around a subset of e-CRM termed partnership relationship management (PRM).

Simplistically, the mission of PRM can be defined as a business strategy and a set of application tools designed to increase the long-term value of a firm's channel network by assisting companies to select the right sales partners, supporting them by offering timely and accurate information and knowledge management resources to deal successfully with channel customers, collectively searching for ways to improve sales, productivity, and competitiveness, and ensuring that each trading partner contributes to customer satisfaction. It would be very wrong, however, to assume that PRM is merely prospecting and announcing promotions or, as Greenberg so aptly describes it, "PRM is not just SFA and a partner" [30]. One of the key differences between sales and PRM is that in customer-facing activities, companies are dealing directly with the customer. In contrast, managing partners focuses on the indirect automation and optimization of layers of network trading partners. The sheer complexity of many supply chains makes allocation of resources, sourcing, lead generation and review, and sales productivity measurement difficult to track and will require companies to completely rethink their former channel strategies. PRM is fundamental in driving this new strategic viewpoint by providing software toolsets designed to automate and enhance communications, processes, and transactions throughout the supply chain system.

SAP's PartnerEdge PRM Application

In 2006 SAP rolled-out for internal use their own PartnerEdge PRM application to get a better handle on their sales business partners.

"Since its" implementation, "We're already seeing the benefits in areas like funds management, where for the first time we have automated calculation of points and levels for partners," reports Sigrid Ippich, project manager at SAP AG. "Another area that's vastly improved is partner life-cycle management. We can easily monitor our partner pipeline to obtain snapshots of partners that want to join, partners in the entry stage, and so forth. Ths tells us how many resources we need to allocate to turn them into productive members. We're using mySAP CRM to efficiently register partners who are ready to join and to perform the initial business planning activities that used to take so much time."

"For more mature partnerships, we use mySAP CRM to calculate how partners are doing against the performance figures they agreed to in their SAP business plans. This was one of those processes that used to take months - now it can be done in days," says Ippich. "Not only do we save a lot of work, but we can share the results with partners immediately. This allows them to make adjustments in their sales and marketing efforts far sooner, which in turn brings additional business that much quicker -one reason why mySAP CRM is boosting revenues substantially for all parties."

By the time the worldwide rollout is complete, all 1,700 of SAP's partners - and the 15,000 individual users they employ—will be using mySAF' CRM as the basis for all their core partnering activities. SAP also plans to use mySAP CRM to support further partnering processes by adding functionality to the partner portal every quarter.

Source: SAP, *"Customer* Success Story — High Tech," found at http://www.sap. com/uk/solutions/business-suite/crm/ pdf/CS_SAP.pdf

---

The foundations of PRM are not unlike those of CRM itself. PRM started as a means to facilitate channel sales and gather metrics based on the marketing and sales efforts of network trading partners. Today, PRM functionality can be separated into five categories.

1. *Partner Recruitment, Development, and Profiling.* A critical component of PRM is the ability assist in the recruitment and qualification of potential channel partners. Once the personal contact with new partner recruits has been completed, PRM tools can assist in ranking the partner database for ongoing marketing/sales assignments. The essential component of PRM software is the population of a *partner profile.* Such a database is critical in managing the capabilities of each partner from contact information and infrastructure to past sales contribution and general performance. By enabling a method to standardize the partner channel, PRM can better enable companies to manage the life-cycles of their partners by providing visibility to partnership risks and rewards, ongoing contract maintenance, forecast of planned revenues, and availability of metrics bearing on profitability and loss.

2. *Marketing Development.* This component of PRM is concerned with communicating marketplace opportunities to the partner network. Perhaps the most important function is *lead generation.* This tool enables parent companies to match customer leads with partners based on their capability profiles. Procedurally, marketers can use the system to analyze the lead, assign it to the most qualified partner, and then capture partner win/loss results. This area of PRM also includes functions to link channel partners to campaigns and promotions and to measure their results. Finally, it also provides for the allocation and budgeting of cooperative marketing funding, charts the productivity of marketing spending, and illuminates methods for improving accountability of partners and promotional campaigns.

3. *Sales Management.* This component of PRM consists of several function that include team selling, catalog management, needs analysis, and order management. Other toolsets provide for quotation management and configuration capabilities that can activate interactive selling tools to customize partner and marketplace needs. Finally, a PRM system should provide partners a window into channel product availability, order status, and service requests and warranties.

4. *Services Management.* A rapidly growing requirement of PRM is the ability to provide for the ongoing training and certification of partners and activation of support capabilities. For example, a partner can get trained on a certain product line, become certified, and then the PRM system will route that partner leads associated with that line. Also of importance is the ability to provide partners with interactive demos and presentation software that combine content and configurability capabilities that dramatically present product to prospects.

5. *PRM Collaboration.* In addition to basic functions, PRM systems should facilitate channel networks to codevelop marketing programs and joint business plans. Finally, effective collaboration will require the ability to transmit analytics and metrics of customer performance, channel sales forecasts, and general marketplace feedback.

Electronic Bill Presentment and Payment (EBPP)

The introduction of Internet-driven applications have produced a revolution in the way today's business is run, how customers are treated, how products are purchased and produced, and how logistics services are contracted. Unfortunately, despite the flourish of high-tech applications, the management of *financial* processes largely remains a paper-bound function, circumscribed by traditional methods of bill and payment transmission. According to a 2001 Garnter Group survey, only about 17% of all business-to-business payments are handled electronically, most of it going through wire transfers, automated clearinghouses, EDI, and credit cards. Still other tools, such as purchasing cards (p-cards), debit cards, and e-mail billing can be found on the fringe of EBPP. For the most part, invoicing is still a manual process that can take weeks. Costs for handling can mount quickly. According to industry research, the cost to the seller to manually generate an invoice averages about $3.45; it costs the payer about $0.60 to pay it manually [31]. While ease and security of payment through credit card and third party financial cybercompanies, such as PayPal, has greatly facilitated business-to-customer Internet purchasing, business-to-business lags behind but is sure to gain traction as the second decade of the twenty-first century opens.

Clearly, companies will either need to acquire or contract electronic payment systems that enable trading partners to receive bills, authorize payments, match payments to purchase orders, and download the data into ERP and accounting systems in a digital format that will be cheaper than credit cards, EDI, or checks. There is even a Web site (see Figure 5.4) where companies can go to find suppliers who can be used to execute all sorts of electronic payments and online billing. Currently, suppliers offering electronic billing solutions can be broken down into the following four categories [32]:

- *Software Suppliers.* Most ERP and SCM software systems contain functionality for electronic payment. Among the functions can be found EDI payment, giro and autogiro payments, bank transfer, and direct debiting.
- *Financial Service Providers.* In this grouping can be found banks and credit unions. As some of the largest billers in business themselves, these institutions have always been keen to deliver their own billing and statements electronically as well as move payments from their client to electronic formats.
- *Consolidators.* These businesses seek to provide services directed at aggregating a customer's bills into a single payment instrument. Acting as a coordinator, these contracted institutions will centralize billing functions and assume the task of paying bills online.
- *Portals and Exchanges.* For the most part, companies in this group offer Internet consumers access to various bill presentment and payment toolsets. Companies such as Quicken.com and Yahoo! have built early leads in this area.

As e-billing capabilities grow, many companies have begun the process of integrating EBPP into their CRM toolsets. Forward thinking executives have come

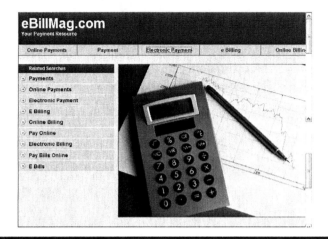

Figure 5.4   e-BillMag.com's Web site.

to understand that the merger of EBPP and CRM provides them with radically new opportunities to develop customer relationships. To begin with, the merger enables companies to offer greater *convenience* when it comes to customers accessing their accounts and answering questions about financial issues. It also provides greater *personalization* that will enable the biller to customize financial transactions and draw the customer closer to the biller's Web site and other services. By utilizing electronic messaging, another channel for *marketing* and cross-selling can be opened, thereby increasing CRM value to the biller. CRM *customer service* functions will be enhanced by the addition of bill and statement information. And finally, customers will gain another avenue for *self-service* and personal management by providing them with the ability to manipulate and analyze financial data, thereby driving down biller services costs, increasing "mind-share," and informing customers in real-time about new policies and procedures.

CRM Analytics

In today's business environment, companies are often suffering, not from a want of information about their markets and customers, but from a glut of too much data. Enterprise systems, marketing and customer service departments, and now the Internet are burying business analysts in a flood of information. The goal of CRM analytics is to provide companies with statistical, modeling, and optimization toolsets that empower organizations to analyze, combine, and stratify their data to better understand the state of their businesses and the status of their customers by group and individual needs. Ultimately, the goal is to provide an information conduit enabling decision makers to architect their organizations in an effort to continually identify and exploit opportunities wherever they may arise in the supply chain.

For the most part, CRM applications have been focused on *operational* functions, such as managing sales and service. While companies have long employed data warehousing and other intelligence toolsets to assist marketers, these applications were usually separate from their CRM program. Today, as analysts search to enhance the capability of their CRM systems, the incorporation of analytical functions that span marketing, sales, and service (operational CRM) and partners and suppliers (collaborative CRM) have become the newest "killer apps."

> What are the engines that drive CRM analytics? According to AMR Research, Analytical CRM uses Online Analytical Processing (OLAP)/ Relational Online Analytical Processing (ROLAP), algorithms, and data mining techniques to provide insight and uncover trends in the data collected by its operational counterparts. The results of this analysis are then fed back into the operational applications to improve the next interaction [33].

As illustrated in Figure 5.5, transactions, clickstream logs, and other customer data are entered through CRM and business system applications and driven into the company's data warehouse. Once the data are assembled, marketers can then apply OLAP, reporting, modeling, and data mining toolsets to identify relationships and patterns in the data that enable predictive analysis. Among the resulting output analysis can be found customer value measurement, risk scoring, campaign measurement, channel analysis, churn analysis and prediction, personalization and collaborative filtering, and revenue analysis. Finally, marketers can utilize the intelligence to drive the development of programs designed to pinpoint individual customer touch points. For example, interactive analytics enable marketers to slice and dice data to carry out what-if scenarios that could be used to create a promotions campaign.

While a significant tool to assist companies architect targeted marketing programs, industry analysts have been careful to point out what today's CRM analytics platforms can really do. According to AMR research, analytical CRM is being used to perform tasks such as profiling customers for focused communications and up-selling/cross-selling opportunities, predicting customer churn and profitability, executing real-time Internet personalization, and assembling the proper mixture of product, price, and channel that maximizes individual customer profitability. At the same time, AMR is careful to detail that analytical CRM is often not fully integrated with operational CRM and is not yet a fully automated set of business files flowing from one system to another [34].

Implementing CRM

As a software application, CRM has been described as an array of technologies for marketplace intelligence gathering and analysis that enables the transformation of customer sales data into metrics detailing sources of competitive advantage,

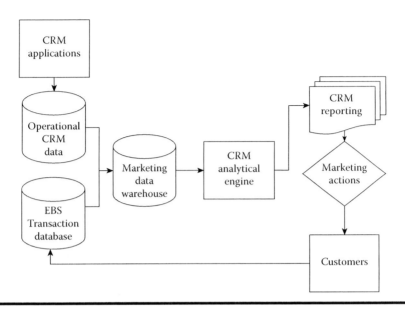

Figure 5.5   Analytical CRM architecture.

customer insight, and high profitability. Originally, CRM was perceived purely as an analytical engine enabling marketing to design strategies to identify high-profile customers and increase profitability. Unfortunately, CRM's massive databases and analytical toolkits have misled marketers into an unfounded belief that CRM, by itself, will reveal optimal combinations of customers and expose problems inhibiting deep relationship building with customers. While CRM began the new millennium with much fanfare, by the mid-2000s, analysts and industry articles had switched their previously enthusiastic endorsement of CRM to one of caution if not opposition.

Practitioners and industry analysts alike were quick to note that CRM had gained a bad reputation for payback, with a distinctive lack of successful implementation and losses in the billions of dollars. According to a 2005 Meta Group study, 55–75% of CRM projects were described as failures; a 2006 Gartner Group reported 55% of CRM projects had failed to produce anticipated results [35]. Charting the reasons for CRM failure has produced an enormous literature. Among the most common causes cited are the following:

■ Too much concentration on the technology and not enough on understanding the customer
■ Requirement to change entrenched, product-based marketing organizations
■ Poorly conceived objectives and goals to be attained by the CRM system
■ Lack of employee and management skills to properly operate the CRM system and deploy functionality to accomplish customer-value generating intelligence

- Insufficient budgeting of training and organizational dollars to successfully configure, test, and implement the chosen solution
- Minimal to no return-on-investment targets that will demonstrate CRM's value to employees, management, shareholders, and customers
- Lack of commitment by senior management for the CRM project

An entire book, *Why CRM Doesn't Work,* was published in 2003 and describes in detail the many faults of the software and the expectations behind them.

Ensuring an effective CRM system requires a comprehensive implementation plan. While the plan requires a detailed set of action steps to guide the implementation process, it also requires companies to clearly define a broader understanding of their CRM system as a business philosophy that extends beyond the software to encompass the entire organization from executives to line workers, all customer–facing processes, and the content of the company's customer focus. The first step in the CRM implementation process is, therefore, to define the projected objectives and benefits of the project. Many companies fail to perform this step. CRM is often seen as a point solution designed to automate tasks and improve efficiencies and not as a clearly articulated strategy complete with metrics to ensure maximum return on investment. Companies often spend large sums to achieve a single CRM function utilized by just a single department and wonder why the return does not have a company-wide affect. In contrast, an effective CRM plan starts with the project's ability to impact corporate strategy. According to Robb Eklund, vice president of CRM product marketing at PeopleSoft, "The Holy Grail of CRM is to move how it is being implemented—as a point solution, to manage sales or run call centers—to an enterprise solution where CRM is integrated and its business processes extended to other areas of a company where customer information may be" [36].

Once a comprehensive strategic plan and ROI objectives have been defined, companies can then begin the process of assembling the CRM suite of products. As discussed earlier in the chapter, a CRM system consists of three integrated functions: *operational* (such as transaction, event, and service management); *collaborative* (such as forecast, process, and information sharing); and *analytical* (such as churn analysis and prediction). Determining which sets of CRM applications to select for implementation should originate in the *requirements* specified in the project definition. This element should clarify the business needs to be solved by the CRM tools. Next, implementers must be careful to match the CRM tool *functionality* to the required solution. Once the above steps have been completed, the specific CRM applications can then be selected. At the culmination of the process, the sum total functionality assembled should map back to the original requirements. Finally, as the project is implemented, CRM analysts must make sure that the proper performance metrics have been closely defined. While it is true CRM costs and lack of maturity of CRM technologies are part of the complaint about the lack of CRM projects showing real ROI, most lackluster efforts are more the direct result of concentrating heavily on the operational side of CRM, coupled with rather weakly defined ROI targets such as customer satisfaction/retention rates, increased sales/revenues, and other metrics.

New Concepts in Customer Management Technologies

Alongside the emergence of the CRM suite of applications has grown a recognition that to thrive and prosper in today's global economy companies need to do more than just deploy analytical tools to generate data from repositories of customer history: they need to become "intimate" with what their customers really want, with increasing the customer experience, if they want to drive customer loyalty and commitment. This microfocus on the customer as an individual has been accentuated by the dramatic changes that have been occurring in communications technologies since the mid-2000s. A new concept called *social networking* has emerged that, by tapping into the connectivity capabilities of the Internet, enables individuals to communicate in real-time with potentially huge peer communities wielding the power to influence the actions of organizations, businesses, and even public opinion. Far from being a temporary phase in the evolution of CRM, *social networking* and *customer intimacy* stand at the forefront of today's customer management environment.

Emergence of Customer Experience Management

As the second decade of the twenty-first century begins, the realization that the customer, and not products or companies, rules the marketplace has become a recognized fact. It is the customer who has assumed the power to direct the design of product and service content, pricing, transaction management, and the medium by which business is conducted. Today's customer expects to be treated as an individual and requires suppliers to provide them with configurable, solutions-oriented bundles of products, services, and information custom designed to meet their unique wants and needs. Finally, customers are simply demanding more control over the buying experience that empowers them to design their own solutions, tap into robust sources of information content, deploy user-friendly tools for order management, self-service follow-up, and financial settlement.

Today, many companies feel that they are effectively responding to their customers by leveraging their CRM systems to architect organizations that are more customer-centered. But while hard components of customer interaction, such as the tracking of demographic and behavior information, Internet-based avenues for transaction management, and brand identification reinforcement to bring into focus critical questions relating to key metrics such as spend, frequency, and cost, are crucial, they constitute only a portion of the effort needed to win and maintain the loyalty of today's customer. Of equal, if not greater importance, are the perceptions and attitudes customers come away with from a buying experience that demonstrates a supplier's value to them. In a word, customers have come to base equal value on both the products and services they receive and on the feelings and expectations, tangible and intangible, which surround their interaction with a supplier, its organization, and its processes. Today's best supply chains are succeeding by unearthing and nurturing this special intimacy with their customers aimed at capturing their loyalty, as well as unmistakable competitive advantage, by listening to the customer experience.

The idea that a superior buying experience is a major contribution factor influencing a customer's future purchasing decisions has always been known by marketers. The steps involved in sourcing, transaction, and postsales functions are imbedded with a range of emotional responses that directly or subliminally impact customers' perception of the exchange and the likelihood that they will again engage in the process in the future. This is not to say that the primary factor in earning customer loyalty is not the product or service. However, according to Strativity Group's 2009 *Customer Experience Management Benchmark Study* [37] companies that invest 10% or more of their revenues on customer experience have significantly lower customer attrition rates, enjoy high levels of referral rates, and are twice as likely to have customer satisfaction scores of 81% or more. These metrics are even more starling when Strativity reports that "40% of loyal customers said they were willing to pay 10% or more to continue buying from companies that delivered great experiences" [38]. Now that the importance of the other side of CRM—customer experience management (CEM)—has been established, what does it mean and how can it be used to grow competitive advantage?

CEM has been defined simply as "the process of strategically managing a customer's entire experience with a product or a company" [39]. A more comprehensive definition has been formulated by the Peppers and Rogers Group. CEM is the

> totality of an individual customer's interactions with a company and its brand over time. It essentially ensures that each department and touch-point—from sales to billing to returns—act collectively in the customer's best interests to generate long-term loyalty. It requires starting from the customer's view first (not the company's), then aligning people, processes, and technology to ensure that interactions are valuable from the customer's vantage point. [40]

Perhaps the best definition of why CEM is the single most important driver of business today is expressed by Todor in his book *Addicted Customers:*

> Commitment and loyalty to a vendor have their roots, not in the product, but in the experience the customer has associated with the product. Experiences are defined by their emotional and psychological consequences. Therefore, businesses that seek sustainable profits and growth must deliver the customer experiences today's customers value. To do

so they must understand what motivates customers, what creates desire and what leads to trusting and valued relationships. [41]

Simply, customers have come to expect the highest quality and the lowest prices available from a wide range of market channel alternatives—the "product" has dropped into second place in the buying decision [42]. What today's customers really value and will reward with loyalty has shifted to how closely the buying experience resonates with both their product as well as their personal psychological and emotional needs.

CEM is a business management strategy and not a "marketing concept." CEM is about how companies and their partners can use their products and services to provide customers with exceptional and fulfilling buying experiences, rather than just an endless parade of features, options, and brand proliferation. CEM seeks to build loyalty and grow profitability by making an emotional connection with the customer that transcends the value of the goods and services they offer. Finally, CEM provides a realistic, measurable approach to creating value built on the strong identification and conscious preference of the customer for a brand based on long-term experience and expectation at the "moment of truth" when the transaction occurs. As marketing and the Internet move into a new age of social networking, companies capable of effectively respond to today's "experiential marketplace" will have to transform their supply chains from product delivery pipelines to customer value creation engines capable not only of superlative service but also of identifying and enriching the customer experience whenever the customer is faced with a buying decision at any channel network touchpoint.

Fulfilling customer expectations

Perhaps the secret to a successful business resides in providing customers with an array of very special values that actually exceed the nature of the product and services being offered to the marketplace. These companies have figured out how to impart to the customer a very unique, and always repeatable experience that, while centered on a quality or unique productlservice envelope, produce in the customer a craving, a psychological need that drives them to the brand even if a clearly better alternative is available.

A perfect example is the almost cult-like devotion of customers to the "Starbucks experience." That people would dutifully stand in line to purchase a very expensive cup of coffee has become a much envied proposition. But it is true that customers' pulse rates climb and their sense of anticipation becomes almost unbearable as they are approaching the shop and is only consummated with the actual purchase, the smell and feel of the warm cup, and sitting down in the atmosphere of the shop among other caffeine addicts. Similar attitudes can be recognized by Wal-Mart, Target, and other bargain shoppers who can not wait to enter their favorite store.

Merging CRM and CEM

As an application technology, CRM enables companies to acquire, grow, and retain profitable customer relationships. Unfortunately, instead of developing meaningful relationships with customers, CRM has been used more as a tool to create databases and generate customer metrics without any real focus on addressing customer experiences. Somehow the real message of CRM's capabilities has gotten lost in the search for profitability and as a support for existing marketing practices. In reality, effectively managing customer experiences is an integral part of CRM. Although some advocates feel that CEM should be split off from CRM and given a special status on its own [43], most marketing professionals consider CRM as both a technology and a CRM tool. CRM defines what the company wants from the customer relationship and gathers the intelligence and insight necessary to unearth what products and services to market and sell into what customer profiles. CEM, on the other hand, is the mechanism by which marketers can clearly identify and build customer loyalty and long-term value. The CEM side of CRM provides an understanding of what creates a positive emotional experience that bonds customers and their favorite suppliers.

Perhaps the best way to perceive CRM and CEM is view them as consisting of two integrated spheres, one focused on ensuring the highest level of customer equity and the other on achieving the deepest, most enduing relationships between a customer and the companies and brands it does business with. Greenberg has come up with an interesting analogy that encompasses these two dimensions of CRM. He calls it "whole-brained CRM" [44]. Similar to the human brain, which is divided into right and left spheres of cognition, so, Greenberg feels, CRM should be considered as consisting of two separate but integrated customer management spheres. The left side of the human brain is described as the "rational" side and governs language, logic, interpretation, and mathematics. The right side of the brain is described as the "intuitional/emotional" side and governs nonverbal processes, visual pattern recognition, perceptions, and interactions. Most of the problem with CRM is that it has been perceived predominantly as a "left-brain" analytical tool concerned with mathematical/statistical functions, with using optimization algorithms to identify and match the right segments of customers with the right products the company has to offer. In this view the utility of CRM is relegated to the hard calculus of corporate cost reduction and profitability increase.

Much of the complaint that has arisen about CRM centers on the failure of organizations to fully understand and embrace the "right-brain" sphere of CRM. These issues boil down to a neglect of human-side factors, such as corporate culture, skills and motives of employees, and attitudes of management. When it comes to dealing with the external part of CRM—*the customer*—the situation is often far worse. Although most CRM projects start off as "full-brained," the human sphere of the customer is often quickly lost. Instead of looking at how the business can provide value to the customer, the focus shifts to how databases and analytics can be leveraged to provide measurements and opportunities for determining how the customer can add value to the company. The left side of CRM looks at customers from the viewpoint of "functional value," while the right side views them from the perspective of the "emotional value" the company provides to them.

Redefining CRM Technologies to Embrace CEM

Today's CRM system requires not only optimization tools for effective customer analytics but also technology tools that enable the timely accumulation and transfer of critical intelligence about the experiences customers are having with a company's products and services. The objective is to not only close the gap between what customers want and what supply chains are actually providing, but also about the intensity of the experience customers are receiving from their interaction with a company's organization, products, and services. CRM systems today have the ability to call on a wide range of technologies to obtain the visibility to data and decision support priorities necessary to closely integrate the customer and the supply chain, manage complexity, and leverage organizational adaptability and demand-sensing capabilities to meet customer experiential expectations.

For example, at Zara, the international apparel manufacturer and retail giant, competitive advantage is based on their ability to electronically link their stores, headquarters, and global production network to design, manufacture, distribute, and retail the latest fashions their customers want before the competition. The process begins with the collection and transmission of customer style intelligence. Once styles have been identified, Zara is able to move from design to the store shelf in 10–15 days. The lightening speed of the process is made possible by linking fabric acquisition from global sources, and then outsourcing patterns, cutting, color treatment, sewing, and assembly. The key is to provide a source of open sharing of information across the supply network that enables customer alignment with channel solutions, productive and distributive agility, timely response, and an experience unavailable at competing suppliers.

While the detail components of today's CRM system have been outlined earlier in the chapter, there are four critical support technologies that enable the linkage of CRM and CEM. The purpose of these technologies is to enable a more cross-functional and cross-enterprise approach to creating the type of experiential value that keeps customers loyal to companies and their brands [45].

■ *Business Enterprise Backbones.* The application of technologies that link CRM and CEM begins with the implementation of business system softwares such as ERP, CRM, and SCM. These applications act as transaction engines and information repositories for the establishment of critical base data, transaction management, information collection, financial accounting, and a bridge to Web-based connectivity tools. Enterprise systems provide companies with an institutional memory about the customer's lifetime relationship to the organization and its brands. IT systems in this area are used to *consolidate all customer information in databases that are shared across the supply chain network.*

■ *Demand Sensing.* If the essence of customer management is the capture of customer demand signals as they occur in the supply chain, then demand sensing tools linking the supply chain together at wherever point the customer experiences products, services, or information is critical. Technology applications in this area receive intelligence concerning the flow of customer demands at they occur at any point in the supply network, which in turn is translated and immediately broadcast to all levels in the supporting supply chain. Technologies in this space are critical to improving the buying experience by putting the customer in the driver's seat and employing the right CRM programs. Technologies in this area are used to *provide a complete view of customer interactions across all channels.*

■ *Operations Optimization.* Once repositories of customer information are available, marketers can network their business system backbones to streamline pipeline supply and delivery processes that exactly match customer requirements. Besides enabling supply chain partners to link and fine-tune their operations to ensure customer expectation fulfillment, these software pieces enable supply points to optimize revenues based on a dynamic assessment of cost versus value trade-offs. IT systems in this area are used to *engage lean concepts and practices to reduce costs and wastes at each channel touchpoint.*

■ *Demand Shaping.* As intelligence about actual demand arising from customer buying events is broadcast through the supply chain, order management and pricing software can be used to open new opportunities for cross-selling and up-selling. In addition, as metrics regarding the actual level of experience customers is captured and passed to each channel node that touches the customer, companies can invent entirely new approaches to the existing product/service mix that are more closely in alignment with customers' expectations. CRM technologies in this area are used to *provide customer-facing functions with focused order management tools and information to service the customer.*

The Advent of Social Networking

The utilization of CRM technologies to gather information about customers and transactions so that marketers can better target sales, promotions, service management, and reduce costs has been termed by Paul Greenberg, the CRM guru, as "CRM 1.0—traditional CRM" [46]. Today, the growth of social marketing tools and Web 2.0 technologies is portending a new age of customer management that extends far beyond the stovepipe approaches to understanding what the customer is thinking via such efforts as focus groups, credit card purchasing behavior, POS detail, and other techniques. Social media in the form of Twitter, LinkedIn, YouTube, Yammer, and other networking services, is threatening to launch what some analysts feel is a revolution in how companies understand their customers and how customers can interact individually with companies through powerful social communities.

The best definition of social networking again comes from Greenberg:

> Social CRM is a philosophy and a business strategy, supported by a technology platform, business rules, processes and social characteristics designed to engage the customer in a collaborative conversation in order to provide mutually beneficial value in a trusted and transparent business environment. It's the company's response to the customer's ownership of the conversation [47].

According to this definition, Social CRM is not a replacement for traditional CRM, but rather is the next step in its evolution focused on perceiving the customer in a radically new way. The operational, transaction-based, and analytical capabilities of CRM still form the building blocks; but now they have to be blended with technology-enabling social features, functions, and processes exposing an entirely new view of the role of the customer. Traditional CRM was about "owning" the customer, about dictating to the marketplace through the corporate Web site or brand power the content of an organization's reputation, products, and value proposition. All customers could do was accept the company message. The big difference today is that customers can not only access the message, they can contribute to it, edit it, and share it with whomever they please [48]. But it is all not slanted to the customer. Social networking also enables companies to "listen to" and direct the stream of chatter and intervene with information/product/service comments as part of the ongoing conversation.

Tapping into the marketing value of social networking is in its beginning stages. Forrester Research has identified five "eras" of social networking [49]. The first, the "era of social relationships," began in the 1990s with the enablement of Internet connectivity. The second, the "era of social functionality," began in 2007. This era is characterized by the use of Internet tools like MySpace, Facebook, and LinkedIn, which has evolved into platforms supporting interactive networks providing increasing power to user communities. The third era, "social colonization" can be said to have evolved in 2009. Utilizing powerful new browser and identity technologies, users in this era are able to transcend social networks and traditional Web sites making every site a social experience. Participants will be able to canvas the input of aggregate communities of members to make buying decisions, communicate with other sites, and see what their friends are doing on the open Web.

By the end of 2010, social networking will enter its fourth era, "social context." Technologies in this era will be capable of site customization based on people's personal identities and their social relationships. In this era users will be able to enable more robust social applications, allow social networks to utilize e-mail features, and serve as a base of operations for everyone's online experience. Consumers could then opt to share their experiences with online communities in exchange for a more relevant Web experience. Finally, sometime in 2011, social networking will emerge into the "era of social commerce." As social networks emerge as the central repository for identities and relationships, they will become more powerful than existing corporate Web sites. Networked communities will be the driving force for innovation. In the end, users will leverage their collaboration networks to define how they want brands to serve them, the content of messages to be broadcast, and what array of network tools can be used to influence companies.

"Twittering away" Brand leadership

"It was a shot heard 'round the blogosphere," comments Lauren McKay, relating to a marketing campaign launched by Johnson & Johnson in mid-November 2008 for their pain-reliever product Motrin. The tongue-in-cheek message was targeted at moms who suffer from pain resulting from baby-carrying devices who were described as "wearing their babies."

The response was swift and negative on the brand. Moms (and their readers) twittered and blogged their anger, accusing J&J of spreading derogatory messages and of just plain being insensitive to the campaign's target audience.

Unfortunately, J&J simply had not been linked-in to the rising crescendo as hundreds of moms voiced their objections, thousands commented on the content, and millions began shaping a negative attitude to the product. "Here was an immediate and instant negative impact on the Motrin brand," says suresh vittal, an analyst at forester research. An expert on brand monitoring or "listening platforms," Vittal feels that if J&J had really been listening they could have blunted the attack in the first hour. Instead, the blogs and twitters continued unchecked, chipping away at the brand for a full day.

The fire storm ended 24 hours later when J&J yanked the campaign and publicly apologized. In the world of social media, punishment is swift and the pain becomes viral as each moment passes by.

Source: Lauren McKay, "Everything's Social (Now)." *CRM*, Vol. 13, No. 6, (June 2009), p. 24.

Defining the Content of Social Networking Technologies

Technologies targeted at the world of social networking must be capable of empowering customers to utilize the following critical networking capabilities [50].

1. *Communications Repository Management.* As the utilization of Net 2.0 technologies matures, organizations must be able to sort through and capture critical data arising from the social buzz about their products and services

(i.e., product value, service responsiveness, openness to the customer, etc.). Software applications must be able to capture and then be searchable to allow marketers to mine histories of customer interactions with the business, organizational responses, and situational/contextual circumstances that could uncover value for future use.

2. *Advanced Networking.* Customer-to-customer or peer-to-peer networking provides an essential window that enables individuals and companies to stay connected to peers, thought-leaders, or just the buzz of the marketplace. By enabling service functions to network with the customer in real-time and with pinpointed context, many problems normally headed to the dreaded call-center can be deflected online. This not only improves the customer experience, it also reduces service operating costs by reducing overheads while expanding support.

3. *Innovation.* As the social networking footprint expands, companies can utilize their social media tools to source reservoirs of knowledge, ideas, and metrics through keyword searches across social networks to locate information relating to just about any research, problem, or opportunity area under review. What is more, gaps in data can be easily filled in by posing key questions to the network, thereby uncovering relevant responses from as of yet untapped social media platforms. The next "best thing" might just arise out of the matrix of ideas for improvements and new capabilities from the collective input of those who know a company's products and services the best—their customers.

4. *Promotion.* There can be little doubt that the customer will increasingly own the message, that the marketing process is about *buying* and not selling. Social networked customers are increasingly ignoring the tradition media devices by which companies sought to shape and direct their customers and turning instead to their own research and the opinions and experiences of their friends and Internet communities in determining product and service choices. Loyal, happy customers will provide a potent positive message that will penetrate and main thoroughfares and isolated backroads stretching across a global terrain impossible even through the most robust viral marketing campaigns.

Realizing the potential of social networking will require important advancements to CRM technologies. To begin with, the software should promote customer-focused communities, such as blogs, wikis, gadgets, Facebook, Amazon. com, Flickr, Ning, Wikipedia, and other sites. The goal is to enhance customers' ability to communicate with each other, share ideas, and create content specific to the products and brands they feel will enhance the buying experience of friends and associates. This means also that a company's social networking capabilities be able to expand beyond its own Web site and online community to the Internet at large. A second critical component is the ability to execute effective user profiling and identify a company's marketplace advocates. Termed *Reputation Management,* this application enables marketers to identify, empower, and influence their most ardent customer advocates as a pathway to positively influence the opinion of entire communities about their products and services.

Another critical technology is the ability to make transparent within existing CRM workflows the ongoing behavior of online communities. This information should be directed to the SFA functions for review or entry into the company's case tracking system. Finally, the technologies must provide data collection and analytical tools so that marketers can effectively see, measure, and manage social network "buzz." According to PricewaterhouseCoopers, the critical social networking metrics enable companies to hear the Internet chatter by focusing on the following:

- *Volume.* This is the number of times something about the company or its products is mentioned rather than just the historic patterns that have been captured.
- *Tone.* Is the chatter positive, negative, or neutral?
- *Coverage.* How large is the number of participants and how many linked communities are discussing a particular issue?
- *Authoritativeness.* What is the reputation of the sources engaged in the conversation? [51]

These metrics will provide a serious source of marketplace feedback that will assist companies to merge the capabilities of traditional CRM with an online window into their customers to assist them in managing the loyalty of their customer communities by keeping them focused on the most critical events, identifying content helpful to the health of their product portfolio, and enabling them to take swift action as negative feedback is actually occurring.

While Social networking is in its very early stages of adoption, their can be little doubt that all classes of organization will be quickly expanding their ability to connect people and information via networks of expertise with minimal technological investment. Social networking provides companies with a new tool to interact with the marketplace in a more effective manner that transforms how customers see a company and what value propositions they favor and what new innovations they want to see in the future.

Summary and Transition

Today's customers have at their disposal an array of technology tools that enable them to interact with their suppliers, view marketing materials, order products, check on delivery status, and pay for goods and services in real-time. This dramatic transformation in the way business is conducted has shifted companies from their traditional preoccupation with simply selling products to a focus on satisfying the needs of individual customers. Instead of a passive view of the customer, who makes decisions purely on branding and market leadership, marketers today are confronted with customers who can actively decide on which companies they wish to do business with. Building customer loyalty today requires businesses to abandon the "one size fits all" strategies of the past and be able to continuously architect organizations, systems, and supply chains agile enough to determine the exact needs of the customer and to propose customized solutions that resonate with the needs of individualized customers.

Establishing a *customer-centric* focus is a multiphased process that involves reshaping the infrastructures of both individual organizations and supply chain partners. To begin with, literally every customer touch point needs to be dismantled and rebuilt around customer service. Second, companies must understand what each customer values and from these metrics design products, services, and communications initiatives that will drive customer loyalty. Third, marketers must stratify their customer databases. Not all customers should be treated equally. By pinpointing what provides individual customer value, companies can define the proper mix of products, services, prices, and other factors to avoid mismatches between their offerings and what the customer truly values. Finally, enterprises must be vigilant in monitoring and measuring their own customer-centricity strengths and weaknesses.

To more effectively respond to the realities of today's marketplace, companies have increasing turned their attention to CRM application toolsets. The mission of CRM is to assemble focused technologies, such as the Internet, SFA, CRM marketing functions, CSM, partner relationship management (PRM), electronic bill presentment and payment (EBPP), and analytical CRM, to structure touch points that continuously enhance the buying experience of individual customers so that they will remain customers for life. CRM is about opening a window into the habits and needs of individual customers so that targeted marketing campaigns can be established, customer profitability can be measured, and mutually beneficial, long-term relationships can be nurtured. Finally, CRM serves as a key foundation in the structuring of integrated, synchronized supply chains that will provide for the seamless satisfaction of the customer across the supply channel network.

Today, the application of CEM and social networking tools are dramatically expanding on CRM's traditional capabilities. CEM enables companies to move beyond the "numbers" side of CRM by requiring marketers to seriously consider the experiential aspect of a customer's individual connection to brands and corporate value propositions. By linking the emotional content why customers buy to the historical analytics of past behavior, CRM can provide a much more robust picture of customers and reveal fresh avenues for competitive opportunities. Social networking provides companies with online visibility to what positive and negative things relating to their images, reputations, and brands are occurring in real-time. Social networking furthermore is bidirectional: not only can companies hear and respond instantaneous to shape or blunt online chatter, it also provides customers with a medium to influence the direction of business policy, marketing and promotions, and product enhancement and development.

Once companies can assemble a view of what their customers value and how they should be managed, the process of architecting the business components that actually build, acquire, and offer the goods and services can effectively be undertaken. Chapter 6 focuses on how today's technology-enabled SCM has altered the way the manufacturing functions of the business are responding to the realities of today's Internet marketplace.

Notes

1. These observations can be found in Newell, Frederick, *Why Doesn't CRM Work: How to Win by Letting Customers Manage the Relationship.* (Princeton, NJ: Bloomberg Press, 2003), p. 4.
2. Greenberg, Paul, *CRM at The Speed of Light: Capturing and Keeping Customers in Internet Real Time,* McGraw-Hill, Berkley, CA, 2001, *xviii.* Greenberg also devotes 33 pages of his first chapter to detailing a variety of comprehensive definitions coming from a number of CEOs and COOs from companies such as PeopleSoft and Onyx Software. In his 4th edition of *CRM at the Speed of Light* (New York: McGraw-Hill, 2010) Green separates CRM into CRM 1.0 and Social CRM 2.0 each with their own definitions. He defines tradition CRM as "a philosophy and a business strategy supported by a system and a technology designed to improve human interactions in a business environment." The definition Social CRM will be discussed at the end of this chapter.
3. Dyche, Jill, *The CRM Handbook: A Business Guide to Customer Relationship Management,* Addison-Wesley, Boston, MA, 2002, 4.
4. Renner, Dale H., "Closer to the Customer: Customer Relationship Management and the Supply Chain," in *Achieving Supply Chain Excellence Through Technology,* 1, Anderson, David L., ed., Montgomery Research, San Francisco, 1999, 108.
5. Quoted in Hess, Ed, "The ABCs of CRM," *Integrated Solutions,* 5, 2, 2001, 41–48.
6. Dyche, 16.
7. These three regions were identified in Ross, David F., *The Intimate Supply Chain: Leveraging the Supply Chain to Manage the Customer Experience.* (Boca Raton, Fl: CRC Press, 2008), p. 13.
8. Morris, Betsy, "Can Michael Dell Escape the Box?," *Fortune,* Oct 16, 2000.
9. Authoritative discussions of the LCV metric can be found in Kotler, Philip, *Marketing Management, 11th ed.,* (Upper Saddle River, NJ: Prentice Hall, 2003), pp. 75–76, Peppers, Don and Rogers, Martha, *Return on Customer: A revolutionary Way to Measure and Strengthen Your Business.* (New York: Currency Doubleday, 2005), pp. 221–227, and Greenberg, Paul, *CRM at the Speed of Light: Essential Customer Strategies for the 21st Century, 3rd ed.* (New York: McGraw Hill, 2004), pp. 655–662.
10. Murphy, Jean V., "Moving the Focus from Customers to Relationships," *Global Logistics and Supply Chain Strategies,* 5, 3, 2001, 56–59.
11. Cokins, Gary, "Are All of Your Trading Partners 'Worth It' to You?" in *Achieving Supply Chain Excellence Through Technology,* 1, Anderson, David, ed., L. Montgomery Research, San Francisco, 1999, 12.
12. The Buy.com story is related in Cooper, Ginger, "The Quest for Customer Centricity," *Customer Relationship Management,* 5, 9, 2001, 35.

13. See the Buy.com Web site accessed September 10, 2009 at http://www.buy.com/retail/toc_feature.asp?loc=15831.
14. Manring, Audrey, "Profiling the Chief Customer Officer," *Customer Relationship Management*, 4, 11, 2000, 84–95. There is also a website dedicated to the vital company role accessed September 11, 2009 at http://www.chiefcustomerofficer.com complete with a CCO Council, yearly executive summits, and supporting literature.
15. Arnold-Ialongo, Donna, "Building Customer Loyalty," *Customer Relationship Management*, 5, 3, 2001, 25–26.
16. Sawhney, Mohan and Zabin, Jeff, *The Seven Steps to Nirvana: Strategic Insights into e-Business Transformation*, McGraw-Hill, New York, 2001, pp. 175–181 have been most helpful in writing the above two paragraphs.
17. Chaffey, Dave, *E-Business and E-Commerce Management, 3rd ed.* (New York: Prentice Hall, 2007, p. 393.
18. Poirier, Charles C. and Bauer, Michael J., *E-Supply Chain: Using the Internet to Revolutionize Your Business,"* Berrett-Koehler Publishers, Inc., San Francisco, 2000, 154.
19. Sawhney and Zabin, 181.
20. These points have been adopted from Greenberg, 1st edition, 55–56.
21. Dyche, 80.
22. *Ibid.,* 85.
23. Fingar, Peter, Kumar, Harsha, and Sharma, Tarun, *Enterprise E-Commerce: The Software Component Breakthrough for Business-to-Business Commerce.* (Tampa, FL: Meghan-Kiffer Press, 2000), pp. 89–90.
24. Taylor, David and Terhune, Alyse D, *Doing e-Business: Strategies for Thriving in an Electronic Marketplace.* (New York: John Wiley & Sons, 2001), pp. 61–62.
25. *Ibid.,* 79.
26. Voth, Danna, "Making Your Mark in the Information Age," *Customer Relationship Management,* 4, 6, 2000, 61–70.
27. These metrics can be found in Giffler, Joe, "Capturing Customers for Life," *Decision Magazine*, May 1998 and Pechi, Tony, "Sublime Service," *Customer Relationship Magazine,"* 5, 8, 25–26.
28. Fingar, Kumar, and Sharma, p. 108.
29. Jason Compton, "Service . . . With a Smile." *Customer Relationship Magazine,* 5, 1, 2001, 40.
30. Greenberg, *CRM at Light Speed, 3rd ed.,* p. 151.
31. Guerrisi, Joseph, "Making Money Move Faster," *Supply Chain Management Review,* 5, 1, 2001, 17–18.
32. See the discussion in Hill, Kimberly, "The Direction of the Industry," *e.bill,* 3, 5, 2001, 20–22.
33. Scott, Kevin, "Analytical Customer Relationship Management: Myth or Reality," *AMR Research Report,* March, 2001, 5.
34. *Ibid.* 6–7.
35. These reports were referenced in Crandall, Richard E., "A Fresh Face for CRM," *APICS Magazine*, (October 2006), 20–22.
36. Weil, Mary, "A Measure of Vision," *Software Strategies*, 6, 9, 2001, 38–41.
37. This study can be found at accessed September 11, 2009 at http://www.strativity.com/pdf/2009CEMStudyExecSummary-final.pdf
38. See the comments in "Why Customer Loyalty Matters" accessed September 12, 2009 http://www.baldrige.com/criteria_customerfocus/why-customer-loyalty-matters/.

39. Schmitt, Bernd H. Schmitt, *Customer Experience Management*. (Hoboken, NJ: John Wiley & Sons, 2003), p. 17.
40. Peppers & Rogers Group, "Turning Customer Experiences into Competitive Edge: Nikon's Journey to Leadership," *Carlson Marketing White Paper*, (2006), p. 3.
41. Todor, John I., *Addicted Customers: How to Get Them Hooked on Your Company.* (Martinez, CA: Silverado Press, 2007), p. xv.
42. According to Customer Experience Maturity Monitor, "the customer's experience is becoming the new differentiator. Consider that more U.S. adults cite "good customer service" (52 percent) as an extremely important consideration versus "good prices" (38 percent) in engendering loyalty to a company, and four out of five will never purchase again from a company that delivers a bad experience." "The State of Customer Experience Capabilities and Competencies," p. 5, accessed Septemeber 20, 2009 at http://www.sas.com/offices/europe/denmark/pdf/20090323customer_intelligence_103820_0209.pdf.
43. Schmitt, pp 15–17.
44. Greenberg, *CRM at Light Speed, 3rd ed.,* pp. 52–64.
45. These points have been summarized from Ross, *The Intimate Supply Chain*, pp. 189–190.
46. Greenberg, Paul, "Social CRM Comes of Age," accessed September 20, 2009 at http://www.computerworlduk.com/cmsdata/whitepapers/3203130/Social%20CRM.pdf, p. 1.
47. Ibid., p. 8.
48. See the comments in McKay, Lauren, "Everything's Social (Now)," *CRM*, Vol. 13, No. 6, (June 2009), pp. 24–28.
49. Owyang, Jeremiah, K., "The Future of the Social Web," accessed September 23, 2009 at http://www.forrester.com/rb/Research/future_of_social_web/q/id/46970/t/2.
50. Some of these comments were summarized from Ashcroft, Jeff, "Social Media in the Supply Chain," *Inside Supply Management,* Vol. 20, No. 10, (October 2009), p. 12.
51. Quoted by Greenberg, "Social CRM Comes of Age," p. 14.

Chapter 6

Manufacturing and Supply Chain Planning: Linking Product Design, Manufacturing, and Planning to Increase Productivities

Today's technology enabled supply chain management (SCM) model seeks to architect collaborative supply chain networks that closely integrate customers, suppliers, business partners, logistics providers, and other functions that have coalesced into various forms of channel configurations. The objective is to engineer the real-time transfer of information anywhere, anytime within the network and provide the connectivity necessary to coordinate and optimize the cost-effective flow of materials, products, and services. One of the most critical components standing at the heart of this convergence of business functions is manufacturing. In fact, it can truly be said that all channel value starts with the conversion of raw materials and components through the production process into products. Once goods have been produced, they then enter the value delivery network where various support functions and services, from sales and marketing to distribution and delivery, augment and complete the transfer to the end consumer.

The competitive value of the supply chain can therefore be said to start with the ability of companies to optimize their productive resources. Effectively managing productive processes encompasses a range of functions that begins with *product life cycle management* (PLM), and progresses through supplier sourcing, production planning, scheduling and shop floor management, finished goods receipt, and financial close-out and performance review. Underlying this management process is, first of all, a model of the plant itself that describes characteristics such as capacity, cost, cycle times, and constraints. This model, in turn, enables the particular configuration by which the plant executes productive processes to achieve targeted objectives such as quality, order due date completion, quantity, and cost. Realizing plant process output, however, is not automatic. Effectively optimizing productive functions is a dynamic management process, and as the variables associated with product and process life cycles, quality, reliance on outside resources, and other factors increases, so does the complexity of the models and methods necessary to manage them. Without effective planning and control tools even the best structured process can not efficiently work and will be poorly utilized, and without the necessary process output, the supply chain pipeline will slowly dry up and the profitability, indeed the very existence, of the entire channel network will be threatened.

This chapter is concerned with an examination of today's best business practices and technology toolsets used to effectively manage the manufacturing function. The discussion begins by reviewing the role of manufacturing in the "age of the global enterprise." Perhaps the keynote of this age is the tremendous changes occurring in the traditional objectives and methodologies of manufacturing. Once consisting of large, inflexible vertical organizations, today's manufacturer has had to evolve against a backdrop of increased globalization, extreme recessionary forces, increased outsourcing, the dramatic shortening of product life cycles and growing requirements for rapid design and release of new products to market, and the creation of new performance metrics to replace dependencies on efficiencies and utilization as benchmarks of manufacturing productivity. Of particular importance is the introduction of a bewildering array of technology tools available to manufacturing to assist in the management of almost every aspect of the business from transaction control to Internet-enabled B2B exchanges.

Beyond the operations technologies and process methodologies available to the twenty-first century manufacturer, the chapter also discusses one of today's most important drivers of productivity—the ability of manufacturing firms to architect collaborative relationships through networking technologies with business partners providing for the synchronization of all aspects of product design, sourcing, and demand-driven planning. Today, manufacturing firms are engaged in what can be called *design for the supply chain*, signifying that the ability to build and distribute products is the focus of not just individual firms but of whole supply networks. Finally, the chapter concludes with an analysis of today's advanced manufacturing planning functions that seek to apply the latest optimization and Web-based applications to interconnect and make visible the demand and replenishment needs of whole supply network systems in the pursuit of competitive advantage.

Manufacturing in the Age of the Global Enterprise

Manufacturing has traditionally been represented as residing at the center of the value chain. The manufacturing function has often been described as the "800-pound gorilla" of the channel network. As illustrated in Figure 6.1, manufacturing owns, interfaces, or is influenced by almost every function inside the organization and outside in the supply channel. According to Staid and Matthews from Accenture [1],

> a single manufacturing location may represent hundreds of millions of dollars of investment, employ tens of thousands of suppliers, thousands of customers, and supply products that are sold for billions of dollars worth of revenue. As such, manufacturing operations typically represent the bulk of cost/value added within a company's supply chain. Achieving actual sellable output in the face of such complexity and interdependency represents a daily logistical triumph.

Because it is such a critical driver of the supply chain, any changes in manufacturing processes are bound to reverberate throughout the entire channel ecosystem. As manufacturing philosophies change to respond to new product requirements and cost and quality improvement efforts, evolve through management and technology breakthroughs, and adapt to meet the needs of suppliers and customers, supply chain strategists must continually rethink the role of manufacturing and how

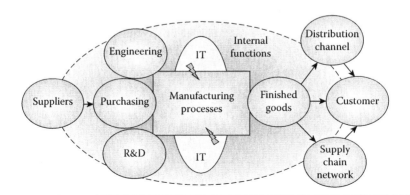

Figure 6.1   Role of manufacturing.

operational, technological, and management changes will impact the entire supply channel galaxy.

There can be little doubt that the management of manufacturing has dramatically changed over the past decade. In the past, acquiring new productive resources and making capital improvements to expand capacities and lower costs differentiated manufacturers and defined industry leadership. Today, manufacturers have come under intense pressure to reduce investment in plant and equipment and reengineer the core processes that remain to be as agile as possible. Part of this shift in the fundamentals of manufacturing strategy is the result of ensuring that companies meet ROA expectations. Value tied up in equipment and inventories peripheral to the business or have poor utilization act as an enormous rock that sinks ROA. Another factor is the declining shelf life of products. Obsolete products and the processes that make them simply add to the asset and not the income side of ROA.

Under the pressure of shrinking margins and global competition, manufacturers during the first decade of twenty-first century have been frantically restructuring the missions and the architectures of their businesses. Many have begun in earnest the process of "deverticalizing" their productive strategies. Companies that once built their own assemblies, offered in-house services, and distributed goods are now outsourcing these functions to channel partners who have much stronger core competencies in these areas. The most radical example of this philosophy is the decision of some former manufacturers to completely outsource all manufacturing and logistics functions and become a "virtual enterprise" focused perhaps on strong internal competencies in brand management or engineering.

For most of us who have grown up in the belief that manufacturing *was* the core competency of any company this change has been quite dramatic, if not bewildering. "Not all companies," writes Don Swann, "have pushed manufacturing to the point of being a contracted commodity service. But the trend is to view manufacturing as an asset-intensive necessary evil rather than the core competency that defines a company's personality, culture, and reputation" [2]. What this means is that manufacturing is now faced with the necessity of redefining its role and meeting the new challenges brought about by today's marketplace realities. There can be little doubt that today's manufacturer faces a host of complex and interrelated challenges unparalleled in history. Everything is changing, and the pace of change is accelerating exponentially. The ability to leverage technological automation, exploit real-time information from demand and supply sources, and closely integrate customer and supplier relationships rather than managing physical assets, such as equipment and inventories, have become today's most critical business dimensions.

Surviving and thriving in this brave new world requires manufacturers to continuously develop solutions to the following four themes. Collectively, these themes present a dramatic departure from past management modes, each requiring radically new solutions. Understanding today's customer is the first theme.

The key to this theme for manufacturers is architecting demand-driven organizations capable of rapid structuring combinations of configurable products, services, and information that will provide the customer with unique value, fast-flow delivery, and a solution to their buying needs. The second theme, managing manufacturing infrastructures, requires manufacturers to develop processes and organizations resilient enough to thrive in an era of global competition. Successfully adapting to the need for collaborative manufacturing partnerships and purposeful outsourcing to meet the realities of today's business environment comprises the third theme. Companies are looking to their supply chain partners to assist in product design, shoulder the task for the actual manufacturing through partnership agreements, and participate in highly complex global networks where value creation is considered a collaborative process. Finally, changes to the importance of manufacturing to the firm have given rise to a new set of performance standards for manufacturing. In this fourth dynamic some of the traditional manufacturing performance benchmarks, such as efficiencies and asset management, will be revisited.

Demand-Driven Manufacturing

Throughout this book, the expanding power of the customer has been described as perhaps the key driver in today's global business environment. Instead of acquiescing in standardized products and one-size-fits-all strategies, customers today are demanding configurable goods and services that provide them with value-added extras, order change flexibility, and a unique, personalized buying experience. In the process they want exceptional convenience, reliability, speed, and self-directed control. In addition, customers have grown accustomed to continuous changes in product design and technologies that shorten product life cycles while simultaneously elevating continuous innovation expectations. Finally, communications and global logistics have extended the reach of customers who are always ready to abandon old loyalties to pursue alternative levels of quality, functionality, and service from competitors located anywhere in the world. For today's manufacturer, it seems that every process or product improvement that is made is quickly eclipsed by even higher levels of marketplace expectation.

Unfortunately, many of today's manufacturing companies are still not customer-centric. Most manufacturers continue to develop their marketplace strategies around product lines or brands and standard service offerings. The focus of such management structures is on developing product and service mixes that correspond to the needs and desires of a majority of the firm's customers determined by detailed marketing analysis. The goal is to utilize economy-of-scale manufacturing practices to develop processes that minimize the impact of product design changes, shrink production costs, support standardized pricing, permit the recycling of marketing and advertising campaigns, and utilize existing distribution infrastructures.

Today, most manufacturers have been migrating from supply driven strategies to ones that are "demand-driven." Rather than pursuing a traditional "push-through" strategy that depends on a series of formalized steps beginning with forecasting and then proceeds to building products to stock, investing in large inventories, and utilizing promotions, pricing, and brand awareness to generate demand, demand-driven manufacturing requires manufacturers to move toward the planning and production of inventories based directly on customer orders as they cascade through the supply chain. Pursuing demand-driven manufacturing means that both manufacturers and their distributors must possess the appropriate technologies for the broadcast of the demand signal to all supply chain partners simultaneously. It also means possessing the ability to fulfill orders to demand without increasing inventories, incurring additional transportation costs, losing margins, or losing customers.

Changing manufacturing companies to respond to a demand-driven strategy is not easy an easy task. Demand-driven production requires the restructuring of traditional manufacturing methods through the injection of Lean principles and concepts, a highly efficient and collaborative supply chain, digitized, fast-flow order management and fulfillment, the efficient deployment of channel resources, and sustained levels of high-customer satisfaction. Of critical importance is an information system infrastructure that enables manufacturing to not only establish efficient flow-through of demand into the planning and shop floor execution systems, but also the establishment of strategic operational guidelines throughout the supply channel enabling demand-information to make fulfillment decision-making more collaborative, effective, and efficient. Demand-driven manufacturers succeed by developing channel strategic objectives and deploying enterprise business systems (EBS), like ERP and CRM, which enables them to extend their vision beyond the business to gain insight and control of the entire supply chain.

Challenges to the Manufacturing Infrastructure

Responding to the realities of such drivers as global competition, demand-pull customer management, and e-commerce have required manufacturers to make substantive changes to their production and supporting supply chain infrastructures. Demand-driven manufacturing necessitates abandoning traditional production management methods in favor of Lean principles, a highly efficient and collaborative supply chain, rapid order turnaround, flexible plant equipment and labor, agile planning and shop scheduling, and the utilization of networking technologies that converge channel members around the common pursuit of customer excellence. Moving to a more demand-driven environment requires manufacturers to pursue the following competencies:

■ *Demand Planning.* While commodity-based manufacturers can leverage standardized electronic collaboration platforms such as *collaborative planning, forecasting, and replenishment* (CPFR) to share, adjust and update demand forecasts with channel partners, make-to-order manufacturers are faced with managing production in the face of demand variability. Solving this dilemma requires organizations to shift from forecasts to the utilization of e-commerce tools that provide a real-time window into transactions, visibility to orders, shipments, and self-serve functions, alerts to out-of-bounds situations, and connecting internal systems and external Internet service providers systems to create new types of values chains.

■ *Management Control and Organizational Structure.* Supporting demand-driven manufacturing models requires centralizing logistics and supply chain functions under a single executive to remove impediments and improve organizational decision-making and responsiveness. Without strong cross-functional teams capable of converging the demand-signal into a single, unified response supply and customer management objectives will be splintered into separate silos making effective response difficult.

■ *Flexible Manufacturing Infrastructures.* Effective demand-driven manufacturing requires productive infrastructures capable of rapid change to meet changes in the internal or external environment. Manufacturing flexibility encompasses such categories as the ability to adapt the existing production system to quickly accommodate product design change, handle a wide mix of product variants by employing equipment with short setup times, adapt processes to manage changes to production volumes, capability to quickly switch product routings as capacities fluctuate, and ability to rapidly adjust for unexpected variations in material inputs.

■ *Lean Production Methods.* At the heart of demand-driven manufacturing is the deployment of the wide range of process improvement tools associated with Lean manufacturing. According to a 2009 survey performed by the Manufacturing Performance Institute [3], three-quarters of manufacturing plants described themselves as pursuing some kind of Lean program for process improvement (Table 6.1). However, the survey was quick to point out that while Lean techniques had a wide appeal, the intensity of Lean initiatives was lacking in many companies. First, plants reported that Lean methods had been applied to only about 50% of production processes. Second, many of the key Lean manufacturing concepts such as strategy deployment (31%), PDCA problem-solving (27%), waste elimination (9%), Kaizen (9%), and value-stream mapping (8%) were missing. The adoption of demand-driven manufacturing is approached incrementally over time, through a process of continuous improvement. Without an effective Lean manufacturing program companies will find the move to a demand-driven strategy extremely difficult.

■ *Fast-Flow Technologies.* Demand-driven manufacturing requires supply chains to deploy technologies that foster fast-flow product development, production planning and scheduling, and timely product delivery. Technologies that enable manufacturing flexibility can be product development based (PLM),

Table 6.1   2009/2010 Deployment of Improvement Methods in U.S. Plants

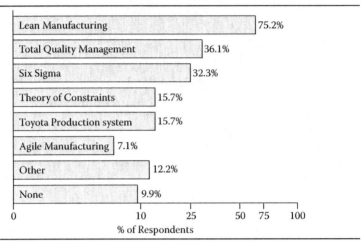

Method	% of Respondents
Lean Manufacturing	75.2%
Total Quality Management	36.1%
Six Sigma	32.3%
Theory of Constraints	15.7%
Toyota Production system	15.7%
Agile Manufacturing	7.1%
Other	12.2%
None	9.9%

% of Respondents

computer-aided design (CAD), processed based (barcode readers, scanners, computerized numerical control (CNC) machines, robotics, manufacturing execution systems (MES), planning and scheduling based materials requirements planing (MRP), finite scheduling tools, operations optimizers, MES, channel partner based (collaborative product development, B2B exchanges), and customer based (e-commerce notification, electronic request for quote, collaborative forecasting, project/contract order visibility, configuration, and Internet fulfillment coordination).

Outsourcing and Supplier Partnerships

Beyond customer-focused infrastructures and technologies, demand-driven manufacturing also require companies to increase their ability to quickly respond to changing marketplace conditions by effectively utilizing the core competencies of their supply chain business partners. Supply chain partners assist manufacturers to overcome barriers to agility by enabling them to pool resources and expertise to reduce cycle times, capture new ideas and innovations, and network capabilities without expensive in-house development. Supply channels fill this gap by providing noncore strategic and operational competencies that reduce manufacturing, distribution, and service costs, improve flexibility, keep companies focused on core competencies, provide access to global networks and superior technology, improve quality and service, reduce capital investment, and increase cash flow.

Over the past 30 years, companies have attempted to leverage their business partners through outsourcing. According to a recent study by Duke University and The Conference Board, the pace of manufacturing outsourcing is expected to quicken [4]. The report stated that between 2005 and 2008, the number of American companies with an off-shoring or outsourcing strategy more than doubled, and, despite the political fallout (especially during the recession years 2008–2009), few companies plan to relocate these activities back to the United States. In fact, 60% of companies currently off-shoring say they have aggressive plans to expand these activities. According to a report performed by PRTM, a management consulting firm, of the 300 companies surveyed it was indicated that they all would be outsourcing more than 50% of their manufacturing [5]. While footwear & apparel were early adopters of outsourcing and off-shoring, Electronics & High-Tech have taken this to a new level in which a global supplier like Cisco reports that 90 of its production is outsourced, giving it a massive global supply chain [6]. Although outsourcing seems to be the panacea to removing costs and growing profitability, the benefits are not extravagant. According to the PRTM survey, respondents reported achieving only 17% cost saving from outsourcing. While labor costs are reduced by 26% and material costs by 18%, management and overhead costs are being trimmed only slightly, which reduces the overall savings.

The decision to outsource noncore manufacturing functions to channel partners can take a strategic or operational form. *Strategically,* outsourcing has been pursued as a method to drive cost reduction, shrink the time necessary for new product innovation, gain quick access to materials and components, and reach global marketplaces. Strategic partners can help their channel partners companies add value to products by improving time to market, decreasing delivery time, or tapping into the success of complimentary products offered by suppliers. Strategic alliances also enable firms to stay focused on developing core competencies and reducing overall risk through sharing agreements,

Operationally, supply partners can assist companies realize demand-pull objectives they would be otherwise incapable of reaching using their own limited competencies. Partner alliances, such as third party logistics (3PL), buyer/seller associations, collaborative product development, retailer–supplier partnerships, and distributor integration, assist enterprises to streamline transaction processing, lower system costs and cycle times, enhance technology systems, and enable resources to be used more efficiently and effectively.

The value of an outsourcing program is based squarely on the ability of business partners to provide the right blend of value, quality, price, and delivery that support demand-pull requirements of the manufacturer. Ensuring that a partner will be able to fill the necessary resource and competency gaps requires the creation of sourcing methods for effectively evaluating, auditing, and developing partners. A successful program requires planners to do the following:

- Validate the desired competencies of the partner by charting key operational metrics such as lead time, inventory turnover, and on-time delivery.
- Confirm the financial health and stability of partners through financial statements and outside analysts.

- Evaluate the level of technical functionality of the partners. This involves a review of the technology tools to be used to process information and to guide the flow of products and services to the customer.
- Determine the "fit" of the partner's culture and values with those of the company.
- Develop a contract that explicitly details expectations on both sides, identifies resources up-front, specifies short- and long-term objectives, and outlines performance metrics, reporting content, calculation, and timing of delivery, states fee structures, and stipulates the nature of legal recourse for nonperformance.
- Agree on continuous improvement efforts in operations, managing change, sharing risk, and expanding the depth of the outsourcing relationship.

Changing Performance Targets

A critical component of responding to the new roles and challenges manufacturers face is revamping the traditional benchmarks that measure the level of manufacturing performance. Past metrics that measured return on investment, asset optimization, and process performance clearly need to be modified and in some cases abandoned altogether and demand-pull strategies gain traction. To be able to compete in a Lean, demand-pull environment manufacturers will have to shift the standards of performance from traditional concerns with efficiencies and utilization to flexibility and information collaboration with channel partners. Many components of plant management will not change: products will still have to be made at optimum quality, costs kept to a minimum, continuous improvement of processes pursued, and integration with engineering, customer, and logistics functions deepened. What will be different is the construction of agile, flexible processes that can be scaled rapidly and at little cost to pursue emerging marketplace opportunities. In fact, the ability to receive and transmit information rather than products will become the central competitive weapon for manufacturers, permitting them to deploy productive resources to meet differing demands from different channel networks and virtual companies.

Manufacturing performance centered on agility, flexibility, and scalability will require significant alteration of traditional ROA and asset utilization metrics. To begin with, manufacturers will have to compress the time normally allocated for ROA. Investments in plant and equipment that exceed 18 months should be examined closely. Risk here involves the possibility of product and process obsolescence, excess capacity, and opportunity to utilize more cost-effective outsourcing alternatives. Another critical area is designing manufacturing processes that are truly customer driven. The goal is to construct productive capabilities that enable companies to make-to-final customer order by synchronizing their networked supply channels to meet order delivery dates while recognizing material, manufacturing, assembly, and logistics constraints. Now known as "the direct model" or the "Dell model," the mechanics of the system seek direct connectivity of the customer with the manufacturer's supply chain that together fulfill the customer's demand. Here, the manufacturer retains the ownership of product design and targeted assembly, while linking in real-time the total production requirement with outsourcing partners who in turn have the capability to pool inventories and leverage core manufacturing processes to supply some or all of the required end-product. A simple way to begin moving manufacturing to this model is to refocus traditional efficiencies from just making products to realizing benchmark utilizations to only making products to the demand schedule. In this scheme process efficiency is credited only when orders are built on time and then shipped to meet the customer delivery date.

While efficiency measurements must change, so must traditional views of plant utilization. Historically, plant managers have been rewarded on high utilizations to ensure products are being made at high efficiency and low cost. Today, the measurement of utilization must revolve around how well productive processes produce to customer demand. While there maybe an objection that high-volume, process manufacturing will have a hard time moving to a customer-centric philosophy, nevertheless, managers even these industries must migrate their planning and control systems to a philosophy that puts the customer at the center of all production decisions. Achieving such a strategy will mean that all aspects of the manufacturing process—product and process design, supporting planning systems, customer management systems, production management, and total plant performance—must continuously move toward greater integration with supply partners and customer-centricity.

The Future of Manufacturing

According to research performed by Capgemini [7], the impact of globalization, integrative information technologies, Lean, demand-driven environments, and closer collaboration with suppliers on manufacturers will be greatly expanding in the next decade. What will manufacturing look like in the year 2020? To begin with, continued globalization and sophistication of developing markets will drive manufacturers to increasingly source, produce, and sell internationally. The Capgemini report projects that manufacturers with foreign presences will expand from around 50% in 2010 to 8% by 2020. This growing global integration will require manufacturing to become more collaborative with companies, involving customers and suppliers at all stages in the manufacturing process. When it comes to the supply chain, half the respondents expect a decline in the number of suppliers used, while 40% expect an increase in the use of distributors to reach new marketplaces.

In the area of research and development, the keynote is deeper collaboration with customers and suppliers. On the customer side, collaboration is expected to expand from involving customers with refinement of features to the involvement of customers at product concept and development of major features phases of R&D. Much the same is expected for supplier collaboration. For the majority of survey respondents refining features was singled out as the prime reason why suppliers are brought in at R&D. However, involving suppliers in the development of major functions and even conceptualizing products is expected to gather increasing traction. These movements toward increased collaboration are driven by a wide consensus in the report that product life cycle will be cut in half over the forthcoming decade.

In the area of manufacturing management, several key trends were identified. The first was the expectation that the role of digital technologies in manufacturing is set to increase significantly by 2020, with the technologies adopted depending mainly on *what* is being made rather than *where*. Integration of production systems from head office to shop floor and from design to quality monitoring, encompassing computer-aided design (CAD), CNC, enterprise resource planning (ERP) and PLM, is predicted to be the main emphasis of development. Beyond technologies, manufacturing will continue outsourcing and internationalization with more companies manufacturing for global markets and operating manufacturing plants in a greater number of countries. The biggest change will be a shift from manufacturing in one country to manufacturing in many. In terms of the products produced, it is expected that by 2020, the balance will have tipped in favor of using subcontractors to help localize products for international markets versus single site plants producing a standardized product. This fits with the general trends towards the internationalization of manufacturing.

The movement to demand-driven, collaborative manufacturing practices will also have an impact on the future of the supply chain and logistics. According to the Capgemini report, manufacturers in 2020 will be relying more on overseas suppliers. Today, suppliers tend to be located in the same geography as the manufacturer's plant. This practices is expected to change with the majority of companies surveyed (60%) expecting to be sourcing from more countries in 2020 than they do today. However, on the other side of the supply chain manufacturers are expected to experience a growing reliance on distributors to service an increasingly complex supply chain. This forecast is attributable to what will be a growing competition and the opening up of new emerging markets forcing manufacturers to market to a growing customer base, which implies a downstream expansion as more distributors and resellers are engaged to satisfy this global demand. Regrettably, in the midst of the globalization figures, discussing growth stands the specter of increasing risks as the supply chain is exposed to ever-greater risks from disruption by weather and natural disaster, terrorism and political instability, and delay introduced through complexity.

Impact of Technology on Manufacturing

Beyond the changes driven by the growing power of the customer, increasing collaboration with supply partners, and globalization, the explosion in integrative information technologies during the first decade of the twenty-first century has had an enormous impact on today's manufacturer. Because of the scale, scope, and complexity of manufacturing, companies have for decades sought to utilize the processing power of information management systems, such as MRP II and ERP, to calculate material planning, plan and control manufacturing activity, and integrate the various functions of the business. As discussed in Chapter 3, the purpose of these enterprise business systems (EBS) is to provide for the organization and standardization of all of the data of a manufacturing company and to enable the integration and optimization of an enterprise's *internal* value chain from purchasing and inventory management through sales, production, and financial accounting.

Today, companies and software developers have all but completed the process of integrating into their EBS manufacturing support applications advanced planning toolsets that enhance tried and true MRP functionality. These advanced applications not only enable manufacturers to gain access to faster and more accurate planning and shop floor execution but also, by utilizing the integrative and collaborative capabilities of the Internet permit them to transmit and receive vital data on a real-time basis. The se advanced Web applications provide today's manufacturer with the means to plan, control, and optimize operations by synchronizing production processes not only with other company processes, but also, through SCM systems, with supporting partners out in the supply chain.

Overall, businesses can expect breakthroughs in and novel applications of technology to continue to impact the theory and practice of manufacturing. In addition to enhancing operating and integrative efficiencies, technology will both drive and enable the development and deployment of new applications that will more closely synchronize supply chain demand with productive capabilities. The growing momentum for the utilization of Software-as-a-Service (SaaS) and cloud computing hold out the promise of reduced implementation time and costs, thereby enabling such functionality for the small- to medium-sized manufacturer. Finally, integrative technology will assist manufacturing functions to accelerate the move out of their historically reactive mode to the influx of new product and process changes and customer requirements to a proactive, strategic role in determining the competitive strength of their companies.

Short History of Manufacturing Planning and Control Systems

Because of the sheer size, scope, complexity, and volume of manufacturing data and diversity of management methods, manufacturing has always been considered a prime area for computerization. As is illustrated in Figure 6.2, over the past 60 years manufacturers have sought to apply the newest hardware and software applications to solve the problem of collecting, calculating, reporting, and

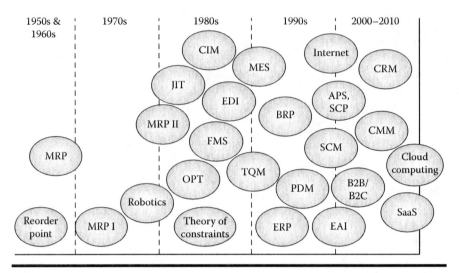

Figure 6.2 Chronology of major computerized manufacturing applications.

utilizing manufacturing information to assist them in making the best decisions for inventory planning, scheduling and controlling the shop floor, and building products to meet customer demand. During the 1950s and 1960s, the first uses of the computer were in the inventory management area. In this era, companies sought to move critical inventory management functions, such as perpetual inventory control, calculation of requirements based on reorder points, and, finally, time phased or MRP bill of material (BOM) explosion, from manual maintenance to the computer.

By the 1970s, advances in computer hardware and manufacturing theory assisted basic MRP to evolve from being purely an ordering system to a set of applications used to integrate company demand with the material plan calculation. Perhaps the most important aspect of what became to be known as closed-loop MRP (MRP I) was the inclusion of sales and operations planning and master scheduling into the system loop to drive the MRP calculation and then to connect the output to shop floor and purchasing release and scheduling (Figure 6.3). As the 1970s closed, manufacturing practice and theory was again advanced with the rise of manufacturing resource planning (MRP II), which enclosed the function of business planning into the closed-loop system, and the appearance of a radical new manufacturing philosophy from Japan—just-in-time (JIT).

The 1980s witnessed a virtual explosion in new computerized tools and management philosophies as the "MRP crusade" gained significant traction. Manufacturers now had the opportunity to supplement weaknesses in the MRP II application suite with the addition of shop floor programmable controllers,

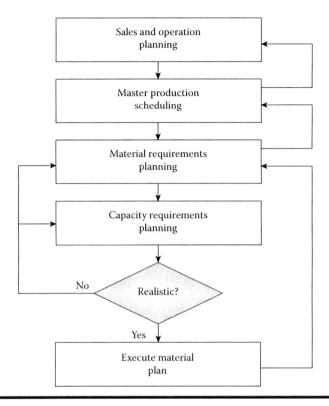

Figure 6.3   Closed-loop MRP.

manufacturing execution systems (MES), enabling tools such as bar coding to assist in shop floor data collection and scheduling, EDI systems enabling the first computer-to-computer linkages, computer integrated manufacturing (CIM) enabling the use of CAD/CAM applications for product design, and new production theories centered around total quality management, constraints management, and PC applications for shop floor optimization and finite loading. These application toolsets were joined in the 1990s by enhancements to MRP II, which became ERP, and product data management (PDM). This decade also was marked by the addition of powerful noncomputerized philosophies, such as business process reengineering (BRP) and SCM, which changed forever the traditional structure and objectives of the enterprise. Today, as will be detailed below, the application of Internet tools has again dramatically changed the nature of manufacturing technologies and moved management from a concern with planning and controlling *internal* productive functions to the opportunity to synchronize plans and pursue collaborative relationships with supply network trading partners.

Functions of an MES	
Core functionality that should come with an MES: • Work-in-process tracking (labor, machines, materials) • Inventory management • Production reporting: finished parts/goods, scrap, material issues and returns, machine setup and downtime • Online work orders, instructions, drawings and quality steps • Lot and serial number genealogy • Quality management: sampling tests and employee certifications • Automation integration: PLC/SCADA/HMI/Historians/SPC • Bar code and RFID data collection • Visibility: real-time and historical reporting And also consider an MES that: • Seamlessly integrates with applications for yard management, supplier enablement, warehouse management and transportation management	• Has pre-built capabilities for common interface methods and integration to ERP systems such as: SAP®, Oracle®, and others • Allows users to adapt workflows and upgrade quickly/cost-effectively without re-applying changes Noteworthy: Most MES do nothing to connect the shop floor with the broader supply chain. An MES should help you take manufacturing to the next level by co-ordinating those operations with the broader supply chain. This will enable you to achieve a smooth flow of inventory and related information among all supply chain participants *Source:* HighJump Software, "The Dirty Little Secrets of Manufacturing Execution Systems," *HighJump White Paper*, (2008), http://www.highjump.com/ResourceCenter/Pages/SpecialReports.aspx

Geography of Today's Manufacturing Systems

The computer toolsets available to today's manufacturer have evolved to respond to the special needs of effectively and efficiently operating twenty-first century manufacturing. As illustrated in Figure 6.4, these applications can be essentially divided into manufacturing planning, production and process management, product design and engineering, plant and quality management, and PLM. Each will be considered in detail.

Manufacturing Planning

The ability to effectively plan, schedule, communicate, and manage the interaction between enterprise departments necessary to execute the timely acquisition of production inventories and finished goods through MRP is perhaps the most mature and recognizable component of today's suite of manufacturing applications. Classically, MRP can be broken down into three separate but integrated functions.

■ *Material Requirements Planning (MRP)*. This function utilizes item planning data to calculate and provide suggested inventory replenishment action to

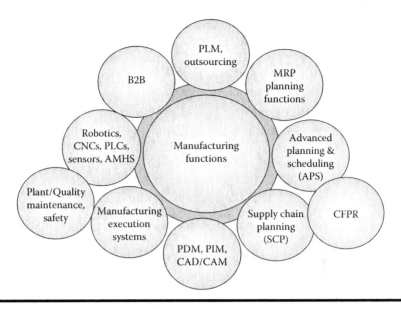

Figure 6.4   Today's manufacturing applications.

meet current demand. MRP's primary mission is to ensure *priority control*: the timely release and ongoing maintenance of open orders to ensure accurate due date completion.

■ *Capacity Requirements Planning (CRP).* This function utilizes the MRP requirements output, converts it into load, and then matches it to available shop work center capacity. By balancing load and capacity, planners can ensure the feasibility of priority plans arising from MRP.

■ *Shop Floor Control.* After priority and capacity plans have been validated, shop orders can then be released to manufacturing. Once on the floor, MRP utilizes tools such as order dispatching and input–output control to ensure jobs are being completed on time.

Despite the tremendous degree of planning and control afforded by MRP, today's requirements for often minute-by-minute update of data and reformulation of plans have rendered the labor-intensive planning and recalculation processes of MRP too cumbersome for real-time shop floor information and execution. Recently, these deficiencies in MRP have been solved with the rise of advanced production planning (APS) and SCM systems. These applications, normally run on PCs and integrated with the MRP backbone, are designed to quickly recalculate shop priorities, level load and optimize resources, provide for inputs into purchasing, and enable planners a window into the material and capacity resources of supply chain partners. Today, these toolsets have been able to tap into the enabling power of the Internet to create real-time linkages with suppliers. Termed collaborative planning, forecasting, and replenishment (CPFR), this collection of business practices is designed to utilize Internet and existing technologies to link the demand and supply capabilities of manufacturers, distributors, retailers, and suppliers in order to integrate channel demand with total network resources, reduce channel inventories, and improve productivities.

Production and Process Management

The continuous search for new manufacturing philosophies and methods to automate shop floor control and optimize scheduling and integrate it more closely with demand planning has been at the core of today's systems approach to manufacturing. Beginning first with the "MRP crusade" of the very early 1980s, shop floor management has been involved in a continuous process of innovation, resulting in a wide range of applications, from scheduling systems and collection devices to machine PLCs and automated handling systems (AMHS). The collective objective of these applications has become mission critical for today's enterprise: how to simultaneously optimize factory productivities, keep costs to a minimum, reduce inventories and cycle times, effectively plan and utilize capacities, and be agile and responsive to the customer.

Fundamental to achieving shop floor management goals is the ability to track production in real-time. For over a decade, MES have attempted to fill in this gap in shop floor management. According to MESA International, a trade association of MES vendors, MES can be defined as a group of applications encompassing order dispatching, operations and detailed scheduling, WIP tracking, labor/machine positing, maintenance, quality management, and document control. The prime function of MES is the control and coordination of work cell and equipment controllers to optimize plant efficiency. Unfortunately, the implementation of MES systems and their integration with EBS backbones and shop devices such as PLCs has been slow in coming. Much of the problem resides in the dynamic nature of MES systems, which makes it difficult to integrate them into MRP and supply chain systems as part of a comprehensive manufacturing model. In addition, most MES systems lack Lean flow capabilities within the four walls, cannot support inventory management outside the four walls, and often require process changes to fit software functionality. Still, forward looking enterprises, such as GE Fanuc, have considered their MES system as the foundation of their integrated approach to manufacturing process management.

Product Design and Engineering

The explosion in integrative networking technologies over the past decade has provided for the creation of computerized tools to assist in the design, development, and rollout of new products. The objective of these applications is to reduce the cost of development and shrink the time from design to product availability. Some of these applications, such as CAD and computer aided manufacturing (CAM) have been available since the 1980s. These applications provide design engineers with automated tools that facilitate product design and negotiate the smooth transition of the product structure into BOMs and process routings the MRP system can use for costing, quality management, planning, and production scheduling. These tools have become particularly important to support the trend toward make-to-order manufacturing. By enabling the quick custom design of modular components into customer-configured products that can be easily transferred to the MRP backbone, CAD/CAM applications can shorten the entire life cycle of design and manufacture.

Plant Maintenance and Quality Management

One of the most important benefits of an integrated factory system is its increased capacity for plant maintenance, quality, and safety. Among computerized toolsets available can be found:

- *Plant Maintenance.* Whether a module in the plant's ERP system or a stand-alone application, the purpose of a computerized maintenance management system (CMMS) is to reduce equipment downtime and maximize production output. The impact of plant maintenance, however, extends beyond the maintenance department. A poor maintenance program results in lost production, poor product quality, increased scrap, missed deliveries, decreased market share, compromised safety, and shortened lifetimes of capital equipment. In today's highly competitive environment, integrating planning, scheduling, production and maintenance makes good business sense. Production wants to reduce downtime and prefers to have equipment maintenance scheduled in a way that maximizes productivity and avoids emergency repairs and equipment failures. Similarly, the maintenance department wishes to reduce downtime through the rapid diagnosis of impending problems through the use of real-time data to be able to predict the probability of failure. The solution is to engineer a CMMS that permits an accurate estimate of equipment failure in sufficient time to assemble the necessary resources from parts/spares control, purchasing, fixed asset management, and maintenance resources and then to schedule production and maintenance together, slotting maintenance work into intermittent periods when production has planned stoppages.

  A critical part of any CMMS is the equipment database. A CMMS should provide engineers with a detailed catalog of each piece of equipment and its place within the plant. Key data would not only consist of equipment process standards, but also information relating to recurring problems, equipment idiosyncrasies, and output quality that, linked together with regularly scheduled inspection and preventive maintenance, can provide an effective trigger alerting maintenance of impending problems. Other technology tools such as equipment sensors to detect impending violations of tolerances, interactive computer maintenance screens facilitating estimating, planning, and scheduling of maintenance, powerful analysis tools generating visual representations from the ERP database, and futuristic modeling capable of automatically reroute production in anticipation of equipment shutdown, assist CMMS to ensure effective maintenance does not significantly impact production.

■ *Quality Management.* For today's manufacturer, unbeatable quality is no longer considered a competitive advantage: it is merely the price of admission to the marketplace. Customers expect that products will either meet or exceed their expectations not most of the time, but all of the time or they simply will migrate to a competitor who can provide it. Most ERP systems, coupled with specialized equipment, can be leveraged to design a range of quality strategies. These systems provide access to methods ranging from periodic product sampling to sophisticated statistical process control (SPC) tools, such as a range of process control charts for plotting process errors, to databases that can be utilized to eliminate *assignable* and continuously reduce *random* causes of process error.

The utilization of software quality management tools can assist a company define the strategic level of quality desired. The primary level of quality management is *inspection*. The purpose of this basic level of quality is to inspect process output for the purpose of plotting the instance of error. However, this method accepts that a certain percentage of output will always have some level of defect and clearly will not assist a company reduce quality error. The next level of quality, *process measurement and improvement*, enables quality management to turn its attention to uncovering and fixing the root causes of error with progress monitored through statistical measurements. The third level of quality management, *process control,* utilizes SPC control chart plotting to monitor processes to keep them under control by continuously eliminating all assignable causes of error and ensuring random causes of error are within proscribed tolerances. The final level of quality management, *design for quality,* requires the utilization of reliability engineering, cooperative design, value engineering, design for manufacturability, and quality functional deployment (QFD) toolsets to engineer into the product design characteristics that will maximally fulfill customer requirements and create quality characteristics, such as durability, performance, reliability, features, and serviceability, unmatched by the competition.

Product Life Cycle Management (PLM)

As the product life cycles continue to shrink and demand for innovation quickens, manufacturers are more than ever faced with the need to deploy methodologies in the quest for shorter product development cycles, flexible manufacturing capabilities, and quicker product rollout capabilities. Today, PLM has been singled out as the solution to solving the bottlenecks characteristic of the product life cycle. PLM can be defined as the quick, concurrent, coordinated, and highly interactive

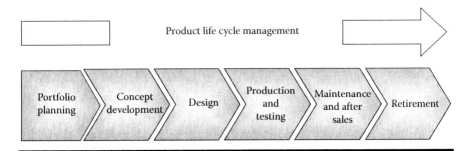

Figure 6.5   Product development life cycle.

development of the product and associated processes through the collaboration of design teams from the manufacturer, the supplier base, and customers. PLM solutions develop a complete digital representation of product design through manufacturing. Conceptual designs are taken from the earliest stages of customer specifications through design development, analysis, simulation, and manufacturing prototyping. Each development step is an enhancement of the previous step, as designs are refined and details added. Figure 6.5 shows this life cycle. An effective PLM program consists of four simultaneously functioning components that can be described as follows [8]:

- *Structured Processes.* PLM is a methodology that seeks to leverage modeling and simulation tools along with cross-functional communication, decision-making, and purposeful action to continuously shrink the design and process times necessary to successfully bring new product to market. PLM is a vehicle to develop a repeatable, structured process that centers on toolsets, such as QFD, to unearth and validate customer requirements and relevant technologies that ensure built-in quality assurance and continuously shrinking PLM efforts.
- *Process Development Tools.* Evolving PLM requires the application of a wide range of technology tools ranging from PDM and CAD through Web-enabled design tools and project management software for billing and reporting tasks. Included in this area are the ongoing deployment of responsive, flexible, modular, and in-line cellular processes and equipment that enable easy configurability while meeting standards for quality and continuous product/process cost reduction. PLM in this area also seeks to effectively utilize computer modeling and simulation for rapid prototyping of products, processes, and systems and the application of programmable processes and equipment that together shrink risk and accentuate systems validation and assurance testing.
- *Supply Chain Partners.* Executing world-class PLM requires companies to expand their access to both internal skilled, multidisciplinary design teams, and the physical capacities of supply chain. The shift to utilizing outsourced

functions has been increasing as companies seek to leverage the core competencies and manufacturing capabilities of their supply networks. The advantages of contract manufacturing are important: OEMs can stay centered on core strengths and outsource marginal functions to companies with specialized expertise; access to new technologies can be easily deployed without the capital expenditure; spikes in demand can be solved without increasing productive assets; and utilizing a partner with global operations can assist OEMs to rapidly ramp up a worldwide market presence.

▪ *Technology Backbone.* Providing connectivity to PLM requires the architecting of enterprise and cross-enterprise computing environments that enable integrated, interoperable, and transparent design, manufacturing, quality, and maintenance systems. These systems include the following: CAD and computer-aided engineering (CAE) systems for design and development; PDM applications providing for product definition, manufacturing process, change control, and documentation management; APS systems that optimize both local factory operations and trading partner supply chains; MES that monitor and control operations at the factory level; SCM systems that apply planning algorithms to activate real-time demand and supply information up and down the channel network; and ERP systems that provide the backbone for transaction posting, performance monitoring, and data repositories for reporting and analysis.

▪ *Ecodesign/Green Products and Regulatory Compliance.* The growth of environmentally conscious manufacturing has pointed to the necessity of into the PLM process requirements to determine the environmental impact of product manufacture across the internal and external supply chain, including after-market service, maintenance and end-of-life disposal or recycling. An effective PLM process should include both suppliers and engineers in the material compliance evaluation process to ensure complete compliance with standards for hazardous substances, FDA regulations, electronic equipment disposal, and automotive end-of-life processing. The future holds the possibility of the requirement of product design that incorporates measurement of energy use and the calculating the size of the carbon footprint as part of the manufacture process.

Impact of Internet Networking Technologies

An emerging component in manufacturing management today is the Internet. While the manufacturing technologies described above seem to be concerned with *internal* systems associated with product design, planning, scheduling, and running the shop floor, increasingly companies have become aware of the need for real-time connectivity with their suppliers and outsourcing partners. As it has had in other parts of the business, the utilization of Internet networking applications means more than simply engaging in a few buying exchanges or setting up a few networking portals connect suppliers and outside designers with internal manufacturing functions. Leveraging the tremendous collaborative power of integrative networking technologies requires a fundamental change in the strategic mission of most manufacturers, which can be detailed as follows:

■ *Manufacturing Process Synchronization.* The application of Web-based technologies has the capability of providing today's product design and manufacturing functions with the ability to receive real-time information from a variety of systems throughout the business and to develop action plans in conjunction with established performance metrics. These applications provide browser-based graphical representations of aggregated shop floor data from multiple sources that facilitate real-time monitoring of productive processes. The goal of Internet integration, however, is more than simply enabling the passage of demand into production planning and monitoring of production: real-time linkage of information will enable companies to be more proactive to synchronize and optimize resources in response to impending changes in demand and supply, An example of such efforts is GE Fanuc Automation North America, Inc. (Charlottesville, Virginia), that coupled its suite of factory automation applications with a Web-based application that provides visibility of real-time data from manufacturing and repair operations throughout the supply chain. Similarly, other companies, such as USDATA Corp. (Richardson, Texas), Intercim, Inc. (Burnsville, Minnesota), Teradyne, and Camstar, Inc. (Campbell, California), are using the Web to push traditional MES capabilities out into the supply chain [9].

■ *B2B Supplier Management.* During the later part of the first decade of the twenty-first century, the utilization of the Internet and B2B for supplier management have been growing in importance for manufacturers. Manufacturers have found the use of the Internet to facilitate product search, order status/tracking, product catalogs, vendor search capability, and supplier/buyer back-end integration particularly critical for raw material and component sourcing. But of greater importance is the use of the Internet to increase collaboration with suppliers. Conventional methods for linking with suppliers, like EDI, are today too cumbersome and slow and focus solely on tactical benefits. Web-based connectivity tools, on the other hand, permit companies to dramatically increase acquisition functions beyond MRO-type purchasing. Internet-enabled tools facilitate custom design through a collaborative-design-chain process involving all elements of the supply network. In addition, B2B connectivity enables manufacturers to link real-time information concerning productive capacity and WIP both within the plant and outside in the supply channel. Such linkages are particularly important in Lean production systems that require tight production and financial integration between trading partners.

■ *Internet-Driven Design Collaboration.* As product life cycles plummet and new products proliferate geometrically, manufacturers have become aware that merely synchronizing sourcing and material flows with supply partners are insufficient to achieve world-class leadership. Responding effectively to today's design environment requires firms to also execute flawlessly *product content synchronization.* Product content can be defined as all the data, such as BOMs, drawings, process execution information including operations and testing, and quality certification, needed to make a product to the correct specification. Accuracy of the product data is absolutely necessary if procurement is to acquire the right components and if manufacturing is to build the right product.

Product content synchronization has become even more important in today's manufacturing environment, as companies continue to outsource production. Outsourcing some or all of the design process has often resulting in out-of-sync product content information among supply chain members who must struggle with conflicting drawings, CAD files, and design documents. Even simple issues such as ECO changes are difficult to transmit, can cause confusion, and often result in component delivery delays. Finally, since channel partners rarely have compatible systems, synchronization of information is difficult and delays in the communication of critical data is common.

In the past, some of the world's largest enterprises, such as Hewlett-Packard and General Motors, developed their own computerized applications that would help coordinate the efforts of both internal and supply chain designers and manufacturers. These toolsets, however, floundered when applied to the proprietary systems often found within the corporation and outside in the supply chain. With the advent of the Internet, these early efforts at design and manufacturing synchronization and collaboration were significantly enhanced. By assembling all the components of a product's content for transmission through a single portal, designers and manufacturers can work with BOMs and process routings through a Web browser. Product ECO changes, documents, descriptive text, costs, all can be transmitted in real-time to associated engineers and production planners. In addition, this information could then be accessed in real-time by suppliers and business partners who could also in real-time begin the synchronization of the flow of components with planned production. For example, 80% of the design of one of Volkswagen's models is executed in cooperation with company suppliers. This cooperative process led leading one vendor to conclude that the whole idea of using exchanges to interact with suppliers has more to do with the design process, rather than just refreshing inventory. Design, the most important function of a manufacturing company, now becomes a B2B fundamental. Instead of an isolated group of designers, product development today is a responsibility of multitiered supply chains of design professionals driven by peer-to-peer (P2P) technologies that enable the construction of interoperable repositories of intellectual property. Instead of disconnected push systems that add time and cost to the product supply process, the Internet permits all participating supply chain partners to participate and collaborate in real-time in the execution of product design and production processes.

Collaborative Product Commerce

Manufacturers have always known that the ability to converge cross-functional development teams consisting of members spanning internal design teams, customers, and suppliers added immense value to the PLM process. Each "age" of product development management had sought to leverage competencies to build teams that could facilitate the design, manufacture, and rollout of new products. During the 1960s and 1970s manufacturers focused on utilizing internal teams to achieve *design for cost* objectives. In the 1980s, *design for quality* became the mantra of product collaboration. By the 1990s, companies began to seriously incorporate their supply partners in the process not only to assist in controlling costs and maintaining the highest level of quality, but also in executing *design for manufacturability* objectives as reengineering and JIT philosophies drove manufacturers to increase outsourcing more of the functions of product management once performed internally. Today, manufacturing firms are engaged in *design for the supply chain* product management. This strategy requires businesses to develop product management processes that will permit them to gauge the impact they will have not only on traditional cost, quality, and features objectives, but also on the production, planning, and distribution requirements imposed upon the supply chain.

This increased focus on integrating and synchronizing the entire supply chain in the pursuit of faster product development, speed-to-market, and shortened time-to-profit is termed collaborative product commerce (CPC). CPC can mean several different things depending on the context of the discussion. CPC can refer to a management philosophy used to leverage a company's supply chain to design and produce products. Then again, it can refer to a group of collaborative PDM techniques. Finally, it can be identified with a group of growing computerized applications.

Defining CPC

The concept of CPC can be described as

> the convergence and rapid deployment of product life cycle management competencies and toolsets found anywhere in the supply chain linked by real-time computer applications focused on the collaborative execution of new products and manufacturing processes to meet the total requirements of the customer.

The difference between CPC and previous product design philosophies and toolsets is dramatic. In the past, all aspects of product design were considered a jealously guarded company secret. Only the most noncompetitive processes were permitted to be discussed with business partners and even these were communicated with the greatest of reluctance. Slowly, companies became aware that by converging resources from all over the corporate development chain new ideas could flourish, costs reduced, products marked by superlative quality and manufacturability increased, and time-to-market intervals shrunk. While some powerful technologies did emerge over time, such as PDM and product information management (PIM), the scope of these design tools limited their use to the four walls of the enterprise.

Today, shrinking product life cycles, the growing predilection of customers for custom, configurable products, globalization, and an increasing dependence on outsourcing have unequivocally driven manufacturers to a position of expanding interdependence on their supply chain partners. Instead of inward-focused product development functions, the critical need has shifted to *inter*organizational collaboration, as companies search to increase outsourcing of their manufacturing and design responsibilities in order to more effectively focus on core capabilities. Guiding this new approach is CPC. With CPC firms can escape the limitations of past philosophies of product development and utilize ideas, capabilities, processes, and technologies regardless of time or place.

The object of CPC is simple: by enabling all stakeholders in the supply chain to have access to design information at any stage in the project, product design teams feel that they can reduce communications costs, delays, and redundant reengineering efforts while improving understanding of actual product performance requirements and characteristics in the early stages of design before significant cost and time have been expended. An example of CPC in action can be seen by examining the process at the Stephen Gould Corporation.

Gould is a pioneering giant in retail and industrial packaging and printing, and is the largest privately owned packaging sales organization in the United States. Almost 99% of Gould's products are highly customizable and often require complex engineering to meet rigid customer specifications. Recently, Gould implemented a CPC system that enables the company to link together in real-time engineers residing in different locations. According to Robert Sherman, the firm's GM, the system provides the ability for engineers to

> look at the same screen at the same time and communicate via [the system's] conferencing application or by telephone. One person at a time controls a pointer on the screen to show exactly where a suggested change needs to be made. The initial drawing is left undisturbed in the central repository while new versions with layered changes are date- and time-stamped and stored as well.
>
> When a drawing is finalized, Gould's customers can access it and communicate changes if the fit is not quite right. Everyone can sign off the prints

in real time. . . . Gould's design process continues with the toolmaker . . . [If a problem emerges], Gould can bring its engineers and customers back online to solve the problem right then and there, instead of going back and forth with the tool shop later during the production run.

Gould also is using their CPC system to facilitate the passage and approval of quality control documents. In addition other functions, such as marketing engineering and production, can utilize the system to communicate necessary changes and concerns in a real-time workspace environment. Finally, the CPC system enables manufacturing to be integrated with possible ECO and BOM changes coming from the customer [10].

Linking Supply Chain Design Capabilities

The definition of CPC detailed above highlights the fact that the technique is primarily a knowledge management exercise focused on populating a common, interoperable repository of product design management data consisting of requirements specifications, design documentation, manufacturing process structures, and postsales product support available in real-time to all concerned parties. Until just recently, however, collaborative manufacturing initiatives between channel network partners were cumbersome and required an enormous management effort to ensure that everything from product content, like BOMs and process routings to demand forecasts, were complete and accurate. As the requirements to satisfy what appears to be an unstoppable trend toward utilizing outsourcing, interoperable real-time access to design databases that provide information about product availability, supply plans, and product content changes from anywhere in the supply chain, CPC has become critical for effective decision-making. The supply chain focus of CPC should be applied to the following processes [11]:

- *Planning and Scheduling.* In conjunction with SCM applications and concepts such as CPFR, CPC requires channel network inventory be visible on a real-time basis, effective forecasts and pull systems accurately describe channel demand and supply requirements, and network productive capacities are accurate and easily accessed.
- *Design.* While each channel node will possess certain core competencies in the design process, CPC toolsets should facilitate the creation of cross-enterprise design teams that are an integral part of all decisions from the point before specifications are finalized through to design for manufacturability.
- *Sourcing and Procurement.* Once product specifications have been completed, an effective CPC system should provide manufacturers with the ability to create intimate relationships with their supply chain. Visibility and collaboration efforts should facilitate vendor management, strategic sourcing, pricing, and linkages to vendor capacities.

- *New Product Introduction.* During the process of moving new products to production, the requirement for increased collaboration is intensified. CPC applications should provide for concurrent access to BOMs, design validation, prototyping, validation testing, and final transfer to volume production.
- *Product Content Management.* Once live production has begun, channel partners must be able to quickly respond to impending design and processing problems. CPC tools should drive product change generation, change impact assessment, change release, and change cut-over/phase-in.
- *Order Management.* Based the product strategy, CPC applications should facilitate order management. Among the key components can be found order configuration functions, available to promise, order tracking, and exception management.

Detailing the Contents of CPC

Today's CPC application suite represents a tremendous enhancement over previous product design software solutions. While many of the toolsets in CPC have their origins in CAD/CAM/PDM, there are several critical differences. The first is that CPC is Internet-driven. Internet-enabled CPC is enabling manufacturers to leap over former barriers of geography and time by empowering product designers anywhere, anyplace on the globe to exchange design data so they can collaborate on a project in real-time. "This is a unique time in the history of technology," says George Ashley, Ingersoll Rand Company's manager of engineering business services.

> There has always been some technology limitation to real-time design collaboration, but now there is not. Design collaboration really works, and the challenge now is letting people learn how many ways they can think collaboratively [12].

The goal of Ingersoll Rand is to use their Web-based software to connect the $13 billion-a-year company so that designers and factories anywhere in the world can work together on new products.

Besides the Internet, today's CPC is markedly different from yesterday's design applications. There are differences in architectures, implementation requirements, and the scope of business functions impacted by the computerized techniques. Compared to PDM solutions, CPC toolsets require less data manipulation and are faster to implement. But most importantly, CPC applications assist manufacturers realize the promise of today's new product development and design environments through the activation of virtual design teams that can utilize the Internet to leverage the core competencies of supply chain partners to achieve shorter design times and quicker product rollout to the marketplace.

Currently, there is no industry consensus on which software applications comprise the CPC suite of products. No one software company today provides a full range of technologies. Finally, exact definition of CPC terminology has yet to be formulated. Still, it may be useful to arrange in a single diagram the suite of toolsets normally associated with CPC. Figure 6.6 attempts such a definition and a brief review of each technology is as follows.

- *CAE/CAM.* Applications in this area contain some of the oldest product design toolsets. Among these applications is found CAE, design, manufacturing, CAD, and other related development tools. The function of these applications is to assist product designers utilize a single data repository to store requirements, specifications, BOMs, CAD drawings, approved suppliers, process descriptions, and other data. Data stored in these databases would be accessible only to internal design team members.
- *CRM.* Today's customer order management systems must provide configuration management functionality that feeds directly into the design engineering engines. The capability to manage these configured structures, so necessary for mass customization, can be found in supporting CAE, CAM, ERP, or PDM systems.
- *CSM/B2B.* Component and supplier management (CSM) systems work with design BOMs. These toolsets provide the logic to generate parts classification and sourcing as a prerequisite to the development of preferred parts lists and

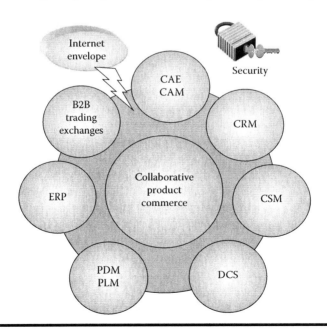

Figure 6.6   Today's collaborative product commerce environment.

approved manufacturers' lists. The most advanced forms of CSM are Web-based portal solutions that enable suppliers to login and access designs and documentation to facilitate update of products changes.

- *DCS.* Design collaboration software (DCS) provides companies with the capability to define design teams that consist of members *outside* of the organization. These software products provide a medium for the synchronous and asynchronous P2P connection of a cross-company design organization. The best of these applications utilize the Internet to facilitate design data transmission and access, virtual manufacturing, life cycle management, and interoperability.

- *ERP.* ERP backbones provide the foundations files for design and manufacturing management and product transactions. Many ERP vendors provide sophisticated product design tools or easily executed interfaces between their ERP systems and third party CPC applications.

- *PDM/PLM.* Software in this area seeks to manage all forms of product design data. The mission of PDM/PLM is to drive requirements specifications, design documents, manufacturing process management, and postrelease product support and evolution documents into a common data repository. Among the associated functionality in this area is electronic work flow, change notification, visualization applications that provide real-time synchronous modeling, CAD support such as versioning and red line markup, Web-based communication, automated e-mail event notification, and other visual tools such as project dashboards and integration to ERP, SCM, and CRM.

- *Security.* Database security is one of the critical components of an effective CPC system. Most manufacturing executives are uncomfortable with CPC security issues. CPC applications must provide companies with functionality to define access privileges, restrict some viewers to read-only status, devise effective encryption solutions to Internet security, and limit designers, particularly from outside the company, access to certain documents and CAD drawings.

While CPC can truly be said to be in its formative stages, the next several years should see its full definition and maturity alongside other time-worn manufacturing tools such as MRP. With new technologies enabling toolsets, such as P2P communications, and the trend toward outsourcing, drive the need for improved collaboration, the pressure for effective, comprehensive CPC solutions can only expect to grow.

Managing Manufacturing Planning Functions

Today's manufacturer is increasingly deploying computerized solutions that enable plants to be driven by actual channel demand that is synchronized with supply chain network resources and constraints. As is illustrated in Figure 6.7, the toolbox

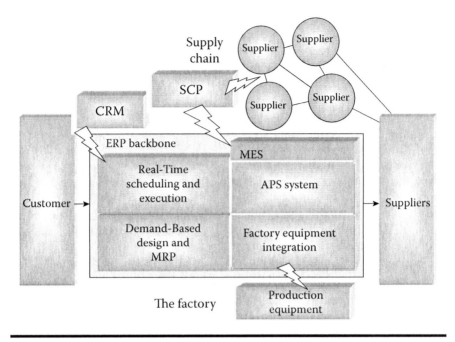

Figure 6.7   Manufacturing planning application suite.

of possible applications includes ERP and MES, as well as new analytical tools, such as APS and SCM, that drive plant information systems and enable management methods such as lean manufacturing that combine planning and factory-level execution. According to Greg Gorbach from ARC Advisory Group, this computerization of plant operations

> must be seen in the context of the extended enterprise and supply chain. What we're talking about are systems for collaborative production management that focus on business processes, and on operating plants within the context of your whole supply chain. [13]

Perhaps the most exciting feature of this movement is the merger of tactical plant operations applications, such as supervisory control, human machine interface (HMI) software, and laboratory information management, with enterprise management tools, applications planning and scheduling analytics, and Web-based portals designed to keep the plant floor synchronized with the supply chain.

In this section the applications associated with the capability of manufactures to merge plant intelligence coming from MES and industrial automation systems with the requirements necessary to effectively drive operational and business information into the supply chain will be explored. This area represents the forefront in manufacturing management functions and is essential for the pursuit of operational excellence through computerized analytical solutions.

Advanced Production and Scheduling Systems

Supply chain management is changing the face of business as it accelerates the flow of products through manufacturing and distribution into the hands of the customer. The ability to execute mass customization and total serviceability while simultaneously reducing cost is possible only when the finite constraints and dependencies of each productive function found across the supply chain can be synchronized and optimized to ensure planners visibility into the total productive capacities of the entire supply chain. The task of advanced planning and scheduling (APS) systems is to address plant-floor constraints and enable the optimization, synchronization, sequencing, and scheduling of demand with individual plant capacity and, ultimately, with total supply chain capacity. When integrated with MES, APS's computerized algorithms can be merged with real-time production floor data to produced optimized schedules that enable effective planning of constraints, capacity, and demand due dates. When integrated with ERP, APS provides the short-term schedule that supports the longer-term resource management tools of ERP used to drive the customer, financial, and distribution applications of the enterprise.

The use of simulation and optimization technologies for APS solutions is not new and has been used in one form or another for the past 40 years. Early adopters of optimization technology tended to be quantitative analysts working in process industries such as chemicals, paper, and steel. These early adopters used general-purpose optimization tools (linear programing packages) either home developed or purchased from software suppliers to develop planning tools that typically ran in a batch mode. Few if any of the available software optimizers dealt with large portions of a company's total supply chain. Over the past decade, business system software suppliers have begun to offer more general-purpose optimization applications that enable linkage of problems, say blending with scheduling. Also important has been the growing power of software suppliers, particularly ERP suites, to embed optimization into their standard solutions seamlessly so they are transparent to the user.

Functionally, an APS system seeks to utilize all materials issued to and resources in a factory to simulate actual delivery capability while illuminating process constraints. To make the APS system work the following elements are necessary:

Accurate Data

The data coming from the typical plant floor is voluminous and dynamic. To effectively structure and filter this information requires that all data elements be as accurate as possible. The data will include such records as individual work center capacities, product structures, process dependencies, lead times, material availability, timely MRP/DRP generations, open purchase orders, and accurate open order operational status. APS data enables planners to realistically model a factory's manufacturing environment.

Planning Timeframe

APS provides for plant simulation over three planning time horizons. The first, *tactical*, utilizes APS simulation logic to make visible constraints over the short- to mid-term horizon to ensure the availability and synchronization of materials and capacities to meet due date commitments. In the second horizon, *operational*, the APS system provides a window for planning the floor on a daily basis. The APS should provide the capability to change the capacity and the load on bottleneck resources and to simulate the effects. In the final horizon, *plant execution*, APS provides planners with tools for the concurrent, real-time management of capacity bottlenecks, adjusting operator shift patterns, balancing manpower allocation, modifying order priorities, and synchronizing materials and process capacities.

Planning Model

The overall mission of an APS system is to achieve an optimal balance between high throughput, minimum inventories, and low-operating costs that results in optimum throughput of product through the production process. Synchronizing orders, operators, machines, and materials are the function of the planning model applied. Basically, there are three techniques available.

- *Mathematical Models.* These models describe the plant environment mathematically and consist of a range of linear and mixed integer programing techniques. This model is best used in stable, repetitive environments.
- *Heuristics.* Models in this group are best used in configure-to-order and make-to-order environments where plant data are not linear and are much too complex for mathematical models. Examples of techniques are theory of constraints (TOC) and process network-based systems that work from the top down—starting from the customer order and working down to the work-center level.
- *Simulation.* This model is characteristic of the earliest type of APS system that evolved into the finite capacity schedulers of the late 1980s. Based around some variation of the queuing theory, simulation models determine order priorities by a bottom-up approach that proceeds from existing orders and demonstrate the optimal sequence of jobs based on the availability of capacity. This model is best applied in capital-intensive industries where the cost of idle machinery and equipment is prohibitive.

Modern APS applications often combine heuristics models with the sequencing functions found in simulation-based systems. The heuristics model assists planners to solve any synchronization issue while the sequencer helps to resolve local sequence conflicts.

Schedule Management

The goal of APS planners is to generate a common schedule of production that can be communicated to the entire factory and consists of the following elements:

- A schedule of optimized sequences for all orders and operations
- Realistic schedule of load priorities for each work center
- Sequenced start and finish times for all orders and resources
- Detailed dispatch list for each work center showing order sequence by operation

Today's APS systems are normally client/server based, utilize memory-resident PC technology so that the complex calculation engines do not have to write to disk, are interfaced to the ERP database, and provide users with interactive scoreboards and drill-down functions that provide drag-and-drop capabilities for working with schedule and material shortage management.

Although APS systems where heralded with great fanfare at the end of the 1990s, the hype has significantly decreased during the last couple of years. Several reasons have been cited to account for the fading interest. To begin with, many companies felt and continue to feel that APS contains more functionality than is required. A significant drawback is the shock to corporate culture represented by an APS system. Planners are expected to forego tried-and-true MRP outputs and spreadsheets for an often complex and involved application that requires extensive training and development of a new mind set. And then and it often conflicted with the existing corporate culture. And then again, much interest in APS was lost during the aftermath of the tech bubble and the ensuing short period of boom and financial bust commencing in 2007. APS all too often was placed on hold and joined other advanced supporting softwares on the shelf.

Besides these issues, another reason for the decline of APS packages was the abandonment at the end of the 2000s of the best-of-breed approach. Many ERP vendors were including APS applications into their expanding suite of products. Because APS is driven by ERP planning data this seemed a logical outcome. Conversely, APS has often been bypassed because a large number of ERP vendors simply do not have it, and companies are unwilling to shoulder the cost and risk of implementing a third party package. Another reason is the confusion between APS and SCM. While the two applications have a lot of similarities, they are focused on solving very different problems. However, many analysts agree that with the migration of APS to Web-based functions, such as browser-based interfaces, support for Internet-based B2B and B2C processes, and availability in an SaaS model, the demand for APS can expect to expand in the 2010s [14].

Supply Chain Optimization Tools

On September 11, 2001, supply chains all over the globe were put under an enormous strain. Focused more on efficiency than on flexibility, supply network capabilities immediately began to break down. Fortunately, while a few companies, such as Toyota, actually had to shut down, supply chains quickly returned to a semblance of normalcy. The lesson was obvious: glitches in the supply chain have not only been proven to torpedo shareholder value, they can also add dramatically to higher costs, lower efficiencies, and reduced flexibility [15]. Countering a possible future crisis the magnitude of 9/11 means that today's manufacturer and distributor must search for alternate ways to construct redundancies in their supply chains through the use of powerful analytical and evaluation tools that allow them to switch channel network capacities on an off similar to Internet technology where, while a node may be down, the entire network remains functional.

Over the past several years manufacturers and distributors have been using constraints-based, optimization technologies resident in APS systems to assist in solving supply chain problems and enabling collaborative planning. Today, these SCM systems have been expanded to include a variety of supply chain functions, including demand planning, supply planning, strategic network optimization, fulfillment scheduling, and CPFR. According to AMR Research, SCM software sales topped $6.68 billion in 2008 and are expected to reach $8 billion by 2010 [16]. The goal of supply chain optimization is to assist planners generate a synergy across the supply chain that better physically positions manufacturing and distribution facilities to achieve the following objectives:

- Networking companies in a supply chain community in order to manage channel complexities by engineering enhanced planning and decision-making capabilities starting with internal ERP systems and extending connectivity to Internet-linked channel trading partners.
- Ensuring that supply channel costs are minimized and that they represent as much as possible the most competitive across geographies and companies.
- Capturing the most profitable customers on a global basis by creating more compelling, value-based relationships than other supply chain networks.
- Securing access to the most value-added suppliers on a global basis by establishing superior Internet-enabled supply chains that offer B2B technology and trading partner relationships.
- Engineering flexible, agile organizations and supply networks that can leverage an array of Internet technologies ranging from CPC to multichannel e-information visibility to capitalize on changes to customer demand and shifts in supply side dynamics.

In many ways applying optimization capabilities to supply channel manage-ment is little more than applying the toolsets and processes of APS. Supply chain planning (SCP) requires first that the SCP environment be described in terms of the actual channel network structure. The channel network structure illustrates the trading partner nodes that add value to the supply chain from raw materials through to finished goods. Similar to APS, SCP enables planning on three levels: *strategic*—concerned with questions relating to where the companies distribution centers should be located or what capacities are need from the supply channel; *tactical*—involves optimizing the flow of goods through a given supply chain configuration over a time horizon and executing sourcing, production, resource deployment, and distribution plans; and *operational*—largely involving schedul-ing, rescheduling, and execution of production and is usually equated with the function of APS.

To make a SCP system work the following components are necessary:

a. *Accurate Data.* SCP systems need to be seamlessly integrated with the com-pany's ERP backbone and distribution, production, and demand planning requirements from all nodes in the supply chain galaxy. For the SCP optimi-zation engines to function properly all data elements should be as accurate as possible. The data will include such records as

- *Supply Chain Structure.* The geography of the supply chain provides the structure for the determination of such elements as demand patterns, forecasts, replenishment, and transportation and their relationship to planning simulation.
- *Product Data.* Data in this component include demand for resources (production, transportation, warehouse space, etc.) and components and materials based on rough-cut routings capacities and item groupings.
- *Available Capacity for All Resources.* There is a wide variety of metrics that could be used including: for *purchasing,* capacity is measured in the number of item units that can be supplied per planning bucket; *transport* is measured in weight or volume units per planning bucket; *production* is based on capacity plans measured in hours per planning bucket; *receiving* is measured in weight or volume units; *dispatch* are the resources that represent equipment/personnel for shipping measured in weight or volume; and *stock area* is the volume or weight for each stock area resource.
- *Costs.* These include total cost for purchasing, transportation, production, and inventory.
- *Penalties.* Penalties assist SCP tools to prioritize market demands when calculating capacity flows. There are two basic types: *missed delivery pen-alties* (prioritization by penalty) and *bucket production penalties* (optimiz-ing production by grouping production into fewer periods to produce in larger quantities).

- *Future Demand for End Products.* This can be expressed either in the form of a simulated demand schedule or sales forecasts for markets and customers.
- *Feedback to ERP/Server.* Time-phased sourcing rules and planning schedule developed during SCP simulation can be transferred to the ERP/Server on request.

b. *Planning Timeframe.* Similar to APS, SCP systems enable planning simulation over three planning time horizons: strategic, tactical, and operational.

c. *Planning Model.* The model chosen for the SCP optimization must be appropriate for the planning level. Planning can precede top down, by starting with data driven from the demand forecasting and order management system and then progressing through production and logistics aggregate optimization planning models and concluding with production and distribution scheduling optimization. Or, the process can precede bottom up by beginning with the MRP and DRP generation and then moving to detailed scheduling and, finally, aggregate planning.

d. *Optimization Techniques.* SCP systems utilize the same optimization techniques as APS.

e. *Schedule Management.* The goal of SCP planners is to generate a common schedule of demand priorities that can be communicated to the entire internal and external supply chain. The software optimization technique will permit planners to consolidate demand and capacity data from all points in the supply chain and then to map and manipulate through simulation various priority alternatives. It is possible to let the mathematical model decide that some demands should not be met because they would result in a loss. This decision is based on profitability (sales price − total costs = profitability). If all demand cannot be satisfied, a predefined prioritization helps to determine which demands to fulfill first. The objective is to meet demand results in the model by pulling product groups to the market while creating a cost-optimized globally synchronized supply chain for end-items and materials. In the process, the model respects all finite capacity constraints and the supply chain network structure.

Planning the supply chain is a complex task since constraints exist on all levels. To ensure that optimization applications provide reasonable, executable solutions several tool sets can be used.

■ Most SCP systems today provide users with graphical user interfaces (GUIs) to facilitate manipulating data and modifying solutions. Typical tools include graphical drag-and-drop planning boards complete with drill down functions. Many graphical planning boards allow users to change a variable (e.g., date or order) and immediately see the impact on the costs, delivery due dates, possible penalties, and to assess new constraint violations. The planner can then make decisions about adjustments and priorities. These decisions automatically update the model's constraints and are sent to the optimizer, which will reoptimize everything based on the latest decisions. The different plans are then compared on the planning board.

■ Users can also control the solution by allowing them to incorporate unique constraints or rules into the model. For example, a planning application might allow a user to specify that customer due dates can be relaxed by 1 or 2 days, or the user may approve the maximum level of overtime.

■ Some software tools permit users to control the progress and performance of the solver method. This is usually accomplished by permitting the planner to set a time limit on the solver and then pause. This lets the planner evaluate the solution as it has been calculated to that point.

■ Some software tools permit planners to define multiple objectives. For example, both customer service and costs can be optimized together—despite the fact that two or more objectives can not really be optimized because they are opposites. These tools provide the ability to strike a balance among competing objectives by a *priority sequence* or through the use of *weights* [17].

■ Today's cutting edge SCP applications enable planners to use the Internet to receive and communicate demand and capacities in real-time from all tiers in the supply chain and to use it in optimization calculations. Termed by Forrester Research extended relationship management (XRM), the use of portal and Web-based technologies enable planners to achieve supply chain connectivity and visibility in a 24/7/365 real-time operational mode. By utilizing supply chain event management (SCEM) alert-messaging functionality planners can see looming constraints in channel trading partners.

SCP optimization applications enable planners to concentrate on making decisions, while the optimizer does the complex and detailed number crunching. The optimizer evaluates all possible solutions and presents the "best" one to the planner. The "best" solution is the one with the lowest total cost for meeting forecasted demand and the one that follows the selected optimization strategy. The result of an effective SCP system is a feasible and optimized plan, which provides the framework for distribution and production planning. All changes and decisions made, such as capacity adjustments, changes to demand and time-phased sourcing rules, are respected by distribution and production planning. SCP optimization assists planners in making controlled decisions in many ways and at different levels of detail. They range from a simple modification of a production capacity in a single planning time bucket to redesigning the entire supply chain structure.

Collaborative Planning, Forecasting, and Replenishment (CPFR)

The complexity of today's supply chain requires manufacturers and distributors to search for new methods to reduce costs, increase efficiencies, reinvent channel models, engineer collaborative relationships, and span functional, cultural, and personal boundaries. While APS and SCM applications provide for the optimization of the supply chain, CPFR seeks to act as a key enabler for the realization of synchronized supply chain forecasting and replenishment. CPFR is the latest generation in a train of channel management philosophies focused on supply chain synchronization. As illustrated in Figure 6.8, CPFR is the maturation of efforts such as quick response (QR), vendor managed inventory (VMI), and efficient customer response (ECR) and can be thought of as the perfect joining of ERP, CRM, and SCP in an Internet-driven supply chain environment dedicated to the integration and synchronization of the entirety of channel demand while reducing total network inventories and costs.

What is CPFR? As the name implies, the mission of CPFR is for all partners in a supply channel network to develop collaborative planning processes based on the timely communication of forecasts and inventory replenishment data to support the synchronization of activities necessary to effectively respond to total supply chain demand. CPFR begins with the development of an agreement between trading partners to develop a consensus forecast that begins at the retail level and makes it way all the way back to the manufacturer. This plan of supply chain demand and replenishment describes what will be sold and when, how it will be merchandized and promoted, in what marketplaces, and during what time period. CPFR interoperable technology permits this data to be freely transmitted up and down the supply channel so that planners at any node in the network can

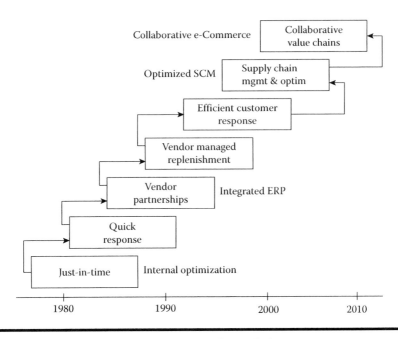

Figure 6.8 Evolution of supply chain planning techniques.

see demand and adjust the plan within certain limits based on possible exception conditions, such as promotions, store openings, or capacity constraints that could impact delivery or sales performance anywhere in the channel. Trading partners would then collaborate to resolve any potential bottlenecks, adjust demand and replenishment plans, and then execute alternative courses of action. The final step in the process, channel replenishment, occurs after consensus on the final forecast [18].

Based on the early experiences of companies such as Heineken USA, Eastman Chemical, Lucent Technologies, Sara Lee, Wal-Mart and others, the Voluntary Inter-industry Commerce Standards Association (VICS) began the task of defining a CPFR standard governing business process, organizational, and technology recommendations in 1996. In June of 1998 VICS released a series of guidelines detailing CPFR industry standards. In the beginning, VICS standards focused on utilizing EDI as the medium for the transmission of collaborative information. Using EDI-enabled companies to leverage existing technology investments to quickly launch CPFR initiatives. By the end of the 1990s, however, the high cost of EDI technologies and the ubiquitous deployment of the Internet enabled even the smallest retailer and manufacturer to leverage the collaborative power of CPFR. In addition, Web-based applications provided business partners to escape from the one-way transmission of data in favor of an interoperable toolset enabling open two-way conversation in real-time supported by formal standards.

In the past, supply chain partners sought to utilize channel inventory management tools such as CRP, VMI, and QR to remove excess assets from the supply network and smooth out demand irregularities. While effective, these toolsets, however, lacked the ability to solve the twin problems at the core of channel replenishment management: forecast inaccuracies and the capability to utilize exception messaging to notify network partners of impending bumps in supply and demand. CPFR provides answers to these two issues by providing for the real-time sharing of sales promotions, point-of-sale (POS) transactions, and total channel inventory positioning that postpones inventory replenishment by linking each level in the supply network with the pull of actual demand. In addition, by systematizing the communication of critical demand and supply data among trading partners, CPFR makes visible all plans and planning variances, thereby assisting companies to improve their forecasting and replenishment decisions to yield the best results. For example, Kimberly-Clark's CPFR manager, Larry Roth, felt that the company's CPFR system has been instrumental in formalizing a structured replenishment approach with all of their suppliers. "Clearly," he says, "CPFR has been a catalyst for re-examining how our various supply chain systems work together and are affected by our customer's plans." There are new examples every month "where our internal plans are improved through both external and internal collaboration" [19].

Summary and Transition

Perhaps the hallmark of the shift in today's economy from the Industrial Age to the Internet Age has been the dramatic changes that have occurred in manufacturing. A few decades ago political economy was measured in industrial output of goods and services by global companies that considered their factories, business processes, equipment, and other assets as the very core of their competitiveness. Today, these same assets are considered as debilitating costs that rob enterprises of flexibility, tie up capital, and blur the focus on core business competencies. Many companies are engaged in a life and death struggle to "deverticalize" the various components of manufacturing, preferring to outsource noncritical functions to channel partners.

All of these dramatic sea changes have required manufacturing to radically redefine its role in order to meet the new challenges brought about by today's marketplace realities. The transformation of global markets and changes to product life cycles, competition, technology, and popular culture are occurring explosively and simultaneously. Responding to these changes requires manufacturers to develop solutions to four critical themes. The first is the absolute necessity of deconstructing past manufacturing management methods that centered on the management of supply and creation of new methods enabling manufacturing functions to be "demand-driven." The key to this theme is the architecting of customer-centric organizations capable of utilizing channel network partners to enable the delivery of combinations of innovation, configurable products, services, and information that provide unique value to the customer. Parallel to the first theme, manufacturers have had to make fundamental changes to infrastructure in their drive to germinate dynamic, high-performance organizations capable of responding rapidly and efficiently to demand for mass-customized products with flexible, cost-effective manufacturing and interactive channel pull systems capable of delivering the necessary product or service to the customer from anywhere in the supply chain.

The ability to effectively respond to the demand-pull of the marketplace can only occur if manufacturers are willing to explore new avenues of supply chain collaboration. In today's fast-paced global environment where product life cycles continue to shrink, demand-driven companies can not hope to succeed without leveraging the competencies of their supply chain business partners. Today's most progressive manufacturers are customer-centric, collaborate with their supply partners, are agile and scalable, and are linked via Web-based architectures to their customers and supply channels. Finally, the concluding theme influencing today's manufacturer is the requirement for a complete revamping of traditional benchmarks of performance. Instead of viewing success based on determining performance on internal ROI and asset optimization, companies need to refocus on agility, flexibility, information collaboration, and service to the customer.

For decades manufacturers have called upon computer systems, such as ERP, to assist in responding to the information management challenges before them. Today, companies have at their disposal an ever-expanding suite of computerized tools for manufacturing planning, production and process management, product design and engineering, plant maintenance and quality management, and PLM that permit them to enhance core competencies by network with their supply channel partners. Of particular importance is the application of the Internet to the suite of today's standard manufacturing technologies. Past enterprise systems, for the most part, consisted of applications used to run *internal* functions in product design, planning, and shop floor scheduling. In contrast, the Internet enables manufacturers to expand design and planning capabilities outside the business to integrate and synchronize real-time demand and procurement data and plans with supply chain partners. As the second decade of the twenty-first century opens, manufacturers have been quick to exploit the potential found in the Internet, a practice that is expected to expand dramatically in the next few years.

Of the technology tools available to manufacturers, the most critical are CPC and supply chain planning. CPC can be described as the convergence and rapid deployment of PLM competencies and applications found anywhere in the supply chain to execute the design, manufacture, and release of new products and processes to the market. The mission of CPC is simple: the assembly of collaborative-design teams from across the supply network utilizing PLM tools linked by Web-based applications focused on understanding customer demand, designing, sourcing and procuring, prototyping and releasing new products to market, and ongoing maintenance. Once products have been designed and released to the marketplace, manufacturers have increasingly come to depend on optimization tools to assist in synchronizing supply chain network demand and individual plant and supply network resources and constraints. The task of APS systems is to address close-in plant-floor constraints and enable the optimization and synchronization of capacities and load. On the strategic and operations level SCM is employed to manage supply channel constraints, search for optimal costs, secure access to the most value-added suppliers, and assemble flexible, agile networks that can leverage the collaborative power of the Internet to make visible in real-time the demand and resources available in the entire supply chain. Finally, manufacturers can apply CPFR functions to supply chain networks to execute the collaborative development of forecasts and synchronization of inventory replenishment. With the use of Internet-based applications, channel partners can freely pass planning information necessary to solve the twin problems of forecast inaccuracies and the capability to utilize exception messaging to notify channel members of possible bumps in supply and demand.

In order to make the wheels of manufacturing hum, the raw materials and finished components consumed in the manufacturing process must be on hand. As inventory and product life cycles continue to shrink and demand for immediate availability of finished products escalates, the pressure on manufacturing and distribution inventories has dramatically increased. In Chapter 7, the function of purchasing and its impact on competitive advantage is examined.

Notes

1. Paul J. Staid and Paul A. Matthews, "The Role of Technology in Manufacturing," in *Achieving Supply Chain Excellence Through Technology,* 1, David L. Anderson, ed., (San Francisco, CA: Montgomery Research, 1999), pp. 78–83.
2. Don Swann, "It's Not Your Father's Manufacturing World," *APICS—The Performance Advantage* 12, no. 1 (2002), pp. 43–45.
3. Manufacturing Performance Institute, "United States 2009/2010 Manufacturers Executive Study," MPI Whitepaper, p. 7, retrieved November 10, 2009. http://storefront.mpi-group.net/p-65-20092010-manufacturers-executive-summary.aspx.
4. Tom Heijment, "Fifth Annual Conference Board/Duke Offshoring Research Network Survey," The Conference Board, August 3, 2009. Summary retrieved Sept. 22, 2009, from http://www.conference-board.org/utilities/pressDetail.cfm?press_ID = 3709.
5. Shoshana Cohen et al., "Global Supply Chain Trends 2008–2010," PRTM, 2008, p. 4. Retrieved Sept. 22, 2009, from http://www.prtm.com/uploadedFiles/Strategic_Viewpoint/Articles/Article_Content/Global_Supply_Chain_Trends_Report_ 2008.pdf.
6. Kevin Harrington and John O'Connor, "How Cisco Succeeds at Global Risk Management," *Supply Chain Management Review* (July 1, 2009). Retrieved Sept. 22, 2009, from www.scmr.com/article/CA6672233.html.
7. Capgemini, "Manufacturing in 2020: Envisioning a Future Characterized by Increased Internationalization, Collaboration and Complexity," retrieved November 15, 2009. http://www.capgemini.com/insights-and-resources/by-publication/tab = 1&businessneed = 0&industry = bf87&alliance-partner = 0&solution = 0&page = 2.
8. These comments have been adapted from James A. Jordan and Frederick J. Michel, *Next Generation Manufacturing: Methods and Techniques* (New York: John Wiley & Sons, 2000), p. 85.
9. These stories can be found in Stephanie Neil, "MES Meets the Supply Chain," *Managing Automation* 16, no. 12 (2001), pp. 18–22.
10. This company story has been adapted from Christopher M. Wright, "They Don't Make 'Em Like They Used To," *APICS—The Performance Advantage* 12, no. 3 (2001), pp. 33–36.
11. These points have been selected from Chris Cookson, "Linking Supply Chains to Support Collaborative Manufacturing," in *Achieving Supply Chain Excellence Through Technology,* 3, David L. Anderson, ed. (San Francisco: Montgomery Research, 2001), pp. 56–58.
12. Steve Konicki, "Groupthink Gets Smart," *InformationWeek*, January 14, 2002, pp. 40–46.
13. Roberto Michel, "Plants Find a New Pace," *Manufacturing Systems* 19, no. 12 (2001), pp. 42–44.
14. For these comments, see Jay McCall, "Is APS Technology Obsolete?," *Integrated Solutions* 5, no. 4 (2001), pp. 72–76, and Jason Krause, "ERP Opens the Door to Collaboration," *Supply Chain Systems* 22, no. 2 (2002), pp. 12–18.
15. Vinod Singhal and Kevin B. Hendricks, "How Supply Chain Glitches Torpedo Shareholder Value," *Supply Chain Management Review* 6, no. 1 (2002), pp. 18–24.

16. Business Software, "Top 10 Supply Chain Management Vendors—2009," Business Software White Paper (2009), 2, retrieved November 17, 2009. http://www. business-software.com/top-10-supply-chain-management-vendors.php?track   = 947&traffic = GoogleSearch&keyword = supply%20chain%20management&gclid = CMDc3_GQjp4CFYJx5Qod3mgToA.

17. Jim Shepherd and Larry Lapide, "Supply Chain Planning Optimization: Just the Facts," in *Achieving Supply Chain Excellence Through Technology*, 1, David L. Anderson, ed. (San Francisco: Montgomery Research, 1999), pp. 166–176, has been most helpful in preparing the preceding paragraphs.

18. An excellent methodology for implementing CPFR can be found in Michelle Lohse and Jeffrey Ranch, "Liking CPFR to SCOR," *Supply Chain Management Review* 5, no. 4 (2001), pp. 56–62.

19. Michael Peck, "CPFR: It Takes 2," *Supply Chain Technology News* 3, no. 2 (2001), pp. 9–10. Syspro retrieved November 17, 2009. http://viewer.bitpipe.com/viewer/view-Document.do?accessId = 10890679.

Chapter 7

Supplier Relationship Management: Integrating Suppliers into the Value Chain

The acquisition of the raw materials, components, and finished goods necessary to service channel network and end-customer demand resides at the very core of supply chain management (SCM). Whether a manufacturer or distributor, the timely acquisition of inventory is fundamental to competitive advantage. Without sufficient inventories the wheels of manufacturing would slowly grind to a halt and distribution pipelines would quickly run dry. Besides providing the goods necessary to meet customer demand, the procurement of inventories also directly affects company financial stability and profitability. Depending on the nature of the business, procurement and services costs can range from 40 to 80% of each sales dollar. To understand the impact of these costs, say, for instance, if a 5% overall reduction in the procurement costs of a typical company could be achieved, it could represent as much as a 50% improvement to the bottom line. To achieve a similar impact, the same company would have to increase sales by 50%, cut overheads by almost 20%, or dramatically cut staff. From such figures it is easy to deduce that the effective management of procurement transcends the traditional mechanics of supplier sourcing, buying, and receiving: procurement is a strategic supply chain function that seeks to integrate and synchronize individual company inventory needs with total channel material flows, trading partner productive capacities, transportation, quality, marketing, finance, and global sources of supply.

The effective management of procurement is, however, more than just buying goods and services. For several decades companies have known that it is not the cost-effective purchasing of inventories but rather the existing relationship between buyer and seller that determines the real value-add component of procurement. As the demands of the customer and the capacities of the supplier are increasingly synchronized, the essential components of procurement are made more efficient, costs decline, the flow of channel inventories are accelerated, and cooperative alliances to improve planning and product information exchange are deepened. In addition, the more integrated the sharing of information becomes the more supply chain partners can fashion truly collaborative relationships where core competencies can be dynamically merged to generate a range of new products, processes, and technologies each partner acting on their own would be incapable of attaining.

In this chapter, the functions of technology-driven purchasing and supplier relationship management (SRM) are explored. The chapter begins with a definition of the basic functions of purchasing. Following, a possible definition of SRM is attempted. Similar to the customer relationship management (CRM) concept, the strategic importance of SRM is to be found in the nurturing of continuously evolving, value-enriching business relationships and is focused on the buy rather than the sell side. While collaborative sharing and merging of procurement competencies dominate the definition, the application of the Internet has opened an entirely new range of SRM networking technologies enabling companies to dramatically cut costs, automate functions such as sourcing, request for quotation (RFQ), and order generation and monitoring, and optimize supply chain partners to achieve the best products and the best prices from anywhere in the supply network. Following, the focus of the chapter then switches to a full discussion of the anatomy of today's SRM system beginning with an outline of enterprise business system (EBS) backbone applications, progressing to technology-enabled SRM service functions such as strategic sourcing and decision support tools. Building on these foundations, the chapter continues detailing SRM processing applications centered around catalog management, RFQ, Purchase order (PO) generation, and logistics, and concludes with a short review of SRM Internet technology services such as Web processing, security, content management, and work flow. The chapter concludes with an exploration of the SRM B2B exchange environment, today's e-marketplace models, and the steps necessary to create and pursue an effective technology-enabled SRM procurement environment.

Defining Purchasing and Supplier Relationship Management (SRM)

The functions of purchasing and supplier management are indivisibly intertwined. Before an effective discussion can occur regarding the impact of the SCM concept on these twin functions, the elements of each must be closely defined.

Defining the Purchasing Function

The acquisition of maintenance, repair, and operating (MRO) and indirect inventories and related services is a fundamental activity performed by all manufacturing, distribution, and retailing companies. According to the *Purchasing Handbook* [1], purchasing can be defined as

> the body of integrated activities that focuses on the purchasing of materials, supplies, and services needed to reach organizational goals. In a narrow sense, purchasing describes the process of buying; in a broader context, purchasing involves determining the need; selecting the supplier; arriving at the appropriate price, terms, and conditions; issuing the contract or order; and following up to ensure delivery.

The purchasing function is normally responsible for the sourcing and acquisition of all products and services used by the enterprise. Broadly speaking, there are three types of purchasing: purchasing for consumption or conversion, purchasing for resale, and purchasing for goods and services consumed in MRO functions. Purchasing for *conversion or consumption* is the concern of industrial buyers and covers a wide spectrum of activities beginning with a determination of what products the firm should produce or outsource, progressing to raw materials and component sourcing, negotiation, purchase order generation and status monitoring, and concluding with materials receipt. Goods purchased for *resale* are the concern of distribution and retail buyers. In this area buyers determine what goods their customers want, search and buy these goods based on targeted levels of quality, delivery, quantity, and price, and sell them at a competitive level based on price, quality, availability, and service. The final type of purchasing, MRO inventories, is concerned with the acquisition of the supply and expensed items and services necessary for the efficient functioning of the business.

The management, planning, and execution of purchasing functions are normally the responsibility of a firm's purchasing department. Overall, the prime responsibility of this function is to communicate the business's purchased inventory requirements to the best suppliers and to ensure timely receipt of materials synchronized to meet the needs of ongoing enterprise operations. The basic activities of purchasing have been arranged below, commencing with functions of strategic importance then progressing to those performed on a daily basis.

- *Sourcing.* This high-value-added activity is concerned with matching business purchasing requirements with sources of supply, ensuring continuity of supply, exploring alternative sources of supply, and validating the level of supplier compliance required to meet or exceed buyer criteria for quality, delivery, quantity, and price. For the past 20 years a critical component of sourcing has been reducing needless redundancies in the supplier base and increasing supplier collaborative partnering.

■ *Value Analysis.* This set of functions is concerned with increasing the value-added elements of the purchasing process. Value analysis can consist of such components as price for quality received, financing, and delivery. An example would be identifying less expensive goods and services that could be substituted at comparable quality and value.

■ *Supplier Development.* In today's environment, increasing collaboration with suppliers has become a requirement for doing business. Pursuing capabilities that promote supplier partnering require buyers to be knowledgeable of vendor capacities, resources, product lines, and delivery and information system capabilities. A key component in the strengthening of this partnership is the development of pricing, technology, and information-sharing agreements that link supplier and buyer together and provide for a continuous win-win environment.

■ *Internal Integration.* Purchasing needs to be closely integrated with other enterprise business areas such as marketing, sales, inventory planning, transportation, and quality management. By providing key information and streamlining the acquisition process, the purchasing function can assist the enterprise to synchronize individual company replenishment requirements with the overall capacities of the supply network. Buyers must also be members of product marketing, research, and engineering development teams if the proper inventory at the best quality, delivery, and cost is to be purchased.

■ *Supplier Scheduling.* One of the keys to effective purchasing is the development of a valid schedule of inventory replenishment. By sharing the time-phased schedule of demand from MRP, firms can provide detailed visibility to future requirements to supply chain partners, who, in turn, can plan the necessary material and capacity resources to support the schedule. In addition, the increased use of purchasing portals and B2B marketplaces have dramatically expanded buyers' ability to search anywhere in the world for sources to meet product and service replenishment needs.

■ *Contracting.* Critical functions in this area consist of the development and analysis of RFQ, negotiation when pricing, volume, length of contract time, or specific designs or specifications are significant issues, and supplier selection and monitoring of performance measurements.

■ *Cost Management.* A critical function of purchasing is the continuous search for ways to reduce administrative costs, purchase prices, and inventory carrying costs while increasing value. The principle activities utilized to accomplish these objectives are purchase cost reduction programs, price change management programs, volume and "stockless" purchasing contracts, cash-flow forecasting, and strategic planning.

- *Purchasing and Receiving.* Activities in this component include order preparation, order entry, order transmission, status reporting, order receiving, quantity checking and stock put-away, invoice and discount review, and order closeout.
- *Performance Measurement.* Monitoring the quality and delivery performance of vendors over time is an integral part of supplier "benchmarking." The ability to measure performance is critical when evaluating the capabilities of competing suppliers and ensuring that costs, delivery, and collaborative targets are being attained. [2]

Defining SRM

Successful supplier management in the twenty-first century mandates that the relationship between buyer and supplier be increasingly conceived as a *collaborative partnership.* As lead times and product life cycles plummet and pipeline flow velocities accelerate, supplier partnering in today global business environment is no longer an option but has become a strategic requirement to maintain competitive advantage. Enhanced by Internet applications that draw buyers and suppliers together in real-time, partnering can assume many forms based on the dynamics of the supply chain. Partnering can be found among allied industries or even competitors and may exist for strategic or operational reasons. Whatever the formal arrangement, partnerships can be described as cooperative alliances formed to exponentially expand the procurement capabilities involved in global sourcing, materials requisition, competitive pricing, operating procedures and efficiencies, and product information exchange.

The increasing focus on the development of synchronized, collaborative relationships between buyers and suppliers has evolved over time and is the product of several marketplace dynamics. As is illustrated in Table 7.1, SRM has undergone dramatic modification and is accented by today's requirement for ever-closer working business alliances. The adversarial nature of yesterday's purchasing arrangement have given way to the structuring of win-win relationships, a mutual commitment to sharing information and resources to achieve common objectives, and the creation of long-term partnerships meant to bind parties in good times and bad. Finally, collaborative partnership often means deconstructing traditional attitudes and practices concerning quality and reliability, delivery, price, responsiveness, trust, the sharing of research and development plans, and financial and business stability. Today's focus on supplier partnerships has grown as a response to the following marketplace realities:

- *Increasing Requirements for Supply Chain Collaboration.* No company in today's marketplace can hope to survive without strong supplier partnerships. As businesses continue to divest themselves of noncore competencies and increasingly turn toward outsourcing, a deepening of partnering

Table 7.1 Traditional Purchasing vs. Collaborative Supplier Management

Traditional Approach	SRM Partnerships
Adversarial relationships	Collaborative partnerships
Many competing suppliers	Small core of supply partners
Contracts focused on price	Contracts focused on long term quality, mutual benefits
Proprietary product information	Collaborative sharing of information
Evaluation by bid	Evaluation by commitment to partnership
Supplier excluded from design process	Real-time communication of designs and specifications
Process improvements intermittent and unilateral	Close computer linkages for design and replenishment planning
Quality defects reside with the supplier	Mutual responsibility for total quality management
Clear boundaries of responsibility	"Virtual" organizations

relationships and mutual dependencies have been eagerly pursued in all industries as fundamental to continuous improvement strategies, total cost management, and competitive advantage.

■ *Changing Nature of the Marketplace.* The dominance of the customer, shortening product life cycles, demands for configurable products, shrinking lead times, global competition, participative product design, and other issues discussed throughout this book have altered forever the nature of sourcing and purchasing and highlighted the importance of supply chain collaboration.

■ *Changing Business Infrastructures.* It has been pointed out earlier in this book that today's enterprise must possess business architectures characterized by extreme agility and scalability while being customer-centric, collaborative, digital, and capable of reliable, convenient, and fast-flow delivery. Although these attributes are normally focused on the sales side of the business, they equally must apply to the supply side. The value chain can be compared to a coin: there is a customer-facing side and a supplier-facing side. The very existence of functions driving one side axiomatically requires the replication of the identical functions driving the other side.

■ *Increased Demand for Cost Control, Quality, and Innovation.* While SCM technologies have been receiving most of the attention, buyers are now more than ever concerned about traditional purchasing values such as quality and reliability. Customers are no longer willing to do business with suppliers who not only can not meet increasingly stringent product and delivery standards, but who also do not possess the capabilities to continuously unearth new product configurations and service management capabilities.

■ *Increased Demand for Risk-Sharing.* True business partnerships mean that the need for trust and risk-sharing be a serious component in any collaborative relationship. As the cost of innovation and operations flexibility grows exponentially and profit levels shrink in the face of competition, partnership agreements that provide for the equal sharing of risk have become a critical method for the management of new product development and controlling spiraling operations costs.

■ *Enabling Power of Internet Technologies.* The Internet Age has had a profound impact on many areas of purchasing, opening new and exciting doors that have provided supply chain partners with the ability to closely integrate demand and replenishment in ways impossible only a few years ago. Applications such as SCM and collaborative planning, forecasting, and replenishment (CPFR) enable whole supply networks to synchronize channel requirements, remove administrative costs, and cut costly lead times out of channel inventory management. In addition, Web-based tools have undercut the need for traditional purchasing functions such as lengthy negotiations, requisitions, and paper-based purchase orders.

■ *Focus on Continuous Improvement.* At the core of SRM can be found a strong commitment to the joint pursuit of continuous improvement as a dynamic process rather than a static business principle. Whereas mutually profitable relations between trading partners might facilitate the achievement of common goals, only those companies pursuing closely integrated collaborative objectives can hope to continually streamline the development and guarantee the availability of superior products and services that consistently leapfrog the competition.

These marketplace realities are directly driving the two central functions of today's SRM applications. The first is responding to tremendous pressures to reduce spend everywhere throughout the supply chain. The second, and seemingly contradictory pressure, is to simultaneously improve supplier relations and collaboration. SRM provides the information to enable purchasing departments to gain full transparency to spend, develop a comprehensive, accurate profile of the supplier base, reveal areas for cost consolidation, identify opportunities for the optimal sourcing of needed materials, equipment, and services. At the same time, SRM provides the reporting and analysis tools to consolidate and prioritize suppliers based on quality, performance and on-time delivery, ensure contract compliance and reduce maverick spending, ensure the quality of purchased items, and ensure appropriate levels of supply. The end result is a procurement function capable of matching business objectives with supplier performance.

While the content of SRM has been broadly discussed above, a detailed definition has yet to be formulated. For example, each SRM software vendor has its own interpretation and can run the gamut covering everything from supplier database analysis to product planning and outbound logistics, including e-procurement, strategic sourcing, auction management, and e-marketplaces. On the other hand, some industry analysts feel that SRM is not a technology at all, but rather a set of business practices that involve the establishment and nurturing of closely intertwined relationships between buyers and suppliers. To complicate matters, the level of acceptance among industry groups is uneven and range from sophisticated adopters of e-business procurement models to those who are firmly committed to manual processes. In addition, SRM is not synonymous with the term B2B or e-business marketplaces. Today's most sophisticated applications of SRM will utilize the Internet to grow supplier collaboration, facilitate processes, and architect new business models, but SRM is not simply a technology. Still, although SRM is in its embryonic stage, enough of the landscape has emerged to venture the following definition:

> SRM is the nurturing of continuously evolving, value-enriching relationships between supply chain buyers and sellers that requires a firm commitment on the part of all trading parties to a mutually agreed upon set of goals and is manifested in the collaborative sharing and timely and cost-effective networking of sourcing and procurement competencies to facilitate the entire material replenishment life cycle from concept to delivery.

SRM enables enterprises to effectively do the following:

- Stratify the supplier base so that procurement can easily identify those suppliers that are most strategic to the organization.
- Construct the governance structure and process guiding how the purchasing organization is to manage buyer/supplier interactions across the life cycle of the supplier relationship.
- Determine the process of supplier development detailing how joint collaboration will enable suppliers to dramatically reduce costs, introduce new services that will resonate with buyers' strategic goals and objectives, and implement new philosophies and technologies that will provide for closer networking and process visibility.
- Establish with suppliers performance measurements targeted at increasing both the overall value of the relationship as well as efficiency and cost reduction.

Components of SRM

The mission of the purchasing function in today's environment can be summarized as the real-time synchronization of the firm's supply requirements with the

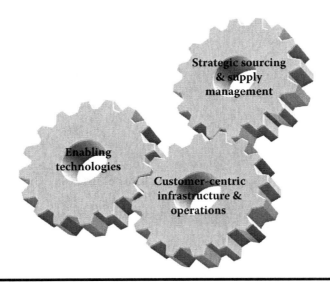

Figure 7.1 Components of SRM.

capabilities of supply channel partners in order to support customers' demand for made-to-order, high-quality, just-in-time goods and services while pursuing reductions in procurement costs and sustainable improvements in performance. According to Hirsch and Barbalho [3], such an approach requires the combination of three critical components as illustrated in Figure 7.1 and detailed below.

Strategic Sourcing and Supply Chain Management

Supplier partnerships require companies to look beyond the everyday purchase of materials to *strategic sourcing*. The goal of strategic sourcing is to find and cement close relationships with those trading partners that account for the majority of a company's purchasing dollars. While strategic sourcing will drive tactical decisions regarding the use of such technology toolsets as Web sites and portals to decide which products to purchase through the Internet and which through traditional mediums, the central focus is on selecting those suppliers who can support the customer-centric objectives of the company. While cost control is a critical element, strategic sourcing is a comprehensive supply management process that involves

> identifying the business requirements that cause you to purchase a good or service in the first place, conducting market analysis to determine typical cost for goods/services within a particular supply system, determining the universe of suppliers that best meet your requirements, determining an overall strategy to procure items in that category, and then selecting the strategic supplier(s). [4]

Depending on the category or type of purchasing to be sourced, other factors, such as the depth of supplier competencies, availability of required services, level of desired product quality, capacity for innovative thinking, and willingness to collaborate, can also be considered key strategic components. Take, for example, the capability of a supplier to support e-commerce functions. The procurement strategy may call for the best suppliers to have in place Web-based catalog applications or interfaces to EBS backbone databases that enable pursuit of cost, quality, and continuous improvement targets.

Applying Technology to the Management of SRM

The effective management of procurement has always depended on the facilitating capabilities of communications and networking technologies. Over the past 50 years, procurement's ability to work with suppliers, communicate requirements, and negotiate quality, pricing, and delivery of goods and services has been driven by technology tools that either match or exceed the velocity of marketplace transactions. The first major technology employed was the telephone. Being able to transcend time and space through the telephone replaced the cumbersome processes of mail correspondence and the necessity of person-to-person contact. The arrival of the fax machine significantly accelerated the processes of negotiating contracts, sending orders, and checking on the status of open orders. Electronic data interchange (EDI) enabled trading partners to interface enterprise resource planning (ERP) systems so that demand, order and shipment transmission, and electronic bill payment could be performed in a paperless environment.

Today, with the application of the Internet to SRM, purchasers have been able to leverage new forms of procurement functions, such as online catalogs, interactive auction sites, radically new opportunities for sourcing and supplier management, and Web-based toolsets that provide for the real-time, simultaneous synchronization of demand and supply from anywhere, anytime in the supply chain network. In addition, the application of a combination of mathematical models and computer software and hardware has enabled businesses to analyze large numbers of supplier bids involving quotes on total spend, volume discounts, transportation, and specification, and business scenarios detailing elements such as constraints or locations to determine the "optimal" sourcing alternative before contract [5].

The application of e-business to the evolving SRM concept can be said to have spawned a new form of procurement management: *e-SRM*. Today, business-to-business (B2B) purchasing transactions continue to grow and are offering firms sustainable and meaningful procurement improvement opportunities from shorter sourcing and negotiation cycles, to reduced costs in ordering and more effective ways to ensure quality and delivery. As will be discussed later in this chapter, the concept of e-SRM has come to coalesce around two Internet-driven functions: *e-procurement*, the utilization of Web-enabled applications to automate the activities association with purchase order generation, order management, and procurement statistics, and *e-sourcing*, the utilization of the Web to develop long-term supplier relationships that will assist in the growth of collaborative approaches to joint product development, negotiation, contract management, and CPFR. While considerable debate still rages as to whether e-sourcing is simply an element of e-procurement, a separate function altogether, which precedes the other, and can the terms be used interchangeably, there is little doubt that companies are utilizing these functions to achieve dramatic breakthroughs in the management of direct and indirect procurement. e-SRM can be defined as the utilization of e-business applications that facilitate the procurement of production and MRO inventories and services. e-SRM provides the mechanism for the structuring of formal and informal supply-side trading relationships that drive the functioning of dynamic value-chains.

SRM-Driven Infrastructures and Operations

SRM requires companies to constantly deconstruct and architect new processes that can be rapidly deployed to meet the shifting of customer requirements while focusing on continuous improvement. Procurement functions unable to respond in a timely fashion to changes in the marketplace with complimentary organizational, technological, and performance management changes will consistently result in suboptimal customer performance. In a way that was virtually impossible in the past, Internet-based procurement toolsets have created an environment where best practices in purchasing can be automated and applied to the acquisition of just about any product and service. This standardization and optimization of the work of the procurement organization extends the expertise of a firm's best purchasing processes through the organization and out into the supply chain.

e-SRM also requires the widening of traditional purchasing functions to include new players in the buy-side economy. As if purchasing professionals have not already been asked to integrate radically new *internal* procurement functions such as private trading exchanges (PTXs) and consortiums, they also have had to come to terms with new *external* trading entities. For example, procurement has had to expand its processes to include working with third party organizations that run e-marketplaces. These *e-market-makers* utilize Internet technologies to connect multiple buyers with multiple suppliers, conduct e-commerce functions, and deploy various forms of Web services such as payment, logistics, credit, and shipping. How individual companies will react to these challenges to existing procurement practices depends greatly on the dynamics of their supply chain network systems. Each business has various types of customers and suppliers that must be served through traditional and e-business methods, along with internal systems that must be integrated.

The Internet-Driven SRM Environment

In today's global environment where connectivity is so easy with any supplier, at any time, at any place, the initial revolutionary fervor as well as the fears that pervaded the implementation of e-commerce after the tech bust has all but disappeared. Instead of "bleeding-edge technology," e-commerce has simply become mainstream for businesses everywhere. For example, in a very difficult period for the airline industry, Delta saved $65 million in the final 3 months of 2001 on its annual $9 billion in overall purchasing costs thanks to a suite of e-SRM tools that went live in October. In the divisions where Proctor & Gamble had implemented e-SRM software, they were able to realize savings of up to 30% of the more than $8 billion in annual spending for indirect goods. Similar savings have been realized by smaller enterprises. ITT Industries, White Plains, New York, has leveraged its new e-business tools to shave as much as 5% on contract items and 13% on noncontract items from its more than $900 million annual purchasing budget [6].

When all the facts, figures, and opinions are compiled, it is clear that the number of B2B e-marketplaces, whether private or independent, can only be expected to expand in the second decade of the twenty-first century. To begin with, traditional buyer–supplier relationships will continue to be transformed into virtual enterprises and industry consortia. As the need for collaboration on all aspects of business accelerates, so will information and transaction management be transferred from manual to digital. Second, as efforts to reduce costs and automate transaction processes are amplified, B2B e-marketplaces will increasingly be seen as critical to achieving operational objectives. Lastly, B2B provides companies that were once considered rivals to jointly participate in the creation of e-marketplaces where they can as a group leverage their collective purchasing power and, in the process, increase efficiency across the entire supply chain. The continuing utilization of B2B e-marketplaces can be expected to achieve the following benefits:

- *Increased Market Supply and Demand Visibility.* B2B e-marketplaces provide customers with an ever-widening range of choices, an exchange point that enables the efficient matching of buyers and product/service mixes, and a larger market for suppliers.
- *Price Benefits from Increased Competition.* Online buying and use of auctions can be employed to increase price competition, thereby resulting in dramatically lower prices.
- *Increased Operational Efficiencies.* B2B applications have the capability to increase the automation and efficiency of procurement processes through decreased cycle times for supplier sourcing, order processing and management, and selling functions.
- *Enhanced Customer Management.* e-Marketplaces assist marketers to accumulate and utilize analytical tools that more sharply define customer

segmentation and develop new product/service value packages that deepen and make more visible customer sales campaigns.

■ *Improved Supply Chain Collaboration.* Today's B2B toolsets enable buyers and sellers to structure enhanced avenues for collaboration for product life cycle management (PLM), marketing campaigns, cross-channel demand and supply planning, and logistics support.

■ *Synchronized Supply Chain Networks.* The ability of e-markets to drive the real-time interoperability of functions anywhere in the supply network focused on merging information and providing for the execution of optimal choices provides supply partners with the capability to realize strategic and operation objectives. Among these can be included shorter cycle times for new product development and delivery, increased inventory turnover, lower WIP inventories, low-cost logistics, and others. [7]

e-SRM Structural Overview

The growing evidence for the realization of the benefits outlined above produced by e-SRM have provoked significant changes in the procurement process and provided radically new toolsets. The challenge during this period has been to search for methods to accelerate the automation and optimization of sourcing and transactional processes through the use of the Internet while at the same time deepening the strategic functions associated with supplier management. As a result of this movement on the part of software developers and procurement strategists, the traditional labor-intensive components of product/service sourcing, RFQ and supplier selection, order release, order receipt, and accounts payable have been greatly standardized and automated through the implementation of ERP systems and team-focused improvements. Today, the application of the Internet to the SRM management process has provided the purchasing organization with the opportunity to venture into unexplored regions of supply chain value using Web-based techniques.

Up until just a few years ago, the procurement process was executed through time-honored techniques. Business requirements were sourced, suppliers signed-up, requirements communicated, and deals negotiated the old-fashioned way through personal meetings, phone calls, faxes, and mail delivery. While most companies did enjoy computerized order processing, order management, and supplier relationship functions through their ERP systems or even EDI, the automation of these back-end functions were inward-facing and did little to enhance the integration and collaborative relationships necessary to speed up the front-end processes that were outside resident in the supply chain. With the application of the Internet to SRM functionality, this gap in the automation of front-end procurement as well as full integration with EBS backbones is rapidly disappearing. Similar to what CRM has done for customer management, SRM is permitting today's cutting-edge companies for the first time to assemble a complete picture of their supply relationships, apply Web technologies to dramatically cut cost and time out of sourcing and negotiating, and utilize real-time data to communicate requirements and make effective choices that result in real competitive breakthroughs.

Figure 7.2 is an attempt to portray the components of today's Internet-enabled SRM system. While earlier in this chapter SRM was described as consisting of two processes: *e-sourcing* and *e-procurement*, it is being proposed that in order to facilitate understanding, SRM should be viewed as being composed of four separate but integrated areas. The first area, *EBS backbone*, comprises the traditional database and execution functions utilized by purchasing to generate orders, perform receiving and transfer to accounts payable, and record supplier statistics. The second area, *e-SRM services*, details the enhancement of traditional buyer functions, such as sourcing and supplier relationships, through the use of Web toolsets. The third

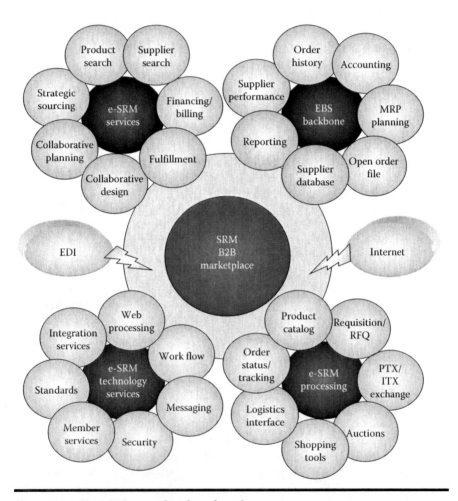

Figure 7.2   SRM B2B e-marketplace functions.

area, *e-SRM processing,* lists the new functions provided by the Internet that facilitate the transaction process. The final area, *e-SRM technology services,* outlines the technical architecture that enables e-SRM front-end and backbone functions to be effectively applied to solve procurement strategies. A detailed description of each of these areas will be discussed below.

EBS Backbone Functions

In this region can be found the procurement functions associated with traditional ERP. The fundamental role of this area is to collect and provide a repository for internal database information to guide purchasing processes. The EBS backbone contains the following critical functions.

- *Procurement History.* The collection of procurement information is fundamental to SRM. Data in this area ranges from static records, such as past transactions, to dynamic information, such as open PO status and active supplier and sourcing files. The accuracy and completeness of this information serves as the foundation for all internal and networked procurement activities.
- *Accounting.* The completion of the purchase order process feeds directly into the firm's EBS backbone for order and price matching, invoice entry and payables, credit management, and any necessary financial reconciliation.
- *Purchasing Planning.* Once total demand has been processed through the MRP processor, the schedule of planned purchase orders can be generated. Depending on the level of communication technologies and collaborative relationships, this statement of planned orders can be used to drive both MRO and indirect materials and production raw materials and component acquisition through the supply chain network.
- *Performance Measurement.* As receiving and payables history is compiled, companies have the capability to generate meaningful reporting and performance measurements indicating the value of their supplier relationships and the degree of success of their continuous improvement initiatives.

The integration of e-SRM toolsets with the EBS backbone is absolutely essential for e-business success. Often overlooked during the frenzy of the dot-com revolution was the fact that e-markets can not hope to deliver expected results unless they are connected to the databases resident in backbone business systems. According to AMR Research's Bob Parker, the crash of indpendent trading exchange (ITXs) in 2001 was not the result of perceived risks on the part of companies of exposing mission-critical processes to competitors, but rather that ITXs "had to be integrated into their members' back-end systems, and very few companies were set up for that type of integration" [8]. Deploying the technology architecture to connect local EBS backbones to the SRM applications is the first step in making Internet-enabled SRM happen.

e-SRM Services Functions

The Internet provides purchasers with fresh new avenues to transform the services traditionally used to execute procurement processes. In the past, purchasers were required to perform laborious and time-consuming searches for sources of new products and services. In contrast, B2B marketplaces provide a level of service features, such as offering an online catalog of products, sales promotions and special pricing, payment processing facilities, and postsales support, impossible in the past. Web-based technologies have made it possible for purchasing functions to significantly streamline this process by utilizing the following B2B marketplace service functions.

1. *Supplier Search.* Historically, the supplier search process suffered from a high degree of fragmentation and discontinuous information flows. The normal process of locating suppliers, performing the mandatory round of RFQ negotiations, and securing contracts was slow moving and often adversarial. Virtual B2B marketplaces, on the other hand, offer large communities of buyers and sellers a completely new channel to reach out to each other in a two-way interactive mode that transcends barriers of time and space. Buyers can leverage Web-based technologies to deepen their existing relationship with preferred suppliers while expanding their search for new suppliers on a global scale. In addition, buyers can explore new dynamic purchasing models, such as online auctions, for sourcing and spot buying.

2. *Product Search.* Instead of cumbersome paper-based catalogs, e-SRM services provide for the creation, aggregation, and Internet access to a wide range of online product and service catalogs that can significantly enhance the sourcing effort. B2B marketplaces host electronic product search for all types of goods and services, including MRO and indirect materials, production, administrative, and capital goods. According to Hoque [9], catalog functionality "can range from a simple keyword search to complex product category classification, parametric search functionality, automatic comparison product offerings, bid-boards for collaborative buying, message boards for posting buyer testimonials, real-time chat for negotiating flexible pricing, and even bidding and auctioning." Effective Web-based applications should enable e-marketplaces to centralize product and service content offerings, permit suppliers to host content on their own sites, and enable buyers to develop customized catalogs.

3. *Strategic Sourcing.* The challenge of e-SRM is to automate and optimize procurement functions, while at the same time improve strategic procurement activities. The difference between the two is important. *e-Procurement* focuses on leveraging Web applications to reduce tactical costs and increase

efficiencies and is primarily focused on nonstrategic, indirect materials. In contrast, *e-sourcing* is focused on the more effective management of vendor sourcing, contract and RFQ, and supplier management during the early stages of strategic, production materials life cycle management. According to AMR [10], strategic sourcing can be defined as a systematic, cross-functional, and cross-enterprise process that seeks to optimize the performance of purchased goods and services through reductions in total cost, sourcing cycle time, and assets. It originates with PLM and asset life cycle management in the make/buy decision process and concludes with contracting and order generation.

The concept of supplier sourcing has been around for decades. However, for the most part it was an internalized, time-consuming, and laborious process centered on cross-functional management of spending categories, corporate aggregation of spending, supplier rationalization, and supplier partnership management. Using today's Web-based technologies, this past concept of strategic sourcing has been completely transformed. In fact, some analysts consider strategic sourcing as the "third generation" of e-procurement, the prior two stages concerned with the automation of the purchase of standard, catalog-based products and the movement to trading exchanges. Termed now *e-sourcing,* software companies, particularly those grounded in PLM technologies, have during the early 2000s developed applications that not only facilitate strategic sourcing but also provide for integration to supply chain planning (SCP) and ERP backbones. These applications can be divided into two categories [11]:

1. *Decision support tools* for creating an effective sourcing strategy that includes these:
   - *Spend Analysis*: historical and forecasted spend by category, supplier, and organizational unit
   - *Item Rationalization*: standardization and elimination of redundant items
   - *Contract Management*: tools assisting in RFQ, bid analysis, negotiation, and contracting that result in lower opportunity, input, and quality costs and shorter product introduction time
   - *Supplier Monitoring and Improvement*: measurement of supplier transaction, quality, and collaborative performance
2. *Negotiation automation tools* that streamline these:
   - *Supplier Databases:* easily accessible databases that reveal supplier capabilities and performance levels to cut the supplier RFQ search effort
   - *e-RFP:* Web applications that provide for electronic request for proposals that link with bid analysis tools
   - *e-Auctions:* Web-based tools that facilitate and fully document auction events

3. *Value-Added Services.* In addition to dramatically enhancing supplier and product search functions, e-SRM leverages the Web to pursue a variety of other critical value-added business services, including the following:

- Financial and billing services such as the use of *payment cards,* or P-cards, credit approval, corporate check payment, clearinghouse functions, and direct electronic billing
- Comparison shopping functions
- Collaborative design and configuration management functions for complex, make-to-order production
- Advertizing, promotions, and dynamic pricing models based on market demand and availability
- Transportation and logistics support to facilitate product fulfillment
- Synchronized supply chain procurement planning
- Establishment of marketplace performance benchmarks and key indicators

While such services are critical in the pursuit of short-term requirements for cost reduction, collaboration of supply chain competencies, and synchronization of channel network inventory plans, they also provide buyers and sellers with the capability to pursue strategic goals beyond pure transaction management that enable them to evolve into e-marketplaces communities.

e-SRM Processing

The goal of e-SRM applications is to streamline the procurement process for the goods and services necessary to produce products and run the enterprise. Originally, the central focus of SRM Web-based applications has been on MRO and indirect inventories procurement. The reason for this was that these goods and services are extremely well-suited to B2B e-markets. They are normally highly standardized commodities, purchased in large volumes, evaluated normally on price alone, require minimal negotiation, and often are acquired through frequent spot purchases. Transferring MRO procurement to the Web is also fairly easy to do, often amounting to little more than creating an online catalog capable of being accessed through Internet-enabled purchase order entry functions. In contrast, applying e-SRM tools to the purchase of production goods, such as raw materials, components, and production equipment, is much more difficult. Procurement in this area oftentimes is subject to highly detailed design constraints and mainly applies to specialized vertical industry suppliers who provide products without sufficient breadth and volume of market demand. What is more, because of the specialized nature of the product/service, actual procurement is often preceded by a complex negotiation process involving RFQ, competitive supplier bidding, long-term contracts, and continued involvement in relationship-specific investments between trading parties. e-SRM in this area is more about greasing the channel with trusted suppliers than opening up a free market bid/ask exchange.

e-SRM requires companies to merge Internet-driven MRO and production procurement under a common umbrella. By consolidating and automating enterprise buying processes companies can capture procurement advantages through economies of scale, more effective negotiations resulting in better pricing, and a deepening of communications and coordination with channel suppliers that will translate into more efficient, collaborative buying. The detailed components constituting this area can be described as follows.

1. *Product Catalog Management.* The management of catalog content has been one of the most critical issues facing e-business from its very inception. The promise of "dynamic e-commerce," defined as the exchange of goods and services via electronic markets where buyers have access to virtual storefronts to search for any product/service mix at the lowest cost, depends on the availability of catalogs containing "dynamic content" that always provides the most current pricing, product information, and product specifications. Meeting this challenge requires content management functions capable of catalog normalization, rationalization, and scalability while providing for rapid supplier data extraction and cleaning, update, aggregation, end-to-end integration, and publishing. For example, Lincoln Electric, a Cleveland, Ohio, manufacturer of welding products, implemented a dynamic catalog management process that enabled it to quickly broadcast product changes and introduce new items to customers. The goal is to engineer a catalog content management system for syndicating product information in formats that customers can receive and use without having to cleanse or reformat it. In addition, the system will allow the company to transmit product updates as changes rather than as a total catalog refresh [12].

2. *Requisitioning.* e-SRM applications seek to facilitate the requisitioning process by integrating product/service catalogs hosted by exchange marketplaces, industry consortia, or third party aggregators located across the Internet into a single "virtual" catalog available through online interfaces. Creating such catalogs requires various levels of effort. Because MRO procurement usually involves highly standardized products and preferred vendors, integrating catalogs for indirect requisitioning is a fairly straightforward affair. On the other hand, developing similar sourcing references for production materials is much more complicated and requires the structuring of catalogs that present buyers with a range of possible suppliers and capabilities robust enough to permit them to perform the depth of value analysis and competitive comparison necessary to ensure alignment of purchasing decisions with strategic procurement targets.

   In addition, e-SRM requisitioning applications must provide buyers with aggregate individual supplier statistics, such as specific contract pricing, service quality performance history, commitment to collaboration, and overall customer care rating. When requirements fail to isolate a preferred supplier, the e-SRM toolset must support such features as browsing, keyword/parametric searching, collaborative filtering, product configuration and other application functions. Finally, the e-SRM application must provide for online document interchange, supplier chat-rooms, open requisition status, and access to the latest budgeting and inventory information for transfer to the firm's EBS backbone [13].

3. *RFQ.* The use of manual forms of competitive pricing for procurement of goods and services normally involves a process that begins with the receipt of sealed bids from prospective suppliers, progresses through supplier selection, moves through negotiation of detailed terms, and ends with a specific contract. This process is usually a labor-intensive burden for both parties, may result in widely different terms from contract to contract, is dependent on the negotiation skills of both parties, and favors the use of existing suppliers with a disregard for past performance. By transferring this process to the Internet, buyers can significantly automate the RFQ process, thereby cutting costs and reducing cycle times. By opening up the bid to a form of real-time auction, buyers can greatly increase marketplace competition and solicit suppliers separated by geography and time to participate in the sourcing process.

   For indirect and MRO bids, the RFQ can simply be passed to the PO generation stage. For production materials or purchases subject to dynamic pricing, the buyer would initiate the RFQ process by either posting the RFQ in a public, online bulletin board for open bidding, or transmitting the RFQ to preferred suppliers by e-mail, fax, or private exchange. As Hoque points out [14], this bid/ask RFQ process will increasingly occur in vertical industry exchanges specializing in specific market segments. In addition to simply hosting bid boards, these exchanges will enable companies to share buying experiences and exchange best-practice techniques.

4. *Shopping Tools.* While still in its infancy, the use of software shopping agents to perform the tasks of Internet browsing and initial gathering of and acting on basic information is expected to expand through time. Basically, shopping agents will augment the work of buyers by performing searches of possible B2B marketplace sites to identify and match targeted products, pricing, quality, delivery, or other desired procurement attributes and execute transactions on behalf of the buyer. In the future, sophisticated buying tools will be able to interact with each other to locate appropriate products or services on the Web and negotiate price, availability, and delivery with a minimum of human interaction. For example, a company's shopping agent would store information related to minimum inventory levels, which

suppliers are used to replenish those items, what level of quality is required for each item, acceptable price ranges, and shipping instructions. The shopping agent would be able to interact with compatible software agents that reside in B2B exchanges, which track hundreds or even thousands of suppliers that have been screened relative to the company's buying criteria. Similar to the computer-to-computer interchange that occurs with EDI, these intelligent shopping agents would automate much of the drudgery of today's buyer [15].

5. *Auctions.* The use of auctions is perhaps one of the most exciting options offered by Internet commerce. Used primarily as a means to buy and sell products whose value is difficult to determine, are commodity-type items, or are custom-designed, the ubiquitous presence and real-time capabilities provided by the Web have enabled this age-old method of purchase to expand beyond the domain of niche markets to reach potential bidders across geographic barriers and traditional industry lines. Almost any product, from airplane tickets to unique products can be offered on an auction site. The use of auctions in e-SRM is, for the most part, confined to non-production-related inventories. The benefits can be significant. For example, Accenture hosted a reverse e-auction for stationery supplies using their own auctioning facility. In one hour eight suppliers watched the contract bid drop from $1.29 million to $0.92 million. The winner of the contract tendered ten bids during the auction, with an average price drop of $46,000 per bid [16].

According to Poirier and Bauer [17], there are five types of e-auctions. The first type, *classical* or *forward auction,* consists of a single seller and multiple buyers who bid on a specific product or lot. The leading bid at the end of the allotted time wins the lot. This method is good for disposing of excess, aged, off-specification, or soon to be obsolete inventories. The second type, *reverse auction,* is the classical auction in reverse in which one buyer and multiple sellers drive the auction. This method is an alternative to the traditional RFQ method by which buyers solicit bids from the marketplace for one-time, high-value purchases. *Dutch auction* is the third type. This auction is characterized by one seller and multiple buyers, but with multiple homogeneous lots available. The lowest successful bid sets the price for the entire collection. This method is applied to products subject to supply shortages or dramatic demand fluctuations. The fourth type, *demand management auction,* differs from the previous models in two ways: there are multiple buyers and sellers, and the market maker plays an active role as the intermediary. This model is used for products that are perishable, characterized by variable or unpredictable demand, and whose prices are marked by extreme elasticity. The final type, *stock market model,* is characterized by multiple sellers and buyers, homogeneity of commodity, and mutual indifference as to the supplier or buyer. This type of auction is limited to true commodities and is normally found in private markets.

6. *Purchase Order Generation and Tracking.* Once the order requisitioning or RFQ process has been approved, a purchase order is generated. POs can be created using EBS functionality and are then transmitted to the supplier through a paper order or electronically via fax, EDI, or the Internet. In addition to serving as the instrument communicating the contract to purchase, the PO can also provide valuable *internal* information. To begin with, the PO record provides purchasing management with information regarding outstanding order data, budgeting, and performance reporting. The progress of the PO can then be tracked and used to provide critical status information needed by manufacturing or distribution planners.

7. *Logistics.* Today's e-SRM order management functions can be significantly enhanced by the utilization of a variety of Web-based logistics services that can be integrated into the procurement process. Logistics partners have the capability to offer Internet enhanced services, such as inventory tracking, carrier selection, supplier management, shipment management, and freight bill management. In addition, logistics service providers can offer advanced functions, such as network planning, dynamic sourcing, and reverse logistics that integrate buyers with supplier e-fulfillment capabilities, dynamic strategies for cross-docking, in-transit merge hubs, postponed assembly, and commingling of loads to optimize shipments.

e-SRM Technology Services

None of the components of e-SRM discussed above would be possible without the necessary supporting technology architecture. Over the past decade, interoperable protocols that enable computer systems to share information, such as Transaction Control Protocol (TCP/IP), Hypertext Markup Language (HTML), and eXtensible Markup Language (XML), have arisen as standards by which companies can conduct business on their own terms and yet be connected to their supply chain network. Also, industry action groups, such as the RosettaNet consortium, have been established to offer companies leadership, influence, and collaboration in the development and deployment of e-business standards for transaction management, exchange protocols, and business processes. Unfortunately, while these toolsets have been growing in sophistication, they also often require a massive integration effort across e-commerce applications, legacy business systems, and best-of-breed models. Depending on the B2B application provider, different standards have been used so that no one provider can be said to offer true end-to-end e-marketplace integration. In any case, e-SRM applications require the following technology support services.

1. *Web Processing.* The ability to drive e-SRM transactions requires a technology focus on data access and transactions as well as optimizing business processes. While the efficient processing of e-commerce transactions stands at the center of the B2B marketplace, the applications should provide for effective and timely decision-making prior to the point when the actual transaction is being made. In addition, the supporting technology should be scalable to handle maximum transaction and data communications volumes. Companies engaged in e-SRM should structure an IT structure equipped to perform load-balancing across multiple servers to ensure adequate performance and high availability of Web-accessed applications.

2. *Security.* In today's global business environment security has arisen to perhaps the key concern of companies engaged in e-market transactions. Security services include such components as information boundary definition, authentication, authorization, encryption, validation keys, and logging of attempted security breaches. The goal is to protect individual files so that confidential information can not be accessed without validation. Many companies have elected not to participate in ITXs or Consortia because of a reluctance to share data for fear of compromising business data security. The predilection for private exchanges is often driven by the desire to severely restrict data interchange to only the most trusted of business partners.

3. *Member Services.* The quest to create Web sites that are characterized by extreme usability, personalization, and customization is perhaps the "holy grail" of e-business management. Winning Web exchanges require marketers to ensure that customers have, first of all, an effective *personal experience*—did each customer's visit validate their expectations and did they leave the Web site with what they wanted, and, second, an *emotional experience*—did each customer develop a positive perception of their interaction with the Web site and do they wish to return for more in the future. Web-based toolsets can assist in developing detailed user profiles and analyzing user browser behavior and shopping preferences so that marketers can customize the customer's next visit to the site.

4. *Content Search and Management.* The essence of e-business is the capability of buyers and sellers to utilize knowledge bases, catalogs, text, graphics, embedded files, and applets to access and transact a broad range of products, services, and information over the Web. The ability to effectively search and pinpoint the desired object is, therefore, one of the most critical elements of e-SRM. Effective searching requires engines that provide access either by *content* (product description, type, business application, classification or category, etc.) or by *parameter* (how the content is organized using hierarchies that, for example, provide drill-down through a search tree or fuzzy logic).

In any case, once content is defined, it should possess the following baseline functionalities: (1) the ability to provide optimal content distribution and content organization to searchers, (2) the ability to transform potentially vast amounts of data resources into a useable format for the searcher, (3) the ability of content/application/system managers to define and organize criteria and rules regarding what may be customized and what potential combinations are valid, and (4) the ability of the content management component to integrate directly with the EBS backbone [18].

5. *Work Flow.* Effective e-SRM requires the delineation of the parameters determining the dependencies that exist between a series of procurement process steps. For example, what business rules govern the process a buyer must execute to move from requisition all the way through to actual purchase and payment? Work flow management provides the vehicle by which this path is detailed and optimized. According to Hoque [19], an e-business work flow component must consist of the following three modules:

■ *Work Flow Definition Module.* The role of this module is to provide a visual model of the application work flow. An effective module should include several templates for basic processes that can be easily modified to match business specific values and logic.

■ *Business Rules Definition Module.* The object of this module is to provide definition of the business rules that govern work flow decisions that are made during the e-business process. An example of a business rule would be determining that if a certain dollar volume was accumulated on the order, the shipping would be free. This rule would be automatically incorporated into the work flow engine.

■ *Work Flow Engine.* The work flow engine receives the user's request and determines the next sequence of screen displays that will match both the process and the business rules definition. As processes are dynamically updated during the course of continuous process engineering, the work flow engine must be capable of responding in real-time in order to transform models into actual working e-business sites.

Anatomy of the e-SRM Marketplace Exchange Environment

The basic types of B2B e-marketplace exchanges are explored in Chapter 2. It was stated that they differ from business-to-consumer (B2C) exchanges in several ways. To begin with, B2B marketplaces are concerned with the transaction of products and services between businesses. In addition, they closely resemble traditional purchasing in that the trading parties involved mostly depended on long-term, symbiotic, and relationship-based collaboration directed toward gain-sharing. It was stated that currently B2B e-marketplaces could be described as belonging to three major types:

- *Independent Trading Exchanges* (ITX) is described as many-to-many marketplaces composed of buyers and sellers networked through an independent intermediary. ITXs can further be divided into *vertical exchanges* focused on providing Internet trading functions to a particular industry and *horizontal exchanges* that facilitate e-business functions for products/services common across multiple industries.
- *Private Trading Exchanges* (PTX) is described as a Web-based trading community hosted by a single company that recommends or requires trading partners participate in as a condition of doing business.
- *Consortia Trading Exchange* (CTX) is described as a some-to-many network consisting of a few powerful companies organized into a consortium and their trading partners.

In this section e-marketplaces will be further discussed. This goal is to provide an anatomy of the B2B exchange environment and how it impacts e-SRM.

Emergence of Today's B2B Marketplace

The rise of Internet commerce over the past decade can be said to have ignited an explosion of strategies in just about every industry designed to utilize the Web as a vehicle for promoting and selling their products and services. Likewise, this rush for new business models has been propelled by a rising tide of corporate buyers who increasingly are turning to B2B exchanges that utilize electronic catalogs, product reviews, market research, and other information databases to access a wider marketplace and accelerate the evaluation and procurement of products and services. In fact, the rapid transition of many buyers from traditional to procurement models that promise low-cost, Internet-enabled marketplaces have in actuality generated an almost bewildering number of B2B buying sites driven by a proliferation of target markets, disparate standards, an expanding array of product offerings, and tangled technology tools ranging from homegrown code to off-the-shelf platforms from software vendors such as Perfect Commerce and Ariba. The application of e-SRM technologies is almost limitless. On the simple side, for example, Hewlett-Packard established a Web-enabled extranet to connect all members of its supply chain from contract manufacturers to component suppliers and plastic molders in an effort to ensure the simultaneous communication of supply and demand requirements. In contrast, the automotive, health care, government, and financial services consortium, Covisint, provides an extremely sophisticated Web-enabled marketplace that will handle anything from online catalogs to electronic document exchange, online tracking of quality, product design, and auction sites.

The development of e-SRM can be said to have emerged over three distinct periods [20]. A short analysis of each period is as follows.

Foundations

The first period can be described as the establishment of the basic B2B model. In this era, e-business was confined to the use of independent portals utilizing online catalog search, facilitating the RFQ process, and providing real-time order transaction and management. The goal of these sites was to sign-up participants fast enough to satisfy capital investors. The first B2B trading sites sought to offer products and services through techniques such as aggregation, performing buyer–seller matching, and hosting auctions. For the most part, these sites were focused on the buying and selling of MRO and nonproduction inventories. For buyers, ROI goals centered on automation and integrated point solutions. Overall, despite the immediate advantages, stage one B2B procurement represented little more than moving catalog operations online and did not offer the marketplace a new business model.

Rise of Collaborative Commerce

The second period in the development of e-SRM materialized during the year 2000. The distinguishing characteristic separating phase two e-marketplaces from their earlier predecessors was their focus on expanding the functions necessary to conduct collaborative procurement and addressing the issue of managing materials and components to be used in manufacturing. During this period, the field of trading exchanges became dramatically overpopulated; in its heyday, at least two exchanges could be found in almost every major industry category. Partly because of the ensuing confusion, but mostly as a way to gain control, companies began to bypass the use of ITXs, preferring to establish private and industry consortia exchanges in an effort to ensure security and engineer the integration of buyer and supplier EBS backbone systems.

Development of Networked Exchanges

The third stage of e-SRM emerged after during the middle years of the decade of 2000. Perhaps the central characteristic of this stage is the transformation of the current field of independent and consortia marketplaces into fully networked exchanges featuring robust functions such as single-data models and joint order management, procurement, financial services, logistics, and network planning that facilitate multibuyer/multiseller interaction and collaboration. Some of these verticals focus on certain business processes, while others focus on industry verticals. Interoperable Web-based technologies today enable true intercompany backbone integration and the seamless utilization of information fostering marketplace-to-marketplace interaction. In addition, B2B sites have gradually weaned themselves away from profitability models based solely on subscription and transaction fees in favor of fees matched to the value delivered to buyers and sellers. For buyers, the narrowing of the exchange marketplace and satisfactory fulfillment of targeted products and services has motivated them to gravitate to closer e-SRM partnerships.

SAS'S SRM solution

Best-of-breed SAS's SRM solution consists of the following software applications:

- **Strategy Alignment & Scorecarding** enables you to quickly determine how the sourcing process functions within a department, division or organization.
- **Opportunity Exploration** enables you to understand your supplier landscape so you can consolidate procurement activities, reduce costs and minimize supply risk.
- **Detailed Analysis and Reporting** offering a point-and-click environment that provides simple to advanced analytics, data manipulation, and reporting capabilities—allowing you to stay on top of your supplier activity and make smart decisions on a daily basis.
- **Procurement Scorecard** allows you to track the performance of the procurement organization and their achievements in support of corporate goals and objectives such as cost savings, contract management and supplier diversity programs.

- **Spend Analysis** lets you identify not only from whom you are buying, but what you are purchasing from each supplier, when it was purchased, and how it was purchased.
- **Ranking** allows you to establish an objective, repeatable, and adaptable measuring system that reliably identifies the best suppliers for your organization.
- **Optimization** by using the power of SAS' analytics you can identify ideal supplier portfolios for any given commodity groups in support of company goals such as reducing your risk exposure, minimizing your purchasing costs, and increasing your leverage for contract negotiations.
- **Data Cleansing** provides powerful data quality tools that enable companies to analyze, standardize, and rationalize supplier and commodity data.

Source: "SAS Supplier Relationship Management" brochure at http://www.sas.com/offices/europe/uk/downloads/sol_srm_brochure.pdf

---

Defining the Trading Exchange

The effective application of e-SRM depends on thoughtful decision-making regarding the optimal procurement strategies and choice of Internet applications. Currently, there are basically three models that companies can imitate: buyer-driven e-marketplaces, vertical e-marketplaces, and horizontal e-marketplaces.

- *Buyer-Driven e-Marketplaces.* This simple B2B model is designed to enable companies to facilitate internal procurement by linking through Internet tools divisions, partners, or companies in order to drive corporate purchasing processes and supplier relationships. These toolsets usually seek to facilitate RFQ and procurement functions by providing aggregate catalogs either on their own systems or the Web sites of service providers that can be used in turn by network trading partners. For example, Chicago-based Quaker Oats's indirect purchasing function was originally composed of 13 distinct, highly decentralized and paper-intensive processes spanning nine disparate systems handling more than 300,000 transactions with 30,000 suppliers. To solve the massive cost inefficiencies, Quaker established a centralized online catalog assembled from a number of MRO catalogs that could be accessed through a portal linked directly to supplier Web sites. The catalog reflects volume-based pricing and rule-based agreements negotiated with suppliers. Requisitions are automatically routed for approval and orders are placed and tracked through the Web portal. Finally, the portal handles paperless invoicing or automatic payment upon receipt.

■ *Vertical e-Marketplaces.* These types of digital marketplaces act as hubs servicing a single industry. Normally this type of e-marketplace exists either because of severe inefficiencies in distribution, or sales or industry fragmentation due to the lack of dominant suppliers or buyers. By automating the exchange process through a combination of technology and deep experience in a particular industry, vertical e-marketplaces focus on reducing industry-specific problems, such as a lack of information flows, high inventory levels, requirements for joint forecasting and planning, or logistics sourcing and contracting. According to Raisch [21], this type of e-marketplace can be divided into three groups:

  1. *Virtual Distributors.* Participants in this group seek to replace or improve a supply network by aggregating a variety of industry-specific product/service catalogs into a single Internet site. Instead of searching across a variety of sources, buyers can search online from a single venue, thereby reducing search and transaction costs while facilitating product and pricing information.

  2. *Exchanges.* This type of e-marketplace focuses on supply networks that are poorly or inefficiently constructed. The mission of exchanges is to facilitate information while reducing costs by permitting network members access to distribution, price, and inventory information of participating companies. Buyers can freely bid and enter orders while viewing the offerings of marketplace sellers.

  3. *Enablers.* Often excess capacities or materials exist in the marketplace. This type of e-marketplace attempts to leverage online tools that distributors or brokers can use to tap into this reservoir of resources and accelerate the matching of potential buyers and sellers. Collabria, for example, has created a market in the commercial printing industry in which they

match idle print capacity as well as price and capabilities to meet buyer needs.

■ *Horizontal e-Marketplaces.* B2B marketplaces in this area range from simple portals to sophisticated collaboration hubs. Perhaps the most important function of these marketplaces is the ability to enable multibuyer/multisupplier interaction and collaboration. By providing a sort of virtual trading "hub" where multiple buyers and sellers can be matched and conduct transactions, these Web sites enable manufacturers, distributors, buying groups, and service providers to develop shared marketplaces that deliver real-time, interactive commerce services through the Internet. Of equal importance is the ability to generate new forms of exchange, such as online sourcing, auctions, and negotiations. Finally, because of their role as a medium, these marketplaces enable trading communities to facilitate the exchange of common information and knowledge. Common horizontal marketplaces can be described as follows:

  – *e-Business Portals* are perhaps the purest form of e-marketplace. They are composed of third party market-makers who provide online buying and selling services to small- and medium-size buyers to create an exchange. The portal offers buyers lower prices, gains in sales and service, and access to exchange members who can create in turn a trading community based on common interest. In exchange, the portal market-maker realizes a range of benefits including an expanded branding awareness and widening exposure to potential customers. Portals include large financial institutions, utilities, telecommunications companies, IT service providers, and commerce service providers [22].

  – *One-to-Many Marketplaces* are typically PTXs that involve one buyer and multiple sellers. Companies hosting these marketplaces normally possess the buying power to force suppliers to participate in the exchange and dictate the terms for participation.

  – *Aggregator Hubs* are third party–led marketplaces that seek to combine the catalogs of several suppliers for display to potential buyers. The more sophistical aggregator hubs provide contracts, authorizations, and other content.

  – *Broker Hubs* normally consist of multiple buyers and sellers presided over by a broker. The central role of the broker is to match buyers and sellers based on product/service requirement and price. According to Hajibashi [23], buyers send their requirements to the hub, which in turn consolidates them to facilitate volume buying and discounting. Sellers then respond to these aggregate requirements. Finally, transactions are typically handled by e-mail versus an automated bidding process, though real-time dynamic transactions are becoming more prevalent.

  – *Collaboration Hubs* enable multiple buyers and sellers to correspond and share key procurement information. For example, buyers may share forecast information to facilitate supplier planning and shipment of products to correspond to required quantities, delivery, and service targets.

- *Translator Hubs* provide similar functions as collaboration hubs, but they facilitate trading functions by offering enterprise application integration (EAI) capabilities that provide true system/data integration between the business system environments of the trading partners. Among the technologies used are fax, EDI, e-mail, and XML.

Future of B2B e-Marketplaces

In the early days of the dot-com revolution, Internet-driven procurement offered to companies what seemed like a radically new way of conducting business. As the second decade of the twenty-first century dawns, the apocalyptic pronouncements of the period, however, seem very archaic as B2B marketplaces have become a standard part of the buyer's toolkit. Much of the calm acceptance of the Internet for purchasing has to do with dramatic advancements in integrative information technologies that have made buyer and selling through the Internet an easy affair. More importantly, B2B has become mainstream today because companies have been effortless able to make the leap to utilizing the Internet to create effective strategic sourcing processes. Not only MRO but production inventories are readily planned for and purchased. B2B applications are now part of the standard suite of most large software supplier such as SAP and Oracle.

Despite the technology breakthroughs, changing Internet-driven SRM from a preoccupation with the original trading exchange model (which emphasized MRO/indirect procurement) to the much more strategic e-SRM model of collaborative marketplace communities (which focuses on direct material procurement and planning), requires alteration to the way companies have thought about B2B marketplaces and the tools used to manage them. The starting point is to reexamine the very concept of SRM stated earlier in this chapter. SRM is about building and nurturing supplier relationships that foster the growth of collaboration and common destiny. Internet exchanges without the presence of a collaborative community of interest are merely transaction sites. Similar to the CRM model detailed in Chapter 5, SRM provides an environment where a company plans to win with its suppliers by establishing a common value chain for the long-term sourcing and procurement of a buyer's critical production inventories rather than simply commodity transactions. According to Foster [24], today's e-SRM functions seek to significantly expand the reach of collaborative supply communities and possess the following enablers:

- ■ e-SRM expands the scope of procurement functions. Most e-SRM application solutions span the life cycle of several business processes from product design through sourcing and manufacturing and onto procurement execution and performance monitoring.

- e-SRM provides for deep integration of business processes. As technology enablers expand the integration between business processes, companies will be able to leverage higher levels of cooperation between functions, including engineering, purchasing, manufacturing, and supply-chain operations.
- e-SRM facilitates direct collaboration between manufacturers and their suppliers. Similar to CRM, the key word in SRM is *relationship*. While e-SRM will automate and standardize transaction information such as RFQs and negotiations, it also will use relationship building applications including CPFR planning data and information regarding product specifications, design, and quality.
- e-SRM enables increased speed and flexibility. Since the centerpiece of SRM is the tight linkage of trading partners, it can assist in shrinking product design and time-to-market life cycles while facilitating the transmission of ECO changes across the entire supply chain.

These very real, immediately measurable advantages have also been accompanied by a change in the attitudes of once highly skeptical buyers. When the whole concept of B2B commerce appeared, many suppliers expected manufacturers to use the Web as a means to play suppliers off against each other in an effort to drive prices down. Now that e-SRM solutions have actually been implemented and have been driving procurement functions, suppliers' fears that B2B e-commerce would turn all products into commodities and destroy perhaps decades of buyer–seller relationships has been shown to be unfounded. If anything, e-SRM has provided buyers and suppliers the opportunity to enhance their relationships and even the most anxious business executive is becoming more at ease with opportunities for information sharing and collaboration.

In summary, manufacturers and distributors can only expect the use of e-SRM applications to increase with time as companies search for solutions that go far beyond simply automating procurement functions. Today, e-SRM has made the jump from its initial stage as a buy-side technology focused on commodities sourcing to fully collaborative marketplace communities. To continue its progress, e-SRM must, first of all, respond to individual company e-SRM requirements, such as product, supplier, and catalog search capabilities, automated RFQ, Web-based order entry and order status tracking, integration with EBS backbone systems, collaborative planning, transportation management, and other exchange integration functions. Second, e-SRM must enable true collaboration across the supply chain. Such a goal requires the existence of enabling infrastructures and business strategies that streamline procurement functions to the point where buyer and seller application functionality, processes, systems, and organizations are merged.

Implementing e-SRM

While the potential benefits of e-SRM are indeed spectacular, achieving them requires a thoughtful and well-designed implementation process. Previously, several critical barriers to e-SRM were detailed. One of the most commonly stated problems is the investment necessary to achieve an e-SRM initiative. Another was the requirement that firms must first reengineer their businesses to align procurement processes with technology capabilities. Technology-wise, effective e-SRM requires companies to integrate their SRM applications with backbone and CRM systems. This also means that companies must have the technology skills to not only successfully install the necessary software, but also to complete any integration requirements. In addition, e-SRM requires companies to take risks. Conducting business on the Internet, particularly open bidding, requires revealing proprietary information to eyes other than those of the parties involved. Then there is the apprehension that open exchanges require price transparency. Open auctions on a global basis could lower profits to the point that companies would not be able to stay in business. Finally, there is the anxiety that engaging in e-business will hurt long-standing relationships with existing suppliers, permanently damaging years of patient negotiation and mutual efforts toward collaboration on specific issues such as product quality and delivery.

Responding effectively to these and other challenges requires a comprehensive e-SRM strategy. An e-SRM strategy must posses the flowing attributes: *comprehensive* (all critical opportunities have been reviewed and the impact on all stakeholders analyzed); *complete* (no area has been left out of the plan and the results of the analysis are meaningful and have weight for the company); and *thoughtful* (the decisions about software, relationships with suppliers, and expected value-added to the supply chain are well documented and capable of discursive analysis). The following critical drivers need to be carefully considered when designing an e-SRM strategy.

e-SRM Value Discovery

Perhaps the first step that needs to be performed is the drafting of a statement of immediate economic benefits and long-term supply chain value to be achieved by implementation of Internet-enabled SRM toolsets. The goal here is to formulate a compelling case that details and positions each of the organization's most critical procurement requirements (i.e., design, sourcing, plan, transact, move, and dispose) with the procurement technologies to be implemented to include ROI, total cost/benefit of ownership, and the risk of not engaging in an e-SRM solution. At the conclusion of the process, the matching of the procurement requirement with the e-marketplace solution should include metrics detailing potential *cost savings* through increased buying economies or improved leverage, enhanced *process efficiencies* attained through decreased time spent on procurement activities, *inventory optimization* achieved through better planning, vendor-managed inventories, or improved supply chain visibility, and *lower development costs* through collaborative design and increased standardization [25]. Finally, companies need to fully understand the magnitude of the project they are embarking on. An e-SRM project is simply not an IT project: It is a strategic enterprise project that will impact the entire organization and business partners out in the supply chain.

Infrastructure Analysis

The next step in the development of an effective e-SRM implementation strategy is performing an assessment of current purchasing practices and organizational capabilities. The goal of this step is to determine the readiness of the organization to utilize Web-based toolsets and e-marketplaces and to decide which Internet-driven strategies possess the highest potential. According to Smeltzer and Carter [26], this process requires the examination of four organizational characteristics. The first, *organizational structure,* seeks to determine the degree to which purchasing is centralized, the position of the purchasing function within the firm, and the level of communication existing between purchasing staffs across the enterprise. The second characteristic, *information technology,* seeks to detail the sophistication of company technologies deployed promoting purchasing automation, current use of e-markets and industry portals, and availability of decision support systems. The third characteristic, *employee capabilities,* seeks to determine the qualification of purchasing professionals to understand and work with e-SRM tools. And, finally, *current purchasing practices,* which reveal the level of the procurement function's level of sophistication in regard to such practices as strategic sourcing, use of P-cards and supplier contracts, and application of supplier performance metrics.

Preparing for Organizational Change

Preparing the organization to pursue e-SRM opportunities requires a considerable degree of organizational readiness. A key task is developing an effective change management plan. Migrating to Internet-driven procurement processes will require overt and subtle changes to the way people have traditionally worked. The change management plan must begin by evaluating existing procurement processes, mapping flows and decision points, designing new processes supported by the technologies, and selecting methods to bridge current with new processes. Instrumental in managing this change is effective education and training. The mission here is to ensure that people know about the concepts and applications of the technologies ensuring their understanding and ability to operate effectively the new Internet-driven applications. The education should, finally, articulate the value proposition behind the e-SRM implementation and motivate people to search for new opportunities to leverage the system for cost reduction and collaborative enhancement.

Spend Analysis

Once internal organizational issues have been detailed, e-SRM project implementers must conduct a thorough analysis of all the goods and services purchased across divisions and strategic business units to determine actual spend levels and degree of supplier fragmentation. The analysis should indicate how much is being spent on individual items as well as product families. Finally, the analysis should identify how much is being purchased by category of goods and type of service from each supplier. The goal of the whole process is to unearth answers to such questions as what is being purchased, from whom, from where, and from what locations.

Item/Service Analysis

Once the initial spend analysis has been completed, the next step is to segment item/service purchases. A critical preliminary is the formulation of a standard for coding and indexing goods and services purchased across the enterprise. This process will further rationalize sourcing, reduce the number of suppliers, and facilitate aggregate demand planning and acquisition. Once completed, the next step is the categorization of purchased goods and services. In this process planners must separate purchases according to their relative risk/exposure and cost/value to the firm. The mission of the segmentation strategy is to determine which purchases are truly commodities with low strategic impact on the organization, which are generic and marked by high dollar expenditures but pose low risk, which are critical and will bring high risk/exposure to the company, and which are strategic in that they provide a distinct competitive advantage. When finally selecting an e-SRM strategy, companies must be careful to architect a procurement system that possesses the capability to integrate applications that can simultaneously leverage automation tools for reducing costs on commodity items while ensuring that strong relations are established with suppliers providing strategic goods and services.

e-SRM Technology Choices

Detailed segmentation of purchased items and services enables planners to more easily identify the required Internet applications that will optimize various supply environments. The process starts by mapping each procurement segment to B2B application enablers, both e-sourcing tools, such as RFQ and catalog search, and e-procurement tools, such as Web-based transaction management. Results may indicate that a portfolio of e-SRM applications will be required for optimum results. Also, it is important to realize that in selecting a technology solution, companies often will have to merge e-sourcing platforms, e-procurement applications, contract management toolsets, supplier collaboration solutions, and content management solutions from several e-SRM software suppliers. The e-SRM choices available can be divided into several possible Web-based strategies. Some of these models are considered extensively in this and other chapters and have been given here only a light analysis.

1. *Software-as-a-Service (SaaS).* A possible strategy to assist in managing the cost of e-SRM technologies is to contract a third party software service supplier. This strategy utilizes software vendors to provide an Internet-based platform in which multiple suppliers or trading partners can engage in cross-enterprise SCM processes that include extended supply chain visibility, work flow, event management, planning, process coordination, and replenishment. The larger software vendors, like SAP and Oracle, provide foundation interoperability toolsets as part of their standard ERP packages. SaaS vendors provide companies without ERP capabilities or who are looking for best-of-breed solutions with connectivity and capability to participate in trading exchanges and Web-based collaborative replenishment programs at a low cost. A SaaS solution permits cash-strapped companies to participate in Web-based technologies without committing to software licensing, extended implementations, and business process reengineering tasks.
2. *Automation Applications.* Much has already been said about the use of Web-based applications designed to automate the procurement process, enhance the productivity of the purchasing function, and facilitate e-sourcing and e-procurement processes. However, while a focus on cost reduction applications will produce short-term, tactical benefits, the biggest advantage will be found in the pursuit of technologies that enhance long-term buying strategies through standardization, aggregation, and collaboration.

3. *Portals.* While portals have many applications and as many definitions, the function of Internet-based portals is to extract and aggregate data from multiple systems, apply certain rules and logic, and present relevant information in a personalized format. According to Poirier and Bauer [27], a procurement portal is a business entity, such as a software provider, a pooled purchasing group, an exchange, or an aggregator, that provides infrastructure and buying/selling services in support of procurement operations. The portal may be focused on a specific vertical industry, commodity-based (MRO supplies), or a hybrid. Portals provide a cost-effective, efficient way for companies to broadcast and link procurement information from EBS backbone systems to trading partners. Portals are relatively easy to implement in comparison to PTXs largely because the requirement for deep system-to-system integration is minimal.

4. *Exchanges and Auctions.* Exchanges involve the use of a neutral third party that operates the exchange, sets the conventions for trading activities, and charges buyers and sellers for its use. An auction site is also normally run by a third party who provides the functionality permitting buyers and sellers to bid on products and services.

5. *PTXs and CTXs.* A decision to participate in a PTX or CTX represents a considerable investment in time and money. These exchanges often require significant technical investment on the part of the host both to establish the exchange hub as well as to assist suppliers to "plug into" the system. These private networks require trading partners to have a series of pass codes that enable them to enter the network and move to the exchange. Also, these exchanges are usually governed by specific agreements that determine the transaction services between partners. These types of e-SRM are strongly oriented around value chain partnering, led by a small group or a single powerful supply chain sponsor who has ownership of the exchange hub.

Performance Measurement

Ultimately, the success or failure of an e-SRM implementation initiative can only be measured against the performance targets that were created at project inception. Dodds and Balchin from Accenture feel that there are essentially two clearly defined but closely related categories of key performance indicators (KPIs) at the foundation of an effective e-SRM strategy [28]. The first category, *implementation success KPIs,* measures the depth of the penetration of e-sourcing and e-procurement into the organization. The second category, *benefit KPIs,* provides the information necessary to determine the extent to which originally identified benefits have been realized. While each area requires different measurement tools and metrics, Dodds and Balchin are quick to point out that these two sets of KPIs must be considered as constituting a single overall measurement. Simply measuring, say the percentage of buying utilizing e-procurement tools, would be meaningless without other indirect metrics such as the proportion of procurement time freed up to focus on value-added strategic activities gained by using an online auctioning application.

Although it can be said that many of the KPIs that are being offered as measurements for e-SRM are essentially the same as those to be found in traditional purchasing management, today's balanced scorecard of e-SRM KPIs require companies to significantly increase the speed by which the data are collected, analyzed, and made available for decision-making. Some e-SRM consulting firms and application suppliers have been developing prepackaged, portal-based solutions and models to benchmark purchasing capabilities that can be plugged directly into the e-SRM project. Among these solutions can be found performance toolsets such as real-time alert notification, data mining and associated data presentation, flexible work-flow-based business process definitions, flexible and user-defined KPI models, and exception-based management processes that are positioned on top of ERP or SCP and execution business systems.

Today, these purchasing measurements systems are being merged with tools to manage performance across the entire supply chain. Termed total cost management (TCM), the Aberdeen Group describes the emerging framework of application and supporting infrastructure for performance measurement as providing for the development and coordination of new organizational and technology architectures that merge supply chain strategies and product and market intelligence with emerging sourcing, planning, procurement, monitoring, and analytics technologies. To effectively drive TCM companies must build technology-based architectures that support the following:

- *Monitoring tools* that provide for the detailed measurement and enforcement of how closely the business complies with supplier contracts and how well suppliers execute negotiated trading agreements and anticipated performance targets.
- *Collaboration* of internal enterprise functions and supply network partners to promote the integration of all procurement and supply chain processes.
- *Process control* that utilizes all types of data communication from EDI to XML to engineer a central platform for the standardization and enforcement of common processes across individual companies and the entire supply chain.
- *Procurement intelligence* that provides a single source of intelligence for the entire supply chain for all procurement-related data and intelligence.

Summary and Transition

The demands of today's marketplace and the advent of Internet technologies have rendered traditional buy-side procurement solutions obsolete and marked the transition to new procurement management concepts, principles, and computerized technologies that have come to coalesce around a new management term—supply relationship management (SRM). Defining exactly what SRM means even after almost a decade since its inception is currently at best a difficult task. Unlike its obvious counterpart (CRM) on the sell side of the business, SRM is not an established management technique nor does it come complete with a defined suite of software applications. In fact, the term SRM is being used to span a variety of procurement functions from supplier management, negotiation, and sourcing to automated order generation, order monitoring, payment, and performance measurement. In addition, SRM also encompasses strategic objectives associated with the integration of supplier collaboration into the mainstream of SCM thought and practice as well as the use of the Internet in the pursuit of tactical objectives such as purchasing activity automation and cost management. Despite the "fuzziness" of SRM elements at this point in time, the following definition of SRM was offered in the chapter:

> SRM can be defined as the nurturing of continuously evolving, value-enriching relationships between supply chain buyers and sellers that requires a firm commitment on the part of all trading parties to a mutually agreed upon set of goals and is manifested in the collaborative sharing and timely and cost-effective execution of sourcing and procurement competencies to facilitate the material replenishment life cycle from concept to delivery.

The application of the Internet to the evolving SRM concept can be said to have spawned a new form of procurement management: e-SRM. As discussed in the chapter, the concept of e-SRM has come to coalesce around two Internet-driven functions: *e-procurement*, the utilization of Web toolsets to automate the activities association with purchase order generation, order management, and procurement statistics, and *e-sourcing*, the utilization of the Web to develop long-term supplier relationships that will assist in the growth of collaborative approaches to joint product development, negotiation, contract management, and CPFR. While there is still much needed discussion concerning the relationship and exact functioning of these two processes, the chapter proposed viewing e-SRM as consisting of four technology components. The first component, the *EBS backbone*, is composed of the traditional database and execution functions utilized by purchasing to generate orders, perform receiving and transfer to accounts payable, and record supplier statistics. The second component, *e-SRM services*, details the enhancement of traditional buyer functions, such as sourcing and supplier relationships, through the use of Web toolsets. The third component, *e-SRM processing*, lists the new functions provided by the Internet that automate and facilitate the transaction process. The final component, *e-SRM technology services*, outlines the technical architecture that enables e-SRM front-end and backbone functions to be effectively integrated to pursue procurement strategies.

While e-SRM-driven processes have been undergoing evolution, so have the B2B marketplaces where the activities of buy-side and supplier management are carried out. In the first era of e-SRM, companies focused on Web-based applications to automate internal transaction processes and to use independent B2C portals for the acquisition of standardized MRO and indirect products. In the second era, companies began to migrate toward independent and consortia exchanges to address issues relating to security and control. In addition, these Internet solutions were seen as prerequisites for the kind of buyer–supplier collaboration required for the strategic sourcing and the purchase of direct or production inventories. The third stage of e-SRM marketplaces, which has only begun to emerge, can be characterized as the transformation of the current field of private and consortia exchanges into fully networked e-marketplaces. These future marketplaces will possess interoperable technologies that will enable true intercompany backbone integration and the seamless utilization of product design, sourcing, and contracting information fostering true marketplace-to-marketplace interaction.

Leveraging the tremendous advantages and benefits to be gained from an effective e-SRM solution requires companies to design an implementation plan that allows them to avoid the common pitfalls that are ready to entrap the unprepared. To begin with, companies must closely define what immediate economic benefits and long-term supply chain value is to be achieved by the e-SRM implementation. Metrics should include potential cost savings, process efficiencies, inventory optimization goals, and lower development costs as well as the types of supplier relationships they wish to build. Following, an assessment of current purchasing practices and organizational capabilities should be performed. The goal is to review the readiness of the organization, its technologies, its employees, and its procurement practices to pursue e-SRM. Critical to organizational readiness is an effective change management and education process. Next, a detailed spend analysis needs to be performed. This activity will define the product types, quantity, and cost of procurement across the entire enterprise. Once these preliminaries have been completed, implementers can choose the most applicable e-SRM strategy. Results may require the adoption of a portfolio of e-SRM applications, dictating the use of SaaS solutions, the level of automation, and type of marketplace exchange. Finally, implementers must be careful to craft a range of meaningful implementation and expected benefit KPIs to monitor the ongoing success of the e-SRM effort.

The development of Web-based CRM and SRM tools to provide the marketplace with instantaneous access to products and services requires that the fulfillment side of e-SCM be equally as efficient and accessible. In Chapter 8, the application of Internet tools to logistics is examined.

Notes

1. Alvin J. Williams and Kathleen A. Dukes, "The Purchasing Function," in *The Purchasing Handbook,* 5th ed. (New York: McGraw-Hill, 1993), p. 5.

2. These functions of purchasing have been summarized from David F. Ross, *Distribution: Planning and Control* (New York: Chapman & Hall, 1996), pp. 443–445.

3. Chet Hirsch and Marcos Barbalho, "Toward World-Class Procurement," *Supply Chain Management Review* 6, no. 5 (2001), pp. 74–80.

4. Ibid.

5. Larry C. Giunipero and Philip L. Carter, "Strategic Sourcing and the Role of Optimization," *Supply Chain Brain* (August 14, 2009), at retrieved August 14, 2009, http://www.supplychainbrain.com/content/headline-news/single-article/article/the-role-of-optimization-in-strategic-sourcing/.

6. These and other success stories can be found in Steve Konicki, "E-Sourcing's Next Wave," *InformationWeek,* March 18, 2002, pp. 57–62.

7. See the comments in Editors, "An e-Outlook," *Midrange Enterprise* (Oct./Nov. 2001), p. 5, and Philip Burgert, "The Changing Face of B2B Commerce," *Electronic Commerce World* 11, no. 8 (2001), pp. 27–28.

8. These comments have been summarized from C. Edwin Starr, Ajit Kambil, Jonathan D. Whitaker, and Jeffrey D. Brooks, "One Size Does Not Fit All—The Need for an E-Marketplace Portfolio," in *Achieving Supply Chain Excellence Through Technology,* 3, David L. Anderson, ed., (San Francisco, CA: Montgomery Research, 2001), pp. 96–99, and Sidney Hill, "Don't Speak," *Manufacturing Systems* 20, no. 1 (2002), pp. 34–38.

9. Faisal Hoque, *e-Enterprise: Business Models, Architecture, and Components* (Cambridge: UK, University Press, 2000), p. 97.

10. Pierre Mitchell, "Strategic Sourcing Gets an 'E'," AMR Research Report, April 2001.

11. Reference the analysis in Carrie Shea, "The Evolution of Strategic Sourcing," B2B Retail and Consumer Goods Benchmarks Report, AT Kearney, September 2001.

12. Karen Abramic Dilger, "Content Upkeep," *Manufacturing Systems* 20, no. 4 (2002), pp. 56–62.

13. See the analysis of B2B purchase requisitioning found in Hoque, *e-Enterprise,* pp. 107–109.

14. Ibid., p. 110.

15. For more discussion on shopping agents, see David Taylor and Alyse D. Terhune, *Doing E-Business: Strategies for Thriving in an Electronic Marketplace* (New York: John Wiley & Sons, 2001), pp. 117–119.

16. Rachel Whitehouse and Yvette Mangalindan, "Online Auctions: What Is Your Bid?," in *Achieving Supply Chain Excellence Through Technology,* 3, David L. Anderson, ed. (San Francisco, CA: Montgomery Research, 2001), pp. 158–160.

17. Charles C. Poirier and Michael J. Bauer, *E-Supply Chain* (San Francisco, CA: Berrett-Koehler, 2000), pp. 106–109.

18. See the discussion in Hoque, *e-Enterprise,* pp. 215–219.

19. See ibid., pp. 221–222.

20. These phases of B2B can be found in Mohammed Hajibashi, "E-Marketplaces: The Shape of the New Economy," in *Achieving Supply Chain Excellence Through Technology,* 3, David L. Anderson, ed. (San Francisco, CA: Montgomery Research, 2001), pp. 162–166, and Bruce Temkin, "Preparing for the Coming Shake-Out in Online Markets," in *Achieving Supply Chain Excellence Through Technology,* 3, David L. Anderson, ed. (San Francisco, CA: Montgomery Research, 2001), pp. 102–107.

21. Warren D. Raisch, *The E-Marketplace: Strategies for Success in B2B Ecommerce* (New York: McGraw-Hill, 2001), pp. 213–214.

22. Reference ibid., pp. 51–52 and 214.
23. Hajibashi "E-Marketplaces," pp. 162–166.
24. Thomas A. Foster, "With SRM, Everything is Relative," *Supply Chain e-Business* 3, no. 2 (2002), pp. 16–21.
25. See the excellent comments in Don A. Eichmann, "E-Marketplace Participation: Reaching the Bottom Line," in *Achieving Supply Chain Excellence Through Technology*, 3, David L. Anderson, ed. (San Francisco, CA: Montgomery Research, 2001), pp. 153–156.
26. Larry R. Smeltzer and Joseph R. Carter, "How to Build an e-Procurement Strategy," *Supply Chain Management Review* 5, no. 2 (2001), pp. 76–83.
27. Poirier and Bauer, *E-Supply Chain,* pp. 101–102.
28. Stuart Dodds and John Balchin, "E-Procurement Measurement: It's Not Broken—But It Needs to Be Fixed," in *Achieving Supply Chain Excellence Through Technology*, 3, David L. Anderson, ed. (San Francisco, CA: Montgomery Research, 2001), pp. 148–151.

Chapter 8

Logistics Resource Management: Utilizing Technology to Enhance Logistics Competitive Advantage

Of the business functions of the modern enterprise, logistics has had perhaps the most visible impact on the economic condition of society. Historically, it has been the role of logistics to solve the problem of product distribution by providing for the efficient and speedy movement of goods from the point of manufacture to the point of consumption. Today, the task of effectively managing logistics has grown dramatically more difficult by the growing complexity of dealing with a global business environment compounded by the continuous downward spiraling of product cycle times, a relentless acceleration in marketplace demands for demand/supply visibility and quick response, and Internet technologies enabling customer-driven services and real-time connectivity with customers and suppliers. In such an environment, logistics competitiveness has migrated from efficiently running warehouses and shipping product to constructing collaborative partnerships with service providers and implementing technologies that facilitate warehouse and transportation management (TM) and enable close integration to enterprise business systems and Web connectivity with supply chain partners.

In fact, the tremendous challenges and opportunities confronting logistics today requires redefinition of the logistics concept and the evolution of a new term— logistics resource management (LRM). Similar to the recent transformation of customer service and purchasing functions into customer relationship management (CRM) and supplier relationship management (SRM), the use of the term LRM is meant to convey a similar expansion of traditional logistics activities to encompass the multifaceted concept of supply chain management (SCM) with its focus on trading partner collaboration, the removal of channel barriers causing excess costs and reduced cycle times, the espousal of Internet technologies that facilitate information and transaction data collection and flow through the supply pipeline, and the creation of agile, responsive organizations linked together in a single-minded pursuit of superior customer service.

In this chapter, the elements of logistics management in the "age of Internet globalization" are examined. The discussion begins with a review of the function of logistics and its evolution to what can be called LRM. After a detailed definition of the structure and key capabilities of LRM, the chapter proceeds to describe the different categories of LRM available today and the array of possible Web-based toolsets driving logistics performance measurement and warehouse and TM. Afterward, the use of third party logistics (3PL) services is reviewed. The different types of logistics service providers (LSP), the growth of Internet-enabled providers, and the challenges of choosing a logistics partner that matches, if not facilitates, overall company business strategies is explored in depth.

Defining Logistics Resource Management (LRM)

Over the past 50 years, the science of logistics management has evolved from purely an operational function to a competitive weapon providing today's enterprise with the capability to leverage Internet technologies to closely link the farthest regions of the supply channel with market demand found anywhere in the globe. Originally, the mission of logistics was to provide responsive, cost effective warehousing and transportations utilities that enabled companies to deliver products in support of internal marketing and financial objectives. In the age of e-business, modern logistics has become in itself a critical competitive resource, creating value by responding to ever-higher levels of customer expectation, engineering operations that integrate trading partners up and down the channel network while reducing costs and expanding the spectrum of value-added services necessary for electronic commerce.

The sheer size and complexity of logistics functions bears witness to its central position in the global economy. According to the 20th Annual State of Logistics Report [1], the cost of logistics just in the United States for the year 2008 exceeded $1.3 trillion. Of this total inventory carrying cost accounted for $420 billion and transportation for $872 billion of transportation cost. Finally, $8 billion was spent on other shipper related costs. Of total logistics costs, motor carrier costs accounted for nearly 78% of annual US freight transportation expenditures. Altogether, logistics costs were the equivalent of 9.5% of the US gross domestic product measured in nominal dollars.

Simply defined, LRM is the process whereby manufacturers, distributors, and suppliers store and move their products through the supply chain to their customers. Perhaps the most often quoted definition has been formulated by the Council of Supply Chain Management Professionals. Logistics is defines as "that part of SCM that plans, implements, and controls the efficient, effective forward and reverses flow and storage of goods, services, and related information between the point of origin and the point of consumption in order to meet customers' requirements" [2]. This and other definitions imply that LRM creates competitive value by ensuring the optimization of logistics operations costs and productivity, better capacity and resource utilization, inventory reduction, and closer integration with customers and suppliers. Beyond cost management, LRM also assists in the creation of marketplace leadership through customer service and timely product and service delivery. Furthermore, the overall success of these objectives depends upon the close collaboration and integration of all logistics partners that populate each node in the supply channel system and who are responsible for the efficient performance of logistics processes. In addition to traditional operations functions, these processes include strategic decisions relating to channel design and structure, resource allocation, human capital, operations, and finance. LRM creates competitive advantage by flawlessly executing customer service objectives, achieving conformance to quality standards, and increasing marketplace value.

Perhaps the best way to understand the contents of LRM is to divide it into three closely integrated sets of functions as illustrated if Figure 8.1. The first set of functions is centered on *logistics performance measurement.* The enormity of the capital wrapped up in warehousing and TM requires managers to keep a close accounting of the performance of these functions through the deployment of analytic and modeling technologies capable of providing full visibility and track and trace to logistics costs. The second component comprising LRM is associated with *warehouse management.* This segment of an enterprise's logistics functions is responsible for the storage and handling of inventories beginning with supplier receipt and ending at the point of consumption. The third and final component of LRM is *transportation management* (TM). This enormously costly component of logistics can simply be defined as the movement of product from one node in the supply chain to another, ending with delivery to the customer.

While the above definition attempts to view LRM from an organizational perspective, Figure 8.2 attempts to view LRM from a process capability point of view. Adapted from a framework developed by Accenture, the figure breaks LRM down into 21 strategic and tactical objectives distributed within a framework

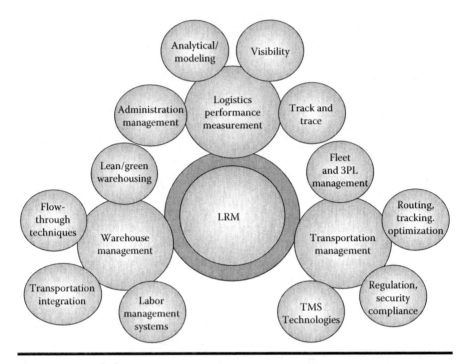

Figure 8.1 Logistics resource management components.

spanning three critical processes: *planning and collaboration, transaction,* and *execution.* In addition, the model contains a *technical infrastructure* component that will determine how information is exchanged and an *infomediary* component that will detail the depth of knowledge and information required to support the capabilities, both internally and externally, to drive the logistics model. While it is critical to acknowledge that not all of the 21 processes will apply to all industries and companies, the LRM capability model does provide a structured approach to understanding the possible components involved in logistics and transportation. In addition, the model provides strategists with a framework for identifying strengths and weaknesses and to develop effective measurements to initiate continuous improvement.

While each of the 21 process capabilities are critical to effective LRM, five of these capabilities can be singled out as absolutely critical, regardless of industry. Mastering these five capabilities provides companies with the most important drivers of total logistics cost reduction, higher customer service, and higher ROI. In addition, these capabilities are the foundation upon which the remaining LRM capabilities can be improved. Developing these and the other capabilities will require strategists to closely define the objectives and processes of each capability and to architect enabling organizations and technologies. These five capabilities can be described as follows.

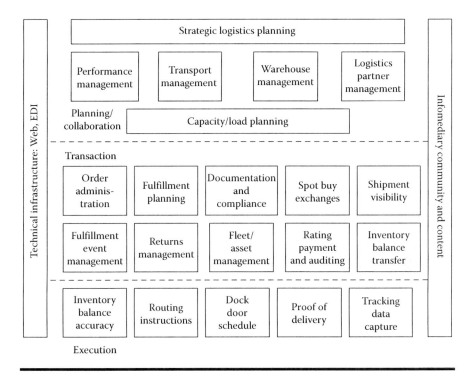

Figure 8.2 The logistics footprint. (Adapted from Lisa Hebert, "The Logistics Footprint," in *Achieving Supply Chain Excellence Through Technology,* ed. Narendra Mulani, Montgomery Research, San Francisco, 2002, pp. 148–181. With permission of Accenture.)

Logistics Performance Management

Perhaps no other business area in the enterprise needs to be measured more closely than logistics. Such a judgment arises in part from the fact that the sheer size and complexity of LRM operations activities (warehouse management, inventory transactions, orders, shipments, carrier negotiations and selection, and so on) touch so many facets not only of individual businesses but also of trading partners on a global scale. Furthermore, since LRM operations drive a significant share of overall enterprise costs, they must be carefully collected and monitored. Accurate measurements assist managers to identify operating inefficiencies and reduce costs. In addition, effective LRM metrics enable strategists to detail how well the company is serving its customers, identify possible avenues for profit by uncovering new value-added services or by differentiating product/service delivery, and architect new channel strategies with suppliers that result in channel inventory reduction and overall improved efficiencies. Finally, as LRM evolves from a tactical to a strategic function, validating the capabilities of logistics has become a critical source of competitive advantage, not only for individual companies, but for the entire supply chain.

LRM functions should be in conformance with the level of performance expected by the customer. Essentially, logistics performance is composed of three key components. The first, *logistics productivity*, is demonstrated by the creation of measurements that can provide meaningful productivity standards, the ability to track and manage logistics costs, the integration of quality management processes, and the broadening of logistics service levels. The second component, *logistics service performance*, is concerned with tracking metrics associated with the ability of logistics functions to meet customer service goals, such as product availability, order cycle time, logistics system flexibility, depth of service information, utilization of technologies, and breadth of postsales service support. A critical issue in this component is ensuring logistics measures are in synch with company fulfillment strategies. For example, if a company's strategy is to be a low-cost provider of products, the logistics systems must be efficient and flexible enough to meet company goals while keeping fulfillment costs to a minimum.

The final LRM performance component, *logistics performance measurement systems*, focuses on what and how performance is to be tracked. Such a task is a difficult one and conceals several traps. Since there are literally hundreds of measurements possible, companies must be careful to select those that paint an accurate portrait of overall performance, such as the sourcing/procurement process, fulfillment process time, cost, quality, and planning, forecasting, and scheduling process accuracies. In addition, companies need to ensure that measurements are not subjective, thereby distorting reality. Finally, LRM information systems should be employed to gather, process, and analyze the tremendous volume of LRM performance activities. Often this means working closely with trading partners to ensure the proper data are being gathered from channel SCM systems, ERP systems, and data warehouses.

Overall, logistics can be said to possess the following seven critical operating performance objectives [3]:

- *Service.* Today's supply chain is about creating *value* as determined by the customer. A *value chain* is about constructing agile, flexible operations that permit customers to configure the optimal mix of product/service solutions they need. Effective channel operations imply that the customer-satisfying elements of manufacturing, product/service availability, fulfillment, and service can be constructed to meet the unique requirements of every customer.
- *Fast-Flow Response.* This performance value refers to the ability of the supply chain to fulfill the delivery requirements of each customer in a timely manner. Rapid response requires the architecting of highly agile and flexible channels where all forms of processing time are collapsed or eliminated. Besides accelerating the order-to-delivery cycle, this attribute also means moving the supply chain away from stagnant pools of buffer inventory driven by forecasts to a demand-driven environment capable of rapidly responding to each customer order on a shipment-by-shipment basis.

- *Reduction of Operating Variance.* In a lean supply chain environment variance in any sphere, from production, to quality, to inventory balances, simply adds to cost and channel inefficiencies. Since logistics productivity is directly increased when variances are minimized, one of the most important duties of logistics management is the establishment of processes for the continuous identification of all forms of process variance, and when they do occur they provide the source for further Kaizen projects.
- *Minimum Inventories.* The levels of inventory necessary to support company sales and revenue objectives are part of the entire supply chain's commitment to customer service. The operating goal of logistics is to continuously reduce inventory to the lowest level as part of the process of achieving lowest overall logistics costs. Achieving this operating goal means pursuing inventory minimization and turn velocity for the entire supply chain and not merely one trading partner.
- *Transportation Reduction.* The movement of products across the channel network constitutes one of the most important supply chain costs. Reduction in transportation costs can be achieved by closer interchannel inventory planning and replenishment, utilization of larger shipments over longer distances to achieve movement economies of scale, and the effective use of third party service providers.
- *Quality Management.* Total quality management (TQM) is not an operating philosophy to be applied just to manufacturing, but rather to the whole supply chain. In fact, it can be argued that the requirement for absolute quality is even more essential for logistics than anywhere else in the company. Logistics transactions often deal with transporting inventories and services over large geographical areas. Once set in motion, the cost of solving a quality problem, ranging from incorrect inventories to invalid orders and late shipments, require lengthy and costly processes to be reversed.
- *Product Life Cycle Support.* As stated above, the continuing evolution of environmental laws and popular opinion has mandated an increase in managing *sustainability* strategies. Some of the critical challenges posed by recalls and returns are that they often involve a complex paper and product trail from worldwide sources, they can mean remanufacturing, repairing, or destroying the return, and they can influence every link within the supply chain. Responding to this operational need requires logistics planners to develop detailed return, recycle, and repair programs that enable the efficient management of product life cycles that can actually provide companies with a wealth of information on product performance, ease of use, defects, and consumer expectation.

Fulfillment Planning and Execution

Whether serving a traditional or an e-commerce channel, one lesson has been made abundantly clear: without effective and timely product fulfillment even the best marketing campaign or the cleverest Web site is destined for failure. In the past, buyer/seller exchange consisted of one-to-one paper-based systems. The fundamental problems revolving around data inaccuracies, lack of connectivity, and lack of information visibility that constrained the fulfillment process were masked by inflated cycle times and channel inventory buffers. In contrast, today's customers have the capability to view marketing materials and promotions and place orders and access shipment information almost instantaneously. Accordingly, customers have become increasingly impatient with the poorly synchronized, inflexible fulfillment processes of the past and requiring in their place integrated supply chains capable of compressing all aspects of fulfillment from order entry to delivery. In such an environment companies need LRM functions that place a premium on fulfillment speed, accuracy, visibility, personalization, and agility, attributes that require companies to deploy the right technologies in support of superbly engineered fulfillment processes that range from tightly synchronized trading partners to strategically located brick-and-mortar warehouses.

Architecting an effective LRM fulfillment plan requires close attention to both operational functionality as well as strategic capabilities. Operationally, fulfillment planning enables companies to convert customer demand into actual shipment schedules, implement lean campaigns that continuously right-size warehousing assets and inventory investment, and optimize TM activities, such as carrier selection and routing. According to Poirier and Bauer [4], the goal of the LRM process is the optimization of supply chain network fulfillment functions. These functions can be separated into the following five operations areas:

- *Freight Cost and Service Management.* The main functions in this area consist of managing inbound/outbound freight, carrier management, total cost control, operations outsourcing decisions, and execution of administrative services. Effective fulfillment planning in this area requires the architecting of fulfillment functions that can optimize inbound materials and outbound product movement, warehousing, and administrative services that utilize the most cost effective yet efficient transportation partners and carriers.
- *Fleet Management.* This area is concerned with the effective utilization of physical transportation assets. Critical areas for fulfillment planning revolve around equipment utilization, equipment maintenance, and total cost. The goal is to determine the optimum use of transportation assets, whether internal or through a third party supplier, without compromising service levels.
- *Load Planning.* Utilizing transportation assets to achieve maximum fulfillment optimization requires detailed load planning. Critical functions in this area revolve around selection of the appropriate transportation mode, load building and consolidation, and possible third party transfer point or cross-docking functions.
- *Routing and Scheduling.* These functions are normally considered as the heart of TM. Areas to be considered are optimization of shipping capacity

utilization, less-than-truckload shipments, and postponement strategies that position actual product at the various nodes in the supply channel. An often overlooked area is shipment documentation and compliance. Key points are concerned with ensuring accurate documentation regarding country quotas, tariffs, import/export regulations, product classification, and letters of credit.

■ *Warehouse Management.* The effective management of inventory in the supply chain requires efficient and well-managed warehousing techniques. Among the components of this area of traditional logistics management can be found order allocation and picking, packaging, receiving, put-away, returns management, and warehouse performance measurements.

Effectively managing each one of these logistics operations areas requires planners to constantly search for methods to automate functions through the use of computerized technologies. Whether through transaction devices, enterprise resource planning (ERP), electronic data interchange (EDI), or the Internet, optimization requires planners to assemble technology toolsets that enable them to leverage channel information flows to make better decisions, thereby increasingly replacing inventory in the supply pipeline with information.

While effective LRM operations will assist in generating incremental improvements to supply chain fulfillment, order of magnitude acceleration of channel value will be achieved when whole channel networks leverage LRM to achieve strategic fulfillment objectives. The goal is to engineer highly integrated supply chain networks that are seamless to the customer, agile in their ability to change to meet marketplace needs, and electronically enabled to provide real-time connectivity. Effective LRM operations planning should provide fresh opportunities for the development of new fulfillment networks and relationships with logistics partners, reengineering of new organizational LRM roles and responsibilities, and the development and implementation of robust, flexible sourcing, warehousing, transportation, and delivery capabilities.

Logistics Partnership Management

Today's fast-paced, Internet-driven marketplace requires businesses to develop new models governing logistics relationships with third party service providers. Often achieving the level of operation optimization detailed above will require the use of a third party logistics (3PL) firm or a lead logistics provider to support fulfillment objectives. According to Hintlian and Churchman [5], today's LRM environment requires logistics service relationships characterized by the following:

■ Increased collaboration between logistics services providers and their customers
■ The establishment of contractual and operations arrangements that foster an environment of win-win between all parties

- A detailed and accurate catalog of core competencies possessed by logistics providers that can be outsourced
- The capability of logistics providers to design support systems that can be used to assist both individual companies and the supply chains to which they belong

LRM support partners have become more valuable as they utilize new technologies to integrate themselves into the supply chain system and create new forms of fulfillment support services. These new fulfillment functions can be divided into three technology-driven enablers. In one region can be found LRM providers who use the Internet to form logistics marketplaces that match buyers and sellers of logistics operational services. In another region can be found infomediaries who harness information technologies to support the synchronization of logistics operational, tactical, and strategic functions for a supply network. In the final region can be found flow management service providers that assist network partners to manage the movement of fulfillment transactions through the supply network.

Shipment Visibility

Shipment visibility today has become much greater than simply order "track and trace." While it is critical that current fulfillment functions provide the capability for customers to access real-time data regarding their shipments utilizing any one of a number of values ranging from stock keeping unit (SKU) number to shipment origin, shipment visibility constitutes the very first layer in LRM. Of increasing importance to logistics managers is visibility to information about inventories and fulfillment capabilities found not just between immediate buyer and supplier, but also among trading partners constituting the entire supply chain. As cycle times and inventories shrink in the supply pipeline, substituting information for inventory has placed a premium on total supply network resource visibility. Assembling effective LRM capabilities has been made more difficult as the complexity of the supply chain has deepened, caused by the increase in contract manufacturing, raw material providers extending several levels upstream, global sourcing efforts requiring overcoming geographic boarders and language barriers, and the establishment of often separate Internet sales functions. Such factors have increasingly rendered just delivering the product to the customer a challenge. Inability to see shipments in the supply chain simply causes time delays, amplifies inventory reserves, and creates a bullwhip effect especially in complex supply channels with many storage nodes where inventory tends to build. Supply chain visibility is about being able to access accurate information about inventory and shipments anywhere, anytime in the supply network so planners can respond quickly and intelligently.

When coupled with event management functionality, the goal of visibility systems is to provide a real-time window into the life cycle of a product transaction beginning with order placement, progressing through picking, packing, and shipment, and concluding with delivery, quality reporting, and performance statistics. Lanier, for example, used a third party shipment solution to eliminate nearly 100,000 phone inquiries about shipment status when it provided dealers Web access to the status of their orders, from pick-and-pack all the way to delivery. Shipment visibility is also critical for inbound shipment management. Target, for example, receives thousands of inbound shipments each week into its distribution centers. In the past, information was handled largely through faxes and phone calls between the company, suppliers, and carriers. Today, Target's supply channel visibility software enables the giant retailer to optimize shipments, consolidate LTL into full truckloads, and better plan the receiving process [6].

As the value of information regarding delivering capability and shipment visibility grows, dependence on various levels of technology can be expected to correspondingly grow and will become a "must-have" for effective LRM. Inside the organization, companies can leverage ERP suites, warehouse management systems (WMS), and transportation management systems (TMS) to view shipment-order linkages and in-transit shipment information, track fill rates, ensure valid available-to-promise information, and control transportation costs. Externally, companies can now leverage supply chain planning, management, and execution applications that provide collaborative planning, forecasting, and replenishment (CPFR) and supply chain event management (SCEM) solutions that provide LRM planners with visibility to the entire supply channel, enabling them to make more strategic logistics solutions, and assist in architecting fulfillment processes that lead to a spiral of continuous improvement and superior service to the customer.

Fulfillment Event Management

Alongside logistics visibility, fulfillment event management (or its better-known name—SCEM) has become one of today's most important buzzwords in supply chain management. Fulfillment event management is one of the most significant and far-reaching of today's software technologies that, when applied to warehouse, order, and TM, can assist in proactively reducing and eliminating common fulfillment errors, such as missed orders, late deliveries, and other shipment and service problems. Event management systems can now feed data into performance management systems so LRM planners can start to understand the root cause of logistics problems such as a transportation delay.. What was absent before was visibility into the reasons why a truck shipment was delayed or why orders were short of inventory. With more granular, extended supply chain visibility, the data and the performance measurement applications to correct the causes of these problems can be mined to shrink lead times or optimize inventories or transportation modes.

Companies such as barnesandnoble.com, the Web-based sales arm of bookseller Barnes & Noble, are using fulfillment event management applications to monitor and send alerts, create work flows, and set up escalation processes to enhance customer service, cut logistics costs, and improve supply chain efficiencies. Once an order is placed via their Web-based entry process, the company keeps customers informed of the stages through which their order is progressing. Fulfillment data, such as when shipments are packed and shipped and when delivery can be expected is e-mailed directly to the customer and is also available on their Web site. AMR Research feels that such capabilities take fulfillment beyond mere track and trace: true SCEM applications should support business processes for the following:

- *Monitoring*: Providing real-time information about supply network events such as the current status of channel inventory levels, open orders, production, and fulfillment
- *Notifying*: Providing real-time exception management through alert messaging that will assist supply channel planners to make effective decisions as conditions change in the supply pipeline
- *Simulating*: Providing tools that permit easy and fast supply channel modeling and what-if scenarios that recommend appropriate remedial action in response to an event or trend analysis
- *Controlling*: Providing channel planners with capabilities to quickly and easily change a previous decision or condition, such as expediting an order or selecting less costly delivery opportunities
- *Measuring*: Providing essential metrics and performance objectives or KPIs to assist supply chain strategists to assess the performance of existing channel relationships and to set realistic expectations for future performance

SCEM can be described as an application integration layer that standardizes and transports fulfillment information as it flows between channel trading partners. SCEM is normally integrated with ERP and supply chain execution systems and has the potential to link to today's emerging forms of trading exchange. SCEM applications currently provide logistics planners with the following functions: order and shipment tracking, work flow, alert messaging/notification, escalation processes, and performance/ compliance management. Basically, the system is engaged when an event, either planned or unplanned, occurs requiring planner intervention. Depending on the impact of the event, the system will trigger a signal, often using Boolean-type logic, to alert planners through a generic work flow process that an occurrence in the fulfillment pipeline has violated predetermined event boundaries. For example, Unilever Home & Personal Care uses their SCEM system to create close Internet links to their third party manufacturers. This permits them to be aware of logistics events such as whether a product is produced, on hold, released, or available for pickup. It also has removed several days in the speed of cycle time fulfillment due to improved information flow through the supply channel [7]. However, while SCEM provides visibility to current events and permits planners to execute operational corrections, the real value in event management is to be found on the strategic level where predefined KPIs, performance scorecards, and dashboards can detail long-term spending and fulfillment performance on shipments and compare the data over a variety of carriers, time periods, and transportation modes to pinpoint critical variances.

Altogether, the benefits of applying the latest computerized techniques that seek to automate logistics functions and increase the accuracy of LRM decisions can produce enormous benefits not only for individual companies, but for entire supply chains. Among the benefits are faster response times so less expensive shipping is required, decreased manufacturing work-in-process (WIP), channel, and in-transit inventories, improved accuracy of order processing and tracking, increased velocity of order management, reduction of returns, and decreases in labor requirements. Additional hard benefits can be found in decreased inventory holding costs, increased inventory turns, and reductions in channel safety stocks.

Dealing with Logistics Uncertainties

As discussed in chapter 4, today's global businesses have been focused on the issues of risk management in the supply chain. The recessions of 2001–2002 and 2008–2009, the events of 9/11, and the continuing wars in Iraq and Afghanistan have arguably issued in an era of unsettledness. An event such as the November 27, 2009, declaration by Dubai that it could not pay its debts sent the Dow crashing 154 points in 1 day. The amounts involved in Dubai were not big enough to destabilize the world economy by themselves, but some investors feared the crisis there foreshadowed other "debt bombs" in countries (such as Greece) that borrowed heavily to weather the recession of 2008–2009.

While it is true that undertaking a supply chain integration project is difficult in itself without threats of global trade disruptions and a soft economy, companies must be careful to avoid the following common misconceptions concerning the application of today's suite of LRM Internet-enabled technologies.

1. *e-LRM is prohibitively expensive, must be debugged by a large in-house IT staff, and will take years before a satisfactory ROI can be achieved.* In reality, participating in e-LRM solutions requires little or no investment from trading partners. While customized private trading communities will require upfront expenses, most providers do not even charge a membership fee for participating in their trading communities. e-Logistics providers charge transaction fees when freight is tendered.
2. *e-LRM will require massive changes in the way the company has historically done business, thereby threatening operations during a period of uncertainty.* Leading-edge companies simply can not wait. Most e-logistics providers possess the technology infrastructures scalable enough to handle enormous transaction volumes, operating redundancies to respond to any emergency condition, and the agility and sophistication to manage even the most complex of supply chains.

3. *e-LRM will require massive changes to existing technology infrastructures.* In reality today's e-logistics providers can easily combine advanced technologies, such as the Internet, XML, and wireless communications, with legacy ERP and TMS solutions. Normally it is the responsibility of the logistics solution provider to provide the necessary connectivity for today's seamless supply chain.

4. *The use of nonproprietary information technologies risks a loss of control and security over sensitive company data.* Virtually all e-logistics marketplaces today are encrypted with sophisticated security technologies that provide safe transmission of sensitive data.

5. *e-Commerce requires a great deal of internal knowledge and it is hard to learn.* Any company with a computer will benefit from e-commerce. Most e-logistics suppliers provide complementary training and 24-hour help desks to support users, and even the most technology-challenged users can acquire sufficient skills to participate in an Internet-driven trading community.

6. *Using e-logistics trading communities will severely damage long-standing relationships with core logistics providers.* Logistics exchanges should be used to enhance, not replace, existing logistics relationships. Logistics managers might want to test the cost of services first on an exchange before turning to core providers. Internet-based logistics providers can also provide service to supply chain areas that have functioned independently of core carriers. Finally, e-logistics providers can add value to existing relationships by supplementing handling and automating financial settlements or special reporting functions that are beyond the competencies of existing logistics partners.

Removing these fears is going to be critical as more and more companies become more comfortable with technology. Many of the key software vendors have developed comprehensive logistics management packages for both warehousing and transportation as have many Web-based 3PLs who have assembled a full range of Internet-enabled services that are flexible enough to meet a variety of customer systems. Online marketplaces are also developing multiple response engines that are more sensitive to their customers' comfort with technology. Today's logistics marketplace is more and more focused on collaborative partnerships where customer needs are identified, logistics process and readiness for e-commerce are mapped, and appropriate Web and EDI connections that accentuate logistics optimization, choice, and execution are discussed at the outset. Such an approach will go far in establishing the necessary trust and comfort and increase understanding of technology benefits.

Defining LRM in the Age of the Global Internet

There can be little doubt that the advent of the Internet has dramatically changed forever the nature and objectives of logistics. In the past, logistics was primarily a back-office function concerned with the day-to-day packing and shipment of products, rate calculations, transportation routing, and inventory chasing. Today, logistics has become a strategic advantage. As delivery times shrink from weeks to days or hours, fulfillment has become a hot topic and the path to survival in today's global marketplace. As technology tools continue to mature, providing companies with logistics systems that can utilize positioning satellites to pinpoint truck, rail car, and shipments in transit, and accurately track transportation assets and the cargo they carry from origin to destination across the globe, competitive logistics depends on effective partnership management and real-time visibility to link all trading partners anywhere in the supply chain. Today's LRM application toolsets are providing better fill-rates, lower levels of channel inventories, better on-time delivery, increased transportation asset utilization, and lower costs while improving customer demand for ever faster and cheaper order fulfillment.

Applying Internet Technologies to LRM

The first applications of Internet technologies to logistics were introduced in the mid-1990s. Despite the complexity of the array of possible logistics functions, the fact that logistics services were basically a commodity and could be easily adapted to the Internet made them an easy candidate for Web-based technologies. Although there was and continues to be much discussion over what is the proper function of e-logistics (should it provide an all-encompassing range of services so that logistics relationships cover end-to-end management of fulfillment or should it be focused on a cafeteria approach where customers engage the discrete service they need), the application of the Internet to LRM has been eagerly adopted by all forms of warehouse and TM and logistics services providers. Initially, companies offering logistics services used the Web to detail information about their organizations, company locations, service and product offerings, and to respond to prospects regarding logistics capabilities and pricing. When the concept of exchanges arose, phase two logistics providers began to use the Internet to sell warehousing and transportation services directly to their customers versus using traditional methods, such as direct sales, phone, or fax. These early Internet-driven e-logistics offerings tended to be horizontal, public exchanges that provided buyers with the opportunity to post loads on a Web bulletin board; in turn logistics companies could then bid on those loads [8].

When viewed in greater detail, these early efforts can be said to consist of two types of logistics exchange: auctions and spot markets [9]. A *logistics auction* features a single party who is soliciting logistics services from or selling logistics services to multiple parties through the Internet. Originally, it was assumed that reverse auctions, where shippers would be able to post requirements on the Web and then entertain bids from logistics providers, would provide fresh opportunity for cost reductions and increased services and dramatically change the nature of the logistics industry. The *spot market* concept would provide lowest-cost advantage to a buyer by broadening the number of bidders; for sellers, the goal was to increase the number of bidders, thereby selling excess capacity at the highest price. The objective of the exchange is to match requirements and provider capabilities and possibly even to expand the use of toolsets such as EDI for managing the transaction or deepening existing buyer–seller relationships.

Unfortunately, these early public exchanges were gripped by the same centripetal forces that plagued the dot-com revolution. To begin with, many logistics providers avoided the exchanges for fear that spot market prices would be construed by buyers as constituting their base services price. Further, logistics services are not the same as commodity products. While the characteristics of commodities can be accurately determined, contracting logistics services requires more than attaining lowest price and must consider additional factors such as transit, reliability, and availability of equipment. Also, the anonymity of the Web was quickly perceived as detrimental to historical relationships that existed between trading partners. Providers wanted to be assured that their customers would promptly pay for services and that there would be no frivolous claims. On their part, customers wanted to be assured that their providers were reputable and that their shipments would arrive on time without damage or pilferage. Finally, since so much business was already covered by logistics contracts (up to 85% of freight moved by truck and more than 90% by ocean shipments), most early exchanges simply failed to generate critical mass and soon died. For companies that had already implemented EDI applications there was little incentive to migrate to the Web.

Today, e-logistics marketplaces have closely followed the experience of other areas of e-business as they migrate from independent to private exchanges and move their focus from cost economies and spot buys to developing collaborative communities. These private logistics exchanges are normally developed by a dominant or group of dominant players who have already established a clear e-LRM strategy and set of logistics partners. This existing relationship provides instant critical mass to the exchange, thereby forcing smaller trading partners in the channel to participate. But again, e-LRM private trading exchanges (PTX) and consortia trading exchanges (CTX) exchanges suffer from the same problems affecting other e-marketplaces. Numerous technical and standards issues must be worked out; while packaged software exists, companies fear being trapped within the boundaries of a certain vendor; many companies are using 3PLs that have their own solutions, negating the use of logistics trading exchanges altogether.

By 2005, the once frantic explosion in the e-logistics marketplace had dissipated and survivors entered a period of shakeout and massive consolidation. Today, the lines separating companies competing in the e-logistics space have become blurred. Some providers offer online load/matching services, some e-logistics software, and others have expanded to include extensive logistics management, financial, and consulting services. e-LRM solutions can be divided into the following six distinct categories [10]:

- *Warehouse Exchanges.* This type of e-LRM provider acts either as a broker seeking to match companies that need temporary warehouse space or as a

VENDORSEEK.COM

Get Free Warehousing/Storage Services Price Quotes

1 **Fill Out Simple Form**
Est. Time - 2 minutes

2 **We Match to Qualified Vendors**
Based on your specific requirements

3 **Compare Quotes & Save!**
No Obligation to Buy

It just takes a minute...

What type of products do you need to store? *
--- Select One --- ▾

Do you require the provider to be in a specific geographic region? *
☐ No Preference   ☐ Northeast
☐ Southeast   ☐ Midwest

Figure 8.3   VendorSeek.com Web site.

full-service warehouse selling services on an as-needed basis. Brokers take on the responsibility of analyzing the buyer's warehouse needs, matching qualified vendors by region or other special requirement, and assembling quotations (Figure 8.3). A full-service warehouse offers a range of services directly to their customers, such as transportation, public warehousing, pick and pack, cross-docking, short-term storage, distribution services, contract manufacturing, and build-to-suit construction (Figure 8.4).

■ *Transportation Exchanges.* This type of e-LRM provider acts as an e-market-place where logistics providers can publicly post and match capacities with customer demands on a spot market basis. An example is Descartes Systems Group, Inc. Descartes' logistics management solutions combine a multimodal network, the Descartes Global Logistics Network, with business applications to provide messaging services between logistics trading partners, brokerage and forwarding management, shipment management services for contract carriers, and private fleet management services.

■ *Transportation Network Infrastructure.* This type of e-LRM provider offers logistics Web-based software applications and communications infrastructures for shippers, carriers, and 3PLs capable of automating global logistics transactions. For example, TransCore, a 70-year-old transportation and communications company, offers a full suite of applications designed especially for the truck transport market, including satellite tracking and in-cab messaging, rail and intermodal services, and operation management. GXS, Inc., offers an EDI, XML, and Web-based suite of applications that address e-LRM requirements for data capture, data quality management, business process visibility, and event management.

■ *Third Party Logistics Provider.* This type of e-LRM business acts as a third party services provider offering transportation, warehousing services,

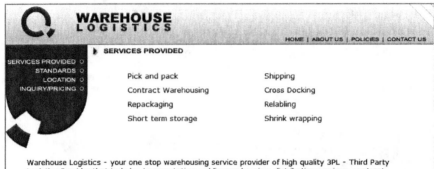

Figure 8.4   Warehouselogisticsinc.com Web site.

shipment consolidation freight payables, reverse logistics, and other functions to clients. Maersk Logistics, for example, offers a full array of document management, warehousing and distribution, ocean transport, air services, customs house brokerage, and miscellaneous services such as on-site management and consulting.

■ *Collaborative Logistics Networks.* This final type of e-LRM provider enables any network member, including shippers and carriers, to connect to any other member in the selection and utilization of logistics functions. For example, part of the Sterling Commerce's *Business Integration Suite, Sterling Collaboration Network* is a hosted service providing secure connectivity and collaboration with customers and business partners.

■ Among the services provided by Sterling Commerce can be found applications to assist companies to build, manage, and expand their business-to-business (B2B) communities; data synchronization tools that simplify the process of data collection, cleansing, registration, validation, and publication; conversion of faxes to EDI or XML documents; e-invoice automation; a *Supply Chain Visibility On Demand* application providing network customers access to inbound and outbound supply chain information; translation and mapping services enhancing communications flexibility by offering high-quality maps, making it easier to exchange and process documents in all formats and standards. *Web Forms* provide a way to exchange documents over the Internet.

While this breakdown can assist in segmenting the types of e-logistics markets, many of today's e-logistics companies have significantly expanded their services to the point that they bridge several of the above categories. For example, Cat Logistics offers a rich array of logistics services. On the strategic side, Cat provides supply chain strategy and decision support to assist client develop effective network design and channel strategies, asset planning, transportation modeling, inventory simulation, facility design and work flow, and process engineering. On the execution side, Cat offers a full slate of services for materials management (demand planning, distributation requiencements planning (DRP), purchasing and expediting, and inventory optimization), manufacturing logistics (value-added processing), transportation services (inbound/outbound, track and trace, freight bill audit and payment, and freight forwarding), distribution center management (warehousing, cross-docking, packaging and labeling, and quality assurance), reverse logistics, and order management (customer service, order entry, claims, and invoicing).

Internet-Enabled LRM Technologies

At the heart of e-LRM can be found several levels of technologies. These computerized toolsets are not only streamlining and enabling fool-proofing warehousing and transportation, but providing networking tools that allow companies to closely integrate their logistics needs with available capacities to be found among their business network partners. Based on the three part division of logistics found in Figure 8.1, the following sections explore the application of information technologies to logistics performance, warehousing and transportation.

Enterprise Performance Measurement

Effective logistics management requires robust analytical solutions capable of providing managers with the capability to evaluate precisely the efficiencies and profitability of supply chain processes. Critical performance metrics should provide analysis, measurement, and communication of the status of key business indicators at both operational and a strategic levels. Operationally, Web-based support tools in this area should provide instant availability to information regarding the status of individual orders and shipments as well as the visibility to track individual products in real-time anywhere, at any time in the supply channel network. Strategically, all network trading partners should be provided with the ability to access in real-time the data they need to monitor the productivity of logistics functions, including an executive dashboard that tracks costs and revenues across the supply chain network.

Fortunately, logistics managers today have a wide variety of technology toolsets available that can assist them to capture, retrieve, and format logistics data so that critical information revealing gaps and trends in performance can be reviewed. Among the tools sets can be found the following:

- *Analytical and modeling tools* that provide mathematical algorithms and dynamic business modeling functions that can assist planners manage transit times, carrier selection, load matching, load optimization, routing, asset optimization and fleet management, and contract management.

■ *Visibility tools* that detail inbound and outbound shipments to a company, its customers, and its trading partners. Included are Web-driven applications providing windows into in-transit inventories focused on customer service as well as sharpening the accuracy of replenishment cycles and reductions in channelwide safety stocks. Visibility to channel resources can also assist companies in selecting the optimum location from which to fill an order based on delivery and inventory costs or other configurable business rules.

■ *Track and trace tools* that provide timely information regarding shipments at anytime, anywhere on the globe. Among the key functions are these:
  - Complete and real-time status and history on all shipments from booking to proof of delivery
  - Electronic transmission and access to bills of lading, shipping labels, waybills, SKU, and, other order details
  - Multimodal tracking for truck, air, ocean, or rail cartage
  - Costs and time requirements for international business
  - Web access to shipments by bill of lading number, PO number, RMA, and other shipper reference numbers
  - Availability of other shipment data such as online reports, document retrieval, claims status, and pickup requests

■ *Administration management tools* providing for the effective reporting and management of logistics financial functions. Among the events to view upon logistics transaction completion are proof of delivery, invoice generation, review and printing of bill of lading, freight bill review, total landed cost calculations, import/export documentation, and any necessary filing/status of loss and damage claims. By providing for Web-based applications that can assist in automating the execution of bill of lading and delivery receipt (POD), audit, and final payment, LRM provides shippers and carriers with the tools to manage fast closure of the order cycle and accelerate time to cash.

Warehouse Management

According to a 2009 report on the state of warehousing [11], Capgemini identified the following five major challenges facing warehousing at the opening of the second decade of the new century:

■ Growing requirements for lean warehousing
■ Reducing of energy and environment costs through sustainable warehousing
■ Increased movement for horizontal collaboration for warehousing
■ Integration between WMS and TMS
■ Increased need to design multichannel warehouses capable of cost-reducing measures in the fulfillment of Internet orders
■ Effective management of labor

While several issues regarding warehousing arise out of these six challenges, all of them involve the increased utilization of warehousing technologies designed to enhance warehouse automation, deepen the planning and control of warehouse functions, and provide for integration with order management and transportation software applications.

The growing application of computerized technologies can be attributed to several factors, the foremost of which are the following:

- Declining costs of warehouse automation
- Increased cost of labor and associated overheads
- Much closer demand for supply chain integration and collaboration
- Value-added philosophies stressing continuous elimination of operational wastes and redundancies
- Requirements for shorter purchasing and customer service cycle times

Two critical areas of technology applications stand out: the expanding use of warehouse automation tools to reduce error, increase productivity, and speed supply chain flow-through and the implementation of software suites for integrated warehouse management, labor and task planning, slotting and carton solutions, and yard management.

The different forms of warehouse automation are illustrated in Figure 8.5. While each area is expressed equally, the biggest impact of technology growth is expected

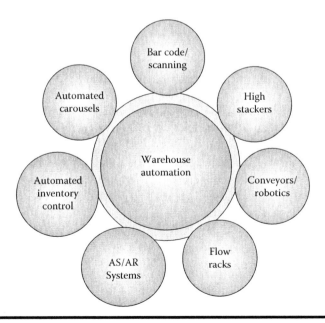

Figure 8.5   Forms of warehouse automation.

not so much in the installation of material handling equipment but in computerized tools that increase workstation capabilities or automate material movement information. These results are in concert with current thinking about the warehouse, emphasizing the prioritization of the enterprise's knowledge base over material handling mechanization. Indeed, the overall goal of automation seems not so much to be on moving product in the warehouse but rather in how to engineer sales, marketing, and supply chain objectives so that the need for physical handling becomes obsolete and storage gives way to direct delivery from producer to consumer.

One particular method that significantly facilitates the movement of goods through the distribution pipeline is *cross-docking*. Cross-docking can be defined as a method of moving products from the receiving dock to shipping without putting it into storage. Typically, arriving merchandize is broken down into case or pallet loads, then quickly moved across the warehouse to the shipping dock where they are then loaded onto trucks bound for the next level in the distribution pipeline. Technology tools such as computers, bar codes, radio frequency identification (RFID), and EDI have facilitated the case and pallet shuffling required to make cross-docking feasible. There are two fundamental goals driving distributors' interest in cross-docking. The first centers on ways to eliminate handling, storage costs, shrinkage, damage, and product obsolescence. The second is to reduce cycle time and get the product to the end user as quickly as possible.

The goal of these changes in the nature of warehouse operations is to keep warehouse size, equipment, personnel, and transportation in line with enterprise demands. This strategy calls for perceiving warehouse assets as modular, meaning warehouse services can be easily disassembled and reassembled with new features with a minimum of cost and time while meeting the targeted level of customer service. To manage short-term problems, the expanded use of public warehousing may be more viable, where possibly a more long-term approach would be the expansion of company-owned facilities. The same could be said of transportation: the use of public carriers, the purchase of equipment, leasing equipment, and the viability of long-term contracts would have to be examined in light of short- and long-term fluctuations in traffic requirements.

The second major area of warehouse information technology is the implementation of WMS suites. WMS software solutions were introduced over 20 years ago and today comprise a billion-dollar market. Over that time period they have evolved from simple back office functions associated with inventory and shipping to today's robust systems. According to an Aberdeen Group survey (February 2009), 59% of the respondents had implemented a WMS while 35% planned to acquire or build a system in the next 12–24 months [12]. Figure 8.6 portrays the higher-level areas managed by a WMS.

In today's rapidly developing technical environment, savvy companies are exploring the use of SaaS for warehouse management software. The economy at the beginning of 2010 has forced businesses to run lean-and-mean, and an on-demand

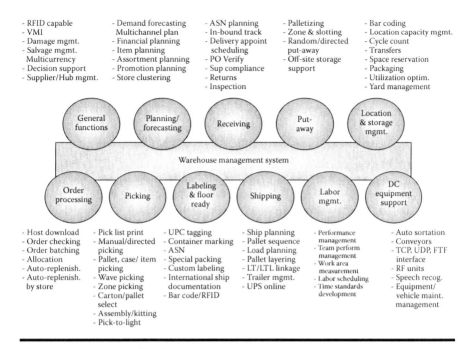

- RFID capable
- VMI
- Damage mgmt.
- Salvage mgmt.
  Multicurrency
- Decision support
- Supplier/Hub mgmt.

- Demand forecasting
  Multichannel plan
- Financial planning
- Item planning
- Assortment planning
- Promotion planning
- Store clustering

- ASN planning
- In-bound track
- Delivery appoint
  scheduling
- PO Verify
- Sup compliance
- Returns
- Inspection

- Palletizing
- Zone & slotting
- Random/directed
  put-away
- Off-site storage
  support

- Bar coding
- Location capacity mgmt.
- Cycle count
- Transfers
- Space reservation
- Packaging
- Utilization optim.
- Yard management

General functions | Planning/forecasting | Receiving | Put-away | Location & storage mgmt.

Warehouse management system

Order processing | Picking | Labeling & floor ready | Shipping | Labor mgmt. | DC equipment support

- Host download
- Order checking
- Order batching
- Allocation
- Auto-replenish.
- Auto-replenish.
  by store

- Pick list print
- Manual/directed
  picking
- Pallet, case/ item
  picking
- Wave picking
- Zone picking
- Carton/pallet
  select
- Assembly/kitting
- Pick-to-light

- UPC tagging
- Container marking
- ASN
- Special packing
- Custom labeling
- International ship
  documentation
- Bar code/RFID

- Ship planning
- Pallet sequence
- Load planning
- Pallet layering
- LT/LTL linkage
- Trailer mgmt.
- UPS online

- Performance
  management
- Team perform
  management
- Work area
  measurement
- Labor scheduling
- Time standards
  development

- Auto sortation
- Conveyors
- TCP, UDP, FTF
  interface
- RF units
- Speech recog.
- Equipment/
  vehicle maint.
  management

Figure 8.6 WMS application components.

solution provides firms with faster time-to-value and rapid deployment, making it easier to do more with less and achieve supply chain excellence quickly by moving away from the high costs and extensive customizations required of on-site software solutions. The bottom line is that the on-demand model makes it possible for customers to eliminate the risk of failure and immediately recognize significant value from their WMS software purchases [13].

Transportation Management

Operating a world-class supply chain requires today's TM function to play a more significant role in the supply channel than just directing modes of transportation and delivering product: it must execute flawlessly, accurately, and timely processes that span the entire order cycle from bid to delivery. It must also drive value through transportation flexibility and optimization, work interactively with third party services, establish and meet sustainability objectives, promote lean practices, and understand its impact on company strategy, finance, sales, and service. Such a strategic view of transportation is essential considering the tremendous challenges facing the function. For example, oil prices reached record highs in July 2008, only drop to 5-year lows a few months later; by 2009, more than 3000 trucking companies had gone out of business, due first to the high cost of fuel and then to the precipitous drop in volume to crippling recession; the US DOT forecasts that over

18.1 billion tons of freight will travel by highway in the year 2020, a 75% increase from 1998. The net impact of these trends is that managing costs, capacity, and congestion (the "three Cs") will become more difficult for shippers and carriers down the road [14].

One of the best ways of responding to the challenges for cost savings and productivity improvements is to utilize a TMS. One of the best definitions of a TMS is from ARC Advisory Group [15]:

> Transportation Management Systems are software solutions that facilitate the procurement of transportation services; the short-term planning and optimization of transportation activities, assets, and resources; and the execution of transportation plans. They address all modes of transportation, including Ocean, Air, Rail, Full Truckload, Less-than-Truckload, Parcel, and Private Fleet. In addition to managing the physical flow of goods, they also manage the flow of transportation-related information, documents, and money. TMS also include performance management and collaboration capabilities.

As illustrated in Figure 8.7, a TMS provides a wide array of transportation solutions that automate and optimize provider selection, multicarrier compliance, rate quotation, routing, manifesting, tracking, cost analysis, and postshipment analysis processes. As trading partners and LSP become linked real-time into the logistics network web, companies can freely retrieve shipping, service, and contact information to identify carriers, transit times, and compliance issues such as Certificate of Origin, customs invoice, global settlement of freight payment and billing, allocating the true cost of transportation based on actual charges, and electronic notice of consignment and statements of revised charges for change of destination of an in-transit shipment.

The scope, functionality, and architecture of TMS applications have changed significantly since their beginnings in the late 1980s. A $1.2 billion industry in 2008, today's software suppliers have been rapidly transforming TMS from a fragmented collection of applications to a unified platform where users can execute role-specific processes via configurable user interfaces, work flows, and Web services. While the transportation automation processes are the centerpiece of the software, the flashpoint of Web-enabled TMS is its ability to serve as the foundation for the networking of information between many external parties, including carriers, suppliers, customers, and LSP across logistics e-marketplaces. For example, Manhattan Associates provides an application called "Logistics Gateway" that provides easy linkage through EDI/XML between buyers and transport sellers. The application enables linkage any logistics supplier to facilitate requisition management, PO management, inspections and quality assurance, fulfillment and shipping and charge-backs. If parties lack Web/EDI toolsets, Manhattan's application provides online portals that open the door to electronic communications for spot

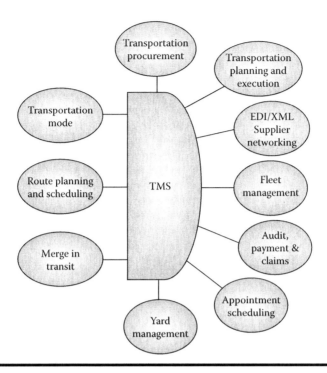

Figure 8.7 TMS applications.

negotiations and to exchange information on status, invoicing, invoice reconciliation and tender acceptance.

Probably the most exciting aspect driving today's TMS, is the ability of enterprises to leverage the SaaS model versus the traditional and past hosted models to acquire TMS capabilities. Until recently, companies not using a TMS often used the excuse that it was too expensive to purchase and implement. Similar to what has occurred in the WMS marketplace, the use of such an excuse is no longer valid or justified. On demand TMS has largely eliminated the cost and resource constraints and shorted the time-to-value companies traditionally faced in implementing new technology and continuous improvement initiatives. Perhaps the key advantage of a SaaS model is that the network of shippers, carriers, and other trading partners are already built into the single-instance TMS provider (this means there is only one software/hardware setup of the TMS that multiple shippers, carriers, and other trading partners use). LeanLogistics, for example, has about 5500 carriers connected to its On Demand TMS network. As part of its *Carrier Procurement* managed service, LeanLogistics identifies carriers within its network that meet a shipper's performance and capacity requirements. Ace Hardware, for example, wanted to cast a "wide net" when it conducted its transportation procurement engagement last year. Lean introduced Ace to a large base of carriers and the RFI was sent to almost 500 carriers. The procurement engagement uncovered a $3.9 million (about 4%) annual savings opportunity for the company [16]. Ultimately, according to ARC Advisory, goal of the SaaS TMS model would be to

> imagine an app store for logistics software, similar to Apple's iPhone App Store, where a transportation manager could browse through various applications (e.g., an app for procurement, appointment scheduling, or freight audit and payment), buy them online using a credit card, download them, run some quick installation wizards, and off they go, on the road to improved productivity and cost savings. [17]

While such a scenario is still somewhat futuristic, there can be little doubt it constitutes the next step in TM.

Jel-Sert, TMS and SaaS

Jel Sert, a 70 year old manufacturer of snack foods and beverages, found itself at the mercy of increased transportation costs, driven largely by the company's inability to scale its manual transportation process with its rapid growth. The company was manually processing over 20,000 orders annually, missing out on load consolidation opportunities, and struggling to manage fuel surcharges and other costs.

"We had examined implementing a TMS in the past," explained Michael Martinez, Director of Distribution at Jel Sert, "but the cost and IT support required prevented us from moving for-ward." As the company continued to grow, however, a TMS solution became inevitable.

The company eventually selected lean Logistics' On-Demand TMS®. A particularly appealing aspect of the "on demand" model for Martinez is that it provides "ongoing service, support, and innovation driven by everyone on the network."

The TMS allowed Jel Sert to standardize and automate their shipment planning and execution processes, as well as implement best practices. Jel Sert brought its freight settlement process back in house, streamlined and automated it, and now has complete visibility to freight bills, overcharges, and payment issues.

Source: Gonzalez, Adrian, "The Economy as a Catalyst for Change: The Role of TMS and Managed Services." *ARC Brief,* (Dedham, MA: ARC advisory group, May, 2009), pp. 8–9.

Understanding the Third Party Logistics Network

The requirements for logistics management in today's business environment have been dramatically accelerated by the pace of global competition and technology. Companies are currently locked in a struggle in which they must constantly search to improve their capabilities to respond to increasingly complex markets characterized by frequent new product introductions, the presence of new Internet-based sales channels, and information technologies that provide for real-time sales management, transaction control, and delivery visibility. Effectively responding to logistics challenges has driven many companies to look at outsourcing their logistics functions to logistics service companies termed 3PL providers. Utilizing 3PLs enable firms to shrink dramatically physical logistics assets, such as delivery systems and computer technologies. Additionally, logistics outsourcing enables shippers to leverage 3PL investments in information, material handling, and operating equipment. Finally, the need to optimize the constantly shifting parameters of supply chain operations requires companies to possess a degree of agility that can only be found in logistics organizations that are totally dedicated to logistics management.

According to industry-wide research, the use of 3PL services has been increasing dramatically over the past 20 years. In addition, while base logistics services are expected to be enhanced, 3PL providers are expanding their offerings to encompass greater functionality in finance, inventory, technology, and data management. In the past, service buyers most frequently outsourced logistics activities that were more transactional, operational, and repetitive, and less frequently those that were more strategic, customer-facing, and IT intensive. However, according to a 2009 study by Langley and Capgemini [18], service buyers are becoming more receptive to outsourcing strategic services from 3PLs. The fact of the matter is that the continuous squeeze on all elements of the supply cycle are simply pushed executives to explore 3PL services to realize added value. According to a study of 3PL value propositions, Mentzer [19] feels that companies today are looking to 3PLs to provide four key sources of logistics value.

- *Trust.* The goal of this value is to find a competent 3PL partner that can relieve the company from the task of managing the supply channel.
- *Information.* The objective of this value is to leverage the technology capabilities of 3PLs to provide logistics information accuracy, quality, and timeliness of the operations they deliver.
- *Capital Utilization.* The reduction of fixed assets in the form of physical plant and equipment is a major source of 3PL value. Less fixed expense can expect to be returned in the form of better working capital.
- *Expense Control.* The overall reduction of logistics channel costs is by far the primary objective of using a 3PL provider. Increase in customer service combined with lower logistics costs is seen by savvy CEOs as a critical path to survival.

Role of the 3PL

Outsourced logistics functions have always been part of the landscape of business. According to the Langley/Capgemini survey, 50% of respondents felt that 3PLs provide them with new and innovative ways to improve logistics effectiveness. Continuously searching to reduce costs, improve inventory throughput, and realize competitive customer service, companies have traditionally used 3PLs to provide a wide variety of transportation and logistics functions beyond the immediate capability of the business. By far the most used services in order are domestic transportation, international transportation, customs brokerage, warehousing and forwarding. Among other core services can be found transportation leasing, small package delivery, cross-docking, product labeling, reverse logistics, freight bill auditing and payment, transportation planning, fleet management, and customer service. In addition to these core functions, the needs of the marketplace fueled by the explosion in technology tools, such as the Internet, have enabled today's top 3PLs to expand their offerings to embrace a variety of advanced services including EDI, Internet order management, and Web-based documentation management.

The addition of these and other services have driven many 3PLs to architect new strategies that are reshaping the type of markets they serve, the operation and technology skill sets required, the investments made, the ROI expected, and the types of partnerships established. While the number of possible permutations is large, today's LSP can be separated into two camps. On the one side can be found LSPs that are focused on offering a limited selection of cost-driven, standardized services through an owned network of transportation and warehouse functions. They provide value to their customers by leveraging their logistics assets to cut SCM costs, managing nonasset cost factors, and architecting process innovation. The overall goal is to attract customers by making their SCM network linkages global and their services ubiquitous. Because they face severe competition due to the commodity nature of their service offerings, these LSP normally are constantly on the lookout for opportunities to apply technologies and assets to enhance core service packages and seize upon new forms of business activities, such as B2B and wireless devices, as they reach critical mass.

In the other camp stands a much more aggressive type of LSP that seek to manage their customers' logistics needs from end-to-end. Termed fourth party logistics (4PL) or lead logistics providers (LLP), LSPs in this category can be defined as supply chain integrators whose strategy is to assemble and manage dynamic organizations composed of a wide range of resources, capabilities, and technologies either within the organization or in partnership with complementary LSPs to deliver a comprehensive, customized logistics solution to the customer through a central point of contact. The goal of the LLP is to manage all aspects of the logistics relationship, beginning with *strategic definition* focused on applying technology and new management concepts to drive reinvention and transformation of the logistics effort, proceeding through *implementation* of business process realignment and systems integration, and concluding with the ongoing *execution* of operational functions associated with daily transportation and warehouse transactions [20]. In summary, an LLP provides a centralized point of contact for the customer with total responsibility for supply chain performance. Most 3PL arrangements fail to deliver benefits beyond one-time operating cost reductions and asset transfers:

LLPs, on the other hand, are dedicated to long-term strategic logistics management and continuous growth in revenues and reductions in operating cost and fixed and working capital.

Internet-Driven LSPs

Today's LSP, whether a narrowly focused provider or a full LLP, has been increasingly assuming the responsibilities for inventory ownership and carrying costs, transportation assets, financial functions, and even component testing and inspection. As technology capabilities have grown, LSPs have also found themselves shouldering the cost for managing information capabilities. Many companies simply do not have the financial and people resources to hook up to today's fast paced technology tools and are increasingly looking to their LSPs to provide the expertise to collect and scrub data and drive it directly into their backbone systems, perform e-commerce functions, provide proactive exception management, and enable participation in the Web-driven supply chain. APL Logistics, for example, provides not only core logistics services, but also advanced technology capabilities to its customers. APL, in fact, has an entire division devoted to developing and integrating the latest in proprietary and industry-standard supply chain technologies for product visibility, exception management, execution systems optimizing shipping and warehouse decisions, Web-based decision tools, and performance reporting. Customers have the capability to then turn up or phase down their use of these services depending upon ongoing needs and level of internal expertise.

Despite the increasing role of LSPs in technology support, the industry is confronted by serious challenges. According to the Langley–Capgemini report on the status of the 3PL industry, with complex legacy applications, aging infrastructure, multiple data repositories, and complex point-to-point integrations consuming substantial resources, 3PLs have fallen behind the cutting edge required by today's fast-paced global environment. The report emphasizes that

> the roots of today's IT challenges lie in yesterday's technology investments. Many 3PL providers operate legacy ERP and operational applications that run on mainframes or mid-range systems. Acquisitions add to the systems complexity. As a result, 3PLs are spending the lion's share of their IT resources keeping it all running. The difficulty of rationalizing and modernizing legacy applications means some 3PLs are maintaining multiple data silos with duplicate and incorrect data. Integration that does occur within the 3PL as well as with partners is often based on proprietary protocols and legacy electronic data interchange (EDI) standards. Shippers, too, face issues with legacy technologies that consume resources and impede integration efforts. [21]

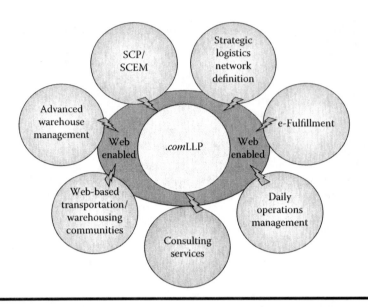

Figure 8.8 *.com*LLP functions.

However, while an IT gap exists, some LSP are moving forward with advanced IT services such as offering their IT platforms on a subscription basis as part of the services contract. The applications most sold this way are WMS and TMS toolsets used for transportation planning, transportation execution, warehouse/distribution center management, visibility, and customer order management. The report found that an average of 10% of the LSPs responding to the survey currently offer these applications on a subscription basis [22].

Earlier in this chapter, it was stated that a variety of LSP services, from inventory tracking and route optimization to carrier selection and freight bill management, are now Internet-driven. This growth in Web-based LRM services caused a couple of logistics experts to dub these LSPs *.com*LLPs [23]. The purpose of the description is to convey the critical competitive advantage available to customers from LLPs who have deep Internet capabilities (Figure 8.8). For example, Ryder Logistics and Transportation utilizes a logistics exchange called Freight Matrix that offers shippers, carriers, and other LSPs a logistics solutions hub to buy and sell transportation, plan cargo requirements, and execute shipment delivery. On its part, the exchange enhances Ryder's ability to access global networked supply solutions, thereby enabling it to maintain its lead as the top LLP provider for Global 1000 companies.

In detail, *.com*LLPs have the ability to provide their customers the following additional information-based services:

■ *Comprehensive Solutions.* Without a doubt, perhaps the greatest advantage of a *.com*LLP is to use technology to reverse the increased propensity for logistics fragmentation as products make their way through the supply channel. *.com*LLPs possess the capability to offer customers a centralized, comprehensive logistics management point that can facilitate the merger of a community of functional logistics providers with the enabling power of a customized, client optimized, uniform technology strategy. Such a technology-based solution has the capability of leveraging networking tools like the Internet to drive the synchronization of real-time logistics planning and execution across trading partners while increasing ever-widening collaborative cooperation.

■ *Improved Information Flows.* The Internet provides LSPs with applications that enable real-time visibility to shipments anywhere, at anytime in the supply chain. In addition, Web-based work flow tools render the transfer of information, such as freight tenders and bookings, a simple process. Also, by significantly accelerating the flow of shipping information and providing electronic funds transfer capabilities the payment cycle can be considerably shortened.

■ *Enhanced Fulfillment Capabilities.* Even today's best companies are being faced with steep challenges when it comes to SCM fulfillment, and particularly fulfillment that is Internet driven. Historically, most companies have developed logistics practices and infrastructure geared to handle the bulk storage and movement of products through the supply network. Increasingly, customers are requiring order picking and fulfillment in smaller, more frequently delivered lot sizes and are turning to their LLPs for the answer. Responding to these fulfillment needs will require nimble LLPs equipped with both the necessary infrastructure and access to Web-based logistics communities that can make delivery of any order size at anytime a reality.

■ *Supply Chain Inventory Management.* Computerized SCM applications provide *.com*LLPs with a range of inventory throughput tools that seek to replace, when possible, information with inventories, enable nimble response to customer requirements, and reduce cycle times everywhere in the supply channel. Among these computerized functions can be found cross-docking, merge-in-transit, remote postponement, and delayed allocation of orders to the latest possible moment before shipment, and the commingling of loads from multiple service providers in order to maximize shipment, provide pricing advantages to shippers, and remove waste from the fulfillment process.

Today's LSP Marketplace Challenges

The use of LSPs, like other business partnerships, has many challenges. On the one side, the need for outsourced logistics services has grown steeply over the years. Other the other hand, LSPs are subject to the same problems that plague other business partnerships. By its very nature, outsourcing requires two often very different organizations to find common ground on business terms, accounting information, operating philosophies, and technologies. While the goal is to construct a long-term relationship, often there are important changes in priorities, levels of urgency, and strategic direction. A lack of agreement on fundamentals, such as requirements and expectations, can serve to slowly erode even the best agreement. Finally, companies often can not resist the opportunity to weigh the promise of exciting new technological, financial, and service enhancements of a new LSP against existing, less-capable LRM partnerships.

Dealing with these and other issues require a special level of care in developing logistics outsourcing. Increasing use of logistics outsourcing requires that companies follow the following benchmarks before launching into a LSP relationship [24].

- *Partnership Flexibility.* The only thing known for sure about a logistics partnership is that the circumstances driving its inception will change through time. While a structured contract is essential for an effective outsourcing arrangement, logistics partners must provide for contract flexibility to account for service changes due to new technologies, remapping of the supply channel, new products, new competitors, change of management, and other issues. Open dialog is critical in ensuring both parties have commonly accepted definitions and terms, detailed performance measurements, and a methodology to adjust logistics functions and expectations to meet current realities. Also, periodic review and affirmation of partnership objectives and measurements used to validate them must occur if the arrangement is to be evolutionary and not stagnant.

- *Defined Cost of Services.* In spite of the best constructed contract, rarely are all the details governing cost and scope of services fully defined. Timeliness, accuracy, thoroughness, and level of service detail can often vary widely from agreed upon contracted standards. Further, although the contract is designed to handle most of the services required, even small deviations and exceptions can stress a relationship. An effective method to cope with these issues is to plan for high, medium, and low costs in the contract. Consistently enforcing the low-cost solution will unsettle even the best relationship.

- *Solving the Human Side of Outsourcing.* The introduction of a LSP into the supply chain equation requires the effective education and concurrence of the people side of an organization. Outsourcing what were once internal functions to an outsider can cause friction and uneasiness among affected employees. Managers, supervisors, and staff should be an integral part of the review and transition process. Their feedback and support will enable them to be integrated into the new environment.

- *Realistic Expectations.* Perhaps the most serious detriment to a relationship is the establishment of false expectations. Customers often overinflate expected savings and assumptions about the effectiveness of outsourced processes. On their part, LSPs often overpromise benefits and services. In reality, both parties must work closely together to establish realistic expectations and then structure a program to guide continuous improvement through time. From the beginning and throughout the life of the relationship, both parties must be prepared to work through practical examples, step-by-step, testing base-line and every possible type of exception process to remove false expectations and uncover measurable benefits.

■ *Technology Misunderstandings.* All too often customers and LSPs exaggerate their technological capabilities as well as downplay what technology expertise is required of each other. Solving this problem requires both parties to come to the table with realistic technology statements of their capabilities. This analysis will then provide the relationship with a base from which to initiate technology improvements that enhance the partnership and drive new benefits.

■ *Partnership Commitment.* Before any negotiations can really begin, it is essential to secure the firm commitment of top and middle management to the proposed outsourced arrangement. The executive team must be apprised of their role as leaders and supporters in the ongoing management of the relationship, both from a change management and from an expanding collaborative partnership perspective.

■ *Retain Core Skills.* Companies seeking outsourced logistics functions must be vigilant to ensure that critical skills and processes are not contracted away. A fine line must be preserved between outsourcing the expense while retaining the skill set to be able to reevaluate the arrangement, recapture the outsourced function, or rethink the entire outsourced strategy. Using a LSP service provider requires companies to thoroughly understand the process and to base their decisions on the cost of the investment and the likelihood of success.

Choosing and Implementing an LSP Solution

There can be little doubt that the trend to outsource logistics functions will only continue to increase for the foreseeable future [25]. In today's high-speed, integrated, and lean supply chains, companies are more than ever obliged to eliminate logistics bottlenecks that can bring even the best run channel network to its knees and dramatically escalate costs. As business and 3PLs alike prepare for the post-recession world of the second decade of the twenty-first century, both are viewing 2010 as a transition point to reevaluate and explore ways to deepen their outsourcing relationships. Newer concepts and technologies are emerging to help both 3PLs and shippers cope with this new, slower-growth world, including horizontal collaboration, new business and resource-sharing models, and more open, standardized IT platforms. By overcoming current limitations and being open to change, 3PLs and shippers can help each other weather the current economic storm and be prepared for tomorrow's volatility.

Critical to any project aimed at architecting an effective logistics outsourcing strategy companies need to careful align their overall business strategies with the logistics strategy it is designed to support. In this final section, the process of strategic logistics design will be discussed. The goal is to provide a broad roadmap to assist today's executive to execute a workable logistics outsourcing plan that not only meets today's competitive challenges but is capable of evolving to meet the needs of tomorrow's marketplace.

LSP Business Models

The trend toward outsourcing noncore competencies has become a widely accepted strategy for most industries today. The outsourcing of functions such as IT, human capital management, warehousing, customer service, and others have become commonplace as companies divest themselves of nonessential functions. Overall, such strategies realize two main goals. To begin with, outsourcing permits businesses to shed costly functions that are actually beyond the normal skill sets and traditional hard asset investments that are at the heart of the company. Secondly, by outsourcing these functions to third party experts, firms can take advantage of world-class resources and processes without developing them in house. However, while the idea of logistics outsourcing seems simple enough, companies need to perform their due diligence in deciding exactly what to outsource and with whom. As the recession of 2008–2009 winds down, the 3PL industry is a graveyard of failed companies and many are facing consolidation in the short-term ahead.

In tackling the construction of a logistics outsourcing strategy, it might be helpful to start by describing the array of possible LSP arrangements that are available. Possible strategies span a wide range of LRM operations models. A summary of each is as follows:

- *Traditional Logistics Model.* On the low end of the spectrum of LRM strategies can be found the traditional model where internal company functions totally control logistics resources. In such environments logistics is not considered a strategic advantage. When LSP activities do occur, they are concerned with the execution of one-time spot buys to solve a temporary under capacities in internal warehousing or transportation capabilities.
- *Classic 3PL Model.* In a classic 3PL model companies seeks to develop a long-term LSP partnership focused on securing logistics capabilities that are on the periphery or are poorly performed by the internal logistics group. An example would be using an LSP to handle international distribution functions that are beyond the core competencies of the business. The use of spot buys with a specific service partner for warehouse and transportations capabilities dominate this model. For the most part, control of strategic logistics functions and design and execution of supply chain strategies is retained by the company.
- *Partial LLP Model.* In contrast to the first two models, which are focused strictly on operational objectives associated with cost and immediate performance, companies pursuing a LLP model would be characterized by a

partial surrender of logistics control to their LLP. While the shipper still acts in the role of logistics integrator and retains control of channel design, LLPs are given responsibility for managing entire portions of the supply chain. Normally, the shipper plays an active part in assisting the LLP to assemble the LLP team, is responsible for enforcing LLP adherence to the channel strategy, and oversees operational logistics function execution.

▪ *Full LLP Model.* In the most robust LSP model, the shipper selects a LLP who assumes full responsibility for logistics management. While the shipper is still an active partner in the architecting and maintenance of both the logistics strategy and the LRM community of providers, full responsibility for the total logistics solution is given to the LLP. The LLP assumes ownership for channel design, 3PL partner selection, and detail operations execution.

Whether a single or a portfolio approach, logistics outsourcing, like any business decision, must take into consideration current and future needs. When beginning strategy formulation companies must first step back and objectively understand the LSP environment, the role of technology, and their own supply chain logistics requirements. To begin with, companies must clearly articulate what types of opportunities can be realized by exploiting the merger of technology and operational excellence made possible by an LSP partnership. Among critical questions requiring answers can be found the following. What is the depth of the firm's current technical environment, assets, and human capital and is it sufficient to pursue targeted technologies? What types of logistics networks are envisioned and how will they support competitive advantage? How customizable are LSP technology-driven services? Can the business take the lead in developing a collaborative logistics community or is that task to be surrendered to a LLP?

Secondly, companies need to integrate new forms of relationships and services into their logistics solution. Today's Internet-based LRM solutions are taxing to the breaking point traditional logistics partnerships, calling upon companies to often architect dramatically new business relationships requiring fundamental process and cultural changes. The *relationship* portion of e-LRM means that logistics strategists must thoroughly understand what their core competencies are before an outsourcing strategy is executed, search for new dimensions of collaboration with their LSPs, ensure that contracts promote a win-win attitude, and construct an outsourcing strategy that promotes the productivity, not only of the company, but also of the entire supply chain. In the final analysis, it is not the technology but rather the ability to create new forms of logistics relationship that is the main driver of fulfillment competitive advantage.

Finally, companies must thoroughly understand the wide spectrum of LSP capabilities that are available when drafting a LRM strategy. In the past, 3PLs provided spot buy services such as emergency shipping, freight forwarding, and small parcel delivery. Today's LSP provides a dramatically wider range of services to meet the needs of an Internet-driven fulfillment environment. Logistics planners can tap into e-marketplaces to search for services, settle a price, or match other buyers and sellers. Making these strategies work requires the type of real-time visibility, optimization, and planning that is only available through Web-enabled applications. In the last analysis, the choice of a LSP strategy again comes back not only to individual company fulfillment execution and scheduling, but also the optimization of these functions across extended, multipartner supply chains.

Steps in LRM Strategy Development

Building a successful long-term LRM strategy requires considering the options available from two separate but interlinked perspectives. The process should begin first with the development of an effective internal analysis targeted at understanding the true logistics needs of the organization and gaining the support of company members. Once completed, the next step is to research the critical criteria necessary to guide logistics management in choosing the best LSP partners. Merging both sets of activities should provide a company with the right information to make the right decisions concerning their LRM outsourcing strategy.

The internal aspects of a LRM project should consist of the following steps.

a. *Logistics Analysis.* Common to all outsourcing initiatives, the formulation of a logistics outsourcing strategy should begin with an analysis of the tradeoffs between using in-house logistics functions and the benefits of outsourcing them to a LSP. There are two parts to this analysis. To begin with, planners must determine how much of the day-to-day logistics functions should be outsourced and what methods are to be used in securing LSP services. Should traditional arrangements with LSPs be pursued, or should Internet solutions be employed as the basis for services selection? Secondly, firms must decide on how much of the logistics solution and supply chain design is to be surrendered to a LLP. This is a potentially dangerous area where entrusting the total management of supply and fulfillment functions to the wrong LLP could spell disaster.

b. *Support for the Customer Strategy.* When performing the analysis of internal logistics strengths and weaknesses, companies must be careful to consider their marketplace cost and service strategies. The LRM strategy implemented must exploit the best combination of in-house and LSP logistics capabilities to realize competitive advantage targets. A number of critical questions come to mind: What new logistics functions will further cement customer loyalty and gain new customers? How deeply can the LRM strategy be integrated into the customer chain and how extensive will be their collaboration?

c. *Select a Technology Solution.* Logistics technologies are evolving rapidly and are becoming increasingly more sophisticated. Companies can take advantage of these tools by either acquiring the expertise or by partnering with a provider who has built the capability. Today, there are many ERP, SCM, and B2B vendors who can provide companies with the ability to connect to Web-driven bulletin boards, portals, public exchanges, and auction sites, or implement their own private exchange or collaborative community. Whatever the path chosen, companies must be careful to select a technology solution that permits them to retain control of logistics information while permitting them the freedom to switch applications and providers to leverage best-of-breed capabilities.

d. *Gain Company Buy-in.* Similar to any strategic project, logistics planners must be careful to gain the support of three critical groups. To begin with, a senior executive must sign-on as the project's advocate on the top management team. This project champion should be a force in the organization, be a supporter of e-business tools, and be prepared to negate any internal resistance to change. The second group is the internal logistics team. Education and training for this group will be necessary to assist them to embrace the change positively as well as have the skills to run the new solution. The final group is the company's IT staff. It will be the responsibility of IT to assist in system integration and leveraging the LSP's technologies and capabilities. All must understand that the LRM outsourcing strategy is a path to improving existing infrastructure to be more competitive.

e. *Start Small.* Consultants and practitioners alike stress the need to build creditability by beginning the LRM outsourcing project by achieving some short-term wins. Achieving, for example, cost reductions by using the e-LRM functionality to manage an already successful transportation lane or for a particular division could help build company confidence. Once achieved, additional functionality can be phased in over time to tackle wider areas of operational efficiency or improved service qualities.

f. *Performance Measurements.* Finally, no logistics strategic effort should be undertaken without charting expected performance measurements. Metrics should enable strategists to determine what benefits are actually being achieved for the expenditure, how well technologies and LSPs are fueling advances in competitive advantage, and how closely results are matching customer expectations.

Searching for LSP partners constitutes the second part of LRM strategy development and is concerned with the following points:

a. *Strength of the LSP.* When searching for a logistics partner, companies should begin by investigating LSP financial strengths, availability of physical assets, investment in technology, capability of infrastructure, and historical commitment to providing logistics services. The goal of this step is to validate LSP partner long-term viability and capability to handle normal as well as increases in volumes. During the exploration, the expertise and experience of the proposed LSP's management team should also be reviewed. The advice is to shy away from virtual .com LSP whose management staffs have weak logistics core competencies.

b. *Selection of a Compatible Technology Solution.* Logistics technologies are evolving rapidly and are becoming increasingly more sophisticated. Companies can take advantage of these tools by either acquiring the expertise or by partnering with a provider who has built the capability. Choosing the latter permits businesses to acquire the benefit of the technology investment without sacrificing their own capital assets. When selecting a partner it is critical to confirm that perspective LSPs possess the necessary technology resources to continuously scale, develop, and implement Internet solutions as strategies change without endangering customer service.

c. *Controlling Information.* The LRM solution selected must enable companies to maintain control over critical logistics channel information flows. Without close control, a company's ability to respond effectively to change can limit its agility to respond to circumstances and the timely choice of alternative options that can result in increased costs and reduced responsiveness. The abundance of software choices and Internet capable LSPs make it possible to architect a solution that will provide just the right amount of flexibility, automation, and collaboration.

d. *Services and Capabilities.* On the services side, LSP partners should be able to provide a competitive level of volume discounting. Choosing a provider with significant buying power in the logistics marketplace will enable them to provide the desired level of service at the lowest price. In addition to discounting, the LSP should provide other value-added services ranging from parcel carrier partnerships, automated materials handling and fleet management, and sorting and labeling to sophisticated e-fulfillment functions such as carrier analysis, network load building, and electronic payment, auditing, and claims management.

e. *Customer Success.* The best LSPs come to table with a history of success. Proper due diligence requires a review of references and their success stories. Existing clients will provide valuable insight into actual provider capabilities, whether they are increasing services use, and the direction in which they see the LSP moving.

Summary and Transition

The evolution of the concept of SCM and the application of the Internet to business has elevated logistics to a position of critical importance in today's marketplace. The mission of SCM is to activate the visioning and productive capabilities of extended communities of businesses to produce collaborative networks whereby products and services can be designed, produced, and distributed in as cost effective and timely a manner as possible. Realizing the possibilities of strategic SCM require the existence of logistics operations that are aligned and structured to meet the quick response needs of high-velocity supply chains. While it is the role of engineering, marketing, production, and sales to respond to the changes of business in the twenty-first century, it is logistics that will be responsible for building the channels where the cycle of innovation, design, production, distribution, and eventual obsolescence and return of materials will be played out.

Responding to the expansion of the role of logistics in today's Internet-enabled environment requires conceiving of it as more than just storing and moving products up and down the supply channel: logistics has become a strategic function. As such, the old term of logistics needs to give way to a new term—LRM. Similar to the recent transformation of customer service and purchasing functions into CRM and SRM, the use of the term LRM is meant to convey a similar expansion of traditional logistics activities to encompass the multifaceted concept of SCM with its focus on trading partner collaboration, the removal of channel barriers causing excess costs and reduced cycle times, the espousal of Internet-enabled technologies that facilitate information, transaction data collection, and flow through the supply pipeline, and the creation of agile, responsive organizations linked together in a single-minded pursuit of superior customer service. In detail, LRM provides channel management strategists with the following five key capabilities: logistics performance management, fulfillment and execution, logistics partnership management, shipment visibility, and event fulfillment management.

While the strategic side of LRM is propelling it into the mainstream of the current trends toward collaboration and Internet-based partnerships, the fact that logistics services are basically a commodity made it an early and easy candidate for Web-based trading technologies. From the beginning, LRM could be seen from two perspectives. Companies could utilize the Internet to architect a comprehensive range of services from carrier search, to rate negotiation, to emergency services selected by logistics managers based on a cafeteria approach. On the other hand, companies could utilize their service partners to provide a robust portfolio of services where all aspects of the logistics channel, including strategy development, technology, and operations, would be handled by a lead logistics provider or LLP. In any case, the range of today's LRM solutions span the following five possible categories: independent *transportation exchanges* designed to match shippers and providers; *transportation network infrastructures* providing hosted Web-based applications to shippers and carriers; *transportation management software* vendors providing Web-enabled logistics software solutions; *3PLs* offering transportation and warehousing services to clients; and *collaborative logistics networks* using Web-based tools to enable communities of shippers and buyers to connect with each other and trade logistics functions.

As the importance of logistics has expanded in today's business climate, so has the propensity of businesses to outsource logistics functions. Historically, companies have always looked to 3PL partners to provide warehousing and transportation functions that were beyond internal capabilities. 3PLs enabled companies to spot buy services, shrink internal expenditures on company physical assets and technologies, and remain agile to meet changes in the logistics environment. The advent of e-commerce and continued acceleration of all forms of business cycle times have engendered a new form of outsourcing partner: the lead logistics provider. The goal of the LLP is to manage all aspects of the logistics relationship, beginning with the *strategic definition* focused on applying technology and new management concepts to drive reinvention and transformation of the logistics effort, proceeding through *implementation* of business process realignment and systems integration, and concluding with the ongoing *execution* of operational functions associated with daily transportation and warehouse transactions.

Drafting a strategy to leverage the various categories of LRM and the various forms of service provider requires companies to take very seriously their search for a service provider and depth of technologies to be used to guide logistics processes. Companies can pursue LRM models than can span a variety of structures ranging from the tradition model where internal functions manage all aspects of logistics, to models that are marked by a progressive surrender of control of operational functions, channel design, and, ultimately, logistics strategy development. Successful selection of the right strategy requires the creation of a continuous process of internal organizational preparation and strategic alignment, and a ceaseless attention to the ever-changing capabilities of LSP in the search for new avenues of competitive advantage.

The presence of enabling technologies not only stands as the basis for the dramatic changes impact the field of logistics, but has served as the central vehicle for the progress of today's business environment. Each chapter of this book has explored how information technology, and particularly the Internet, has reshaped almost every business function, from sales and service to new product development and procurement. In the final chapter in this book, the process of designing strategies to successfully implement today's array of supply chain technologies, from local automation applications to full enterprise business systems are explored.

Notes

1. Rosalyn Wilson, "20th Annual State of Logistics Report," *Council of Supply Chain Management Professionals*, June 19, 2009, at accessed June 19, 2009, http://cscmp.org/memberonly/state.asp.
2. This definition can be found at accessed June 21, 2009, http://cscmp.org/aboutcscmp/definitions.asp.
3. These points have been adapted from David F. Ross, *Distribution Planning and Control: Managing in the Era of Supply Chan Management* (Norwell, MA: Kluwer Academic, 2004), pp. 44–46.
4. Charles C. Poirier and Michael J. Bauer, *E-Supply Chain: Using the Internet to Revolutionize Your Business* (San Francisco, CA: Berrett-Koehler, 2000), pp. 135–136.
5. James T. Hintlian and Phil Churchman, "Integrated Fulfillment: Bring Together the Vision and the Reality," *Supply Chain Management Review Global Supplement* 5, no. 1 (2001), pp. 17–18.
6. The company examples were found in Jean V. Murphy, "Seeing Inventory in Real Time Lets You Have and Have Not," *Global Logistics and Supply Chain Strategies* 6, no. 5 (2002), pp. 34–40.

7. Joy LePree, "Take the Right Path," *Manufacturing Systems* 20, no. 1 (2002), pp. 52–55.

8. For a summary of the origins of e-LRM, see Beth Enslow, "The Virtual Logistics Department: Next Generation Logistics Exchanges," in *Achieving Supply Chain Excellence Through Technology*, 3, David L. Anderson, ed. (San Francisco, CA: Montgomery Research, 2001), pp. 278–281.

9. See the analysis in Theodore Prince, "Transportation E-Commerce: Pressing Toward Portals," in *Achieving Supply Chain Excellence Through Technology*, 4, Narendra Mulani, ed., (San Francisco, CA: Montgomery Research, 2002), pp. 184–187.

10. C. John Langley, "Analyzing Internet Logistics Markets," *Supply Chain e-Business* 2, no. 5 (2001), pp. 22–26.

11. Rob van Doesburg, Patrick Baptist, and Wouter van Heijst, *Warehousing Report 2009*, (Utrecht, GN: Capgemini Consulting, 2009).

12. Brian Wyland and Scott Pezza, "Five Key Steps to Optimizing Warehouse Management," *Aberdeen Group White Paper* (February 2009), p. 22.

13. Michael Tichelaar, "The Time For On-Demand WMS Is—Yesterday," September 11, 2009, accessed September 11, 2009, http://www.supplychainbrain.com/content/nc/technology-solutions/warehouse-management/single-article-page/article/the-time-for-on-demand-wms-is-yesterday.

14. See the comments in Adrian Gonzalez, "The Economy as a Catalyst for Change: The Role of TMS and Managed Services," *ARC Brief* (Dedham, MA: ARC Advisory Group, May 2009), pp. 1–12.

15. Adrian Gonzalez, "The True Value and Meaning of Software-as-a-Service TMS," *ARC Brief* (Dedham, MA: ARC Advisory Group, October 2009), p. 3.

16. Ibid., pp. 5–6.

17. Ibid., p. 9.

18. John Langley and Capgemini, "The State of Logistics Outsourcing: 2009 Third Party Logistics," found at accessed september 11, 2009, http://www.uk.capgemini.com/services/supply-chain/the_state_of_logistics_outsourcing_2009_thirdparty_logistics/, p. 4.

19. Found in Joel Sutherland, CA and Thomas W. Speh, "Using 3PL Service Providers to Create and Deliver Significant Supply Chain Value," in *Achieving Supply Chain Excellence Through Technology*, 4, Narendra Mulani, ed. (San Francisco, CA): Montgomery Research, 2002), pp. 176–178.

20. See the methodology in Douglas Bade, James Mueller, and Bryan Youd, "Technology in the Next Level of Supply Chain Outsourcing: Leveraging the Capabilities of Fourth Party Logistics," in *Achieving Supply Chain Excellence Through Technology*, 1, David L. Anderson, ed. (San Francisco, CA: Montgomery Research, 1999), pp. 260–263. There are several types of 4PLs. In one group can be found consulting firms exclusively specialized in logistics, transportation, and supply chain management such as SCMO (company), BMT Limited, MVA Consulting, TTR, and Intermodality, which offer complete ranges of services, from strategy to implementation. In another group are found consulting generalists Deloitte, PricewaterhouseCoopers, Ernst & Young, and KPMG, as well as Accenture, Arup, Atkins (company), Mott MacDonald, Parsons Brinckerhoff, and AECOM. Finally, firms such as McKinsey & Company, Bain & Company, A. T. Kearney, the Boston Consulting Group, and Booz & Company may also play the role of 4PL with a different value proposition, and are considered to be "pure strategy" firms only. Found at accessed September 13, 2009, http://en.wikipedia.org/wiki/Fourth-party_logistics.

21. Langley and Capgemini, *The State of Logistics Outsourcing*, p. 22.

22. Ibid., p. 25.

23. See Fred A. Kuglin and Barbara A. Rosenbaum, *The Supply Chain Network @ Internet Speed: Preparing Your Company for the E-Commerce Revolution* (New York: AMACOM, 2001), p. 129, and Rob Schryver, "The Trade Tsunami: Traditional 3PLs Expand Roles," *Inbound Logistics* 21, no. 9 (2001), pp. 71–76.

24. These points have been summarized from Robert E. Sabath, "Getting Outsourcing to Work in the Supply Chain," in *Achieving Supply Chain Excellence Through Technology*, 1, David L. Anderson, ed. (San Francisco, CA: Montgomery Research, 1999), pp. 276–280.

25. According to Langley and Capgemini, *The State of Logistics Outsourcing*, p. 4, shippers and 3PL respondents agree that key factors responsible for the widespread use of logistics outsourcing relationships include the following:

openness, transparency and good communication; the ability to create personal relationships on an operational level; the flexibility of 3PLs to accommodate customers' needs; and the ability to achieve cost and service objectives. Shipper respondents devote on average 47% (in North America) to 66% (in Europe) of their total logistics expenditures to outsourcing, and all geographies anticipate increases in these percentages in the next five years.

Chapter 9

Developing SCM Technology Strategies: Creating the Game Plan for the Successful Implementation of SCM Technologies

At the heart of the supply chain management concept can be found the continuous unfolding of a variety of dynamic organizational, technology, and business channel collaborative strategies. Unlike the corporate strategies of the past, which focused narrowly on internal budgets and detailed metrics regarding the performance of company-centric market segments, products, and processes, the concept of supply chain management requires companies to rethink the very nature of their critical success factors and the way they do business in the supply chain in light of the tremendous opportunities brought about by the merger of the competitive power of today's networked ecosystems and Internet-based technologies. Progressive companies of the second decade of the twenty-first century seek to utilize the Web to drive new business models and attain levels of competitive advantage that consistently and continuously achieve order-of-magnitude breakthroughs in the creation of the products and services customers really want by leveraging the capabilities found anywhere in the supply network to engineer seamless value chains capable of satisfying customers with any solution, at any time.

This chapter seeks to explore how companies can build effective market-winning business strategies by actualizing the opportunities to be found in today's technology-driven supply chain. The discussion begins with an investigation of how today's dependence on supply chains has dramatically altered business strategy development. Structuring effective business strategies requires companies to closely integrate the physical capabilities, knowledge competencies, and technology connectivity of their supply chain networks alongside company-centric product, service, and infrastructure architectures. Building such a powerful technology-driven SCM strategy requires that companies first of all energize and inform their organizations about the opportunities for competitive advantage available through the convergence of SCM enablers and networking applications. As the chapter points out, strategists must be careful to craft a comprehensive business vision, assess the depth of current supply channel trading partner connectivity, and identify and prioritize what initiatives must be undertaken to actualize new value-chain partnerships. The chapter concludes with a detailed discussion of a proposed SCM technology strategy development model. The model consists of five critical steps, ranging from the architecting of purposeful supply chain value propositions to assembling performance metrics that can be used to ensure the proposed SCM technology strategy is capable of achieving the desired marketplace advantage.

Changing Views of Enterprise Strategy

Almost a decade after the dismal end of the dot-com era, the deployment of Internet-based technologies has not only arisen to play a significant role in the economies of today's business environment, they also require companies to devote serious attention to the direction their businesses are going to take with regard to Internet-based networking technologies. This means that companies not only must be concerned with the development of business strategies regarding products cycles, productive functions, finance, marketing, and supply chains, they must also architect these strategies with an eye to the role play by and potential of their Internet content. As illustrated in Figure 9.1, Internet technologies are no longer a novelty restricted to businesses like Amazon.com: every area of today's enterprise has an Internet content and technology must be seen as the prime driving force for all channel strategies. Missed opportunities to exploit the power of Internet technologies will result in the inability of today's organization to fully realize the objectives outlines in their corporate strategies.

While the new millennium promised by the dot.com revolution failed to materialize as promised, the revolution in business networking and connective that was the underlying engine of the era has indeed dramatically altered the playing field forever. For decades companies had leveraged technology tools and management

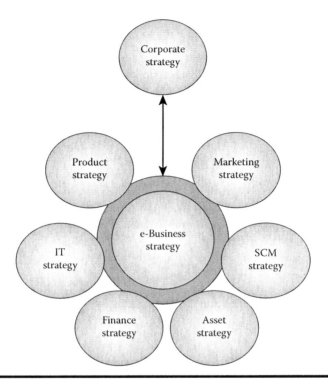

Figure 9.1   Relationship between e-business and business strategy.

philosophies, such as MRP II, just-in-time (JIT), electronic data interchange (EDI), total quality management (TQM), enterprise resource planning (ERP), and BPR, to drain from enterprise processes stagnant pools of productivity, inefficiencies, waste, and lack of customer focus while enabling bold innovation, the continuous generation of agile organizations positioned to deepen customer and supplier relationships, and the capability to leverage core competencies focused on creating new forms of customer value. With the application of the Internet in all of its various forms, enterprises were now poised to merge these inward-focused improvement tools with networking technologies capable of enabling unprecedented levels of productivity, market dominance, and customer responsiveness.

The bottom line is that savvy executives who had the knowledge and experience to see through the exaggerated hype of the era had quickly surmised that the Internet offered enterprises the potential to fundamentally transform themselves much like MRP, JIT, and TQM had provided companies with the capability to revolutionize business in the past. The difference is that the Internet technologies enable companies to escape from old economy business models that view organizations as internally focused and distinctly separate from their customers and suppliers. The application of Web technologies is about firms exploring new value propositions on how to work with their customers. It is also about new methods of defining unexplored regions of collaboration and synchronization with their suppliers to define radically new infrastructures that will expand the dynamics of competition. Charting an effective technology strategy is not merely a matter of buying and bolting-on Web-enabled software; it is about reinventing the corporation around the tremendous opportunities driven by today's networking technologies.

Overview

In the past, business schools taught and executives fashioned elegant corporate strategies designed to ensure that their businesses were able to gain and then sustain a competitive advantage in the marketplaces in which they competed. The object of strategic planning was straightforward: How should a company deploy the advantages it enjoyed in processes, products, services, assets, and brands to beat out the similar offerings of rival firms and capture a specific customer segment? Once market superiority was gained, market winners then strove to solidify their supremacy by continuously improving products and services, reducing process and overhead costs, and investing in new product development and productive processes to keep the engines of competitive advantage moving forward. Because neither markets nor products changed much over time, enterprises that gained initial superiority could leverage considerable resources and process knowledge, mature distribution channels, advertizing and marketing clout, and the newest technologies to maintain that lead.

Today, it is evident not only that there is no such thing as *sustainable advantage*, but that all advantage is *temporary*. As Fine in his concept of business "clockspeed" tells us, no core competency is unassailable, no lead is uncatchable, no kingdom is unbreachable [1]. In fact, the faster the marketing environment, the quicker the profits are amassed, and the more revolutionary the products and processes, the faster the clockspeed, the shorter the reign. And clockspeeds are increasing in every industry, rendering obsolete products, processes, and infrastructures and abrogating power over markets.

What is causing this acceleration in the erosion of competitive advantage? In some cases it can be traced to new technologies and the dramatic explosion in products and services and shrinking product life cycles that have been rapidly diminishing the longevity of corporate cash cows. In other cases, shocks to the business and economic environment, such as the recessions of 2001–2002 and 2008–2009, international instability in the Middle East and Korea, and the ever-present threat of terrorism can suddenly dry up capital investment, drive companies to focus inward, and result in mass unemployment and whispers of a deflating economy. Then there is the juggernaut of globalization and outsourcing that are swiftly shifting economic balances of power. Still more deadly is the sudden growth of new, unorthodox business models that have been quick to challenge long successful leaders for marketplace supremacy by leveraging special competencies that permit them to violate market boundaries and raid targeted customer segments once considered the exclusive preserve of more mature, established companies. Other threats can arise from upstarts utilizing radically new technologies or management styles that not only steal business away but also increasingly render obsolete the intellectual and physical assets of their more senior rivals. Such fast clockspeed businesses seem to be everywhere today: it took Amazon.com just 2 years to achieve the same total annual sales as it took Wal-Mart 12 years to achieve; in 1993, Dell computers had $2.9 billion in sales, and it had $12.3 billion 4 years later.

The Primacy of Value Chains

The basis of business in the twenty-first century rests upon twin pillars: the core competencies possessed by each enterprise and the capabilities of their channel trading partners. The sum total output of a business's core competencies can essential be described as constituting its *value proposition*. The value proposition comprises the way a firm markets its products and services, the processes by which it acquires and manages its assets, and the organizational infrastructure that guides the overall direction of the business and ensures the continuous and effective tactical governance of the firm. The *function* of the enterprise can be summarized as its ability to continuously design, assemble, and deliver what can be said to constitute its *customer value proposition*. The *structure* of the enterprise is the particular configuration in time of its knowledge, capabilities, productive processes, and physical assets that enables fulfillment of the value proposition. And finally, the *business architecture* of the enterprise is the design of how the various parts of its structure are integrated and directed to realizing the goals of its functions.

Historically, most of the literature regarding the development of business strategy has focused on designing the *business architecture* of the individual corporation. However, it is obvious that companies have never been self-sufficient entities and that their business architectures have always implied, even if in a passive manner, the inclusion of networks of business partners whose competencies and resources could be deployed to actualize the value proposition. In fact it can be argued that a truly comprehensive business architecture must not only focus on the core competencies *within* the boundaries of the enterprise but also on the competencies of the firm's *extended* organization. Recently, it has been often repeated that "companies no longer compete—supply chains do." While companies do indeed define the products and services they sell and how they will be managed, competitive differentiation today is determined by how fast, how cost efficiently, and how effectively whole supply networks deliver on customers' demands for products and services.

Assembling an effective *supply chain business architecture* requires companies to move beyond viewing their trading partners purely as passive channel constituents. When supply chain relationships are static, suppliers and customers are perceived as separate, loosely linked business entities whose own strategic goals and operations capabilities stand outside, and sometimes even in conflict with, each others value propositions. In contrast, today's market leaders differentiate themselves from their competitors by their ability to continually identify and invest in the rich array of supporting competencies to be found in their supply network partners. In fact, it may be said that a company's real competitive advantage consists in its capability to design and integrate the capabilities of its supply networks in the pursuit of marketplace advantage, albeit temporary, against the backdrop of accelerating changes in markets and competitive forces. Success in tomorrow's marketplace will go to those enterprises that focus on the strategic capabilities of their entire value chain, rather than simply on the strength of company-centric products, services, and infrastructures.

Structuring a competitive *supply chain business architecture* requires strategic planners to view the supply chain from several perspectives. Dynamic channel architectures understand that trading partners provide a matrix of competitive advantages ranging from specialized competencies to collaborative relationships that provide the enterprise with a depth of customer satisfying values they could not possible hope to assemble acting individually. Dynamic value chains consists of three interdependent dimensions, as portrayed in Figure 9.2.

■ *Supply Chain Structure.* This dimension details the *physical* composition and interconnecting links of the supply chain system. Understanding this dimension requires supply channel planners to first of all map out each network system that composes the entirety of the supply chain galaxy, beginning with the most remote supplier systems and concluding with the last customer to be found out on the rim of the channel network. The second activity of channel planners is to assess the importance of each network trading partner in relation to the value it contributes to the competitive advantage of the entire supply chain galaxy. These two exercises are directed at identifying trading partners that possess the greatest strategic importance and illuminating those

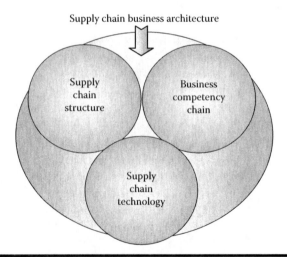

Figure 9.2   Supply chain strategic dimensions.

regions that contain dramatically innovative network nodes that have the power to radically change and possibly destabilize the structure of the supply channel system.

■ *Business Competency Chain.* Drafting a map of the physical structure of a value chain is a comparatively easy task in comparison to the second dimension, establishing a map of channel partner competencies. This process begins with a definition of the core capabilities existing within the organization. Elements such as product design skills, marketing, logistics, and others are examples. Next, strategic planning teams need to understand what key capabilities each member physically constituting the supply chain system contributes to the firm's value proposition. Such an exercise should reveal the interrelationship of each trading node in the channel galaxy and provide answers to such questions as: Which are robust systems expanding at a stable rate that can be counted upon to steadily provide new ideas and new capacities? Which are old systems whose productivities are spent and will eventually darken and die? Which are brilliant novas that emit radically new competencies and challenge the prevailing architecture of the entire channel galaxy but that have fast clockspeeds and are destined to die out after a brief and often unsettling life? The goal of the process is to identify the strengths and weaknesses of trading partners as a means to determine the level and volatility of value enablers from physical resources to innovation capabilities.

■ *Supply Chain Technologies.* This final dimension of supply chain planning references the potential robustness of the connecting links integrating and synchronizing each node in the supply channel galaxy. The goal is to identify the technology tools that determine how databases and processes from each channel node can converge and support one another. Whether it be through the use of analog methods, EDI, or the Internet, it is critical that the technologies employed be reviewed to clarify how deep into the supply chain they provide for database and competency transfer. Some may be single pair connections that stop at the boarders of the business relationship. Others may provide localized integration where isolated systems of suppliers and customers in some region of the network galaxy are closely synchronized. Still others may possess technologies that lie in far off regions, but are capable of opening wormholes in the fabric of the trading galaxy linking suppliers in the nth tier with customers far off on the rim of the supply chain system.

Being able to constantly generate the products and services necessary to win customers and drive home competitive advantage means that today's enterprise must be able to transcend the limits of company-centric products and process competencies by utilizing the capabilities of their supply chain partners and the enabling power of Internet business networking. Viewing the supply chain galaxy from a three-dimensional perspective enables companies positioned within the network system to structure a host of collaborative and synchronized functions ranging from concurrent engineering to forecasting to customer database management that can open immense opportunities to build and enhance market leadership. Focusing and energizing the supply chain business architecture requires that not just individual companies, but the entire supply chain system utilize integrative technologies to access databases and transmit process information in real-time, unencumbered by traditional information silos. Thinking about the application of Internet-based forms of information and transaction management is the next step in the discussion before an effective integrative networked supply chain strategy can be detailed.

Barriers to Internet-Driven SCM Technologies

In the early 1980s, the concept of just-in-time (JIT) burst upon the scene as a revolutionary method of running the enterprise. At first it was thought that the dramatic breakthroughs in productivity being reported by Japanese companies could be duplicated by simply applying JIT to any production system. Papers and books were written detailing formulas and techniques. Executives went on plant tours. Methods and technologies of traditional management, particularly MRP, were deemed obsolete. Despite the hype, however, the results continued to be disappointing. Critics complained that the Japanese had unfair government support; they focused on commodity-type products that easily fit the JIT model; their economic system of slow incremental growth did not fit the US system of investor-driven high return. Even ignorant racial stereotypes about the homogeneity of Japanese society were offered as excuses for lack of success.

Slowly, it began to dawn on executives that the advantages of JIT could not be attained by simply bolting-on the *techniques* of JIT; companies also need to implement the *philosophy* of JIT. It simply did no good to design a shop floor *kanban* card system if the demand side of the business did not also develop a comprehensive pull system so that customer orders and not some form of economic order quantity (EOQ) drove production. Similarly, JIT required changes in purchasing, engineering, finance—well the entire infrastructure of the company! What is more, companies soon found that their JIT initiative did not really work if supply chain partners ran their businesses with methods that were out of synchronization with their JIT initiatives. It took a while for companies to understand that JIT was in reality both a technique *and* a philosophy for running not only individual enterprises but entire supply chains in the pursuit of a common competitive advantage.

Today, as businesses embrace and explore what the Internet means to their businesses and supply chains and how they are going to integrate Web technologies with current business practices, companies are finding themselves facing the same kind operational and philosophical conundrum characteristic of their past efforts to implement JIT. Utilizing the Web seems to be a relatively easy decision to make. The Internet seems to offer companies a relatively low-cost technology that enables them to effectively connect with their customers and suppliers. However, once the Internet initiative begins, implementers are immediately faced with the same kind of problems that once faced JIT implementers. Unless the objective is to simply set up a static Internet marketing information site, a host of operational and strategic critical questions immediately becomes apparent. What impact will e-business have on business operations? How will Web tools work with current information technologies and what will it take to integrate them? How will customers and suppliers react to the new technology? Can e-business open entirely new avenues of competitive advantage and what kind of infrastructure must be in place to make such opportunities happen?

As Sawhney and Zabin point out [2], e-business initiatives *always* result in one or more of the following possible sets of outcomes:

1. Cost reduction
2. Revenue expansion
3. Time reduction
4. Relationship enhancement

Deciding which of the above e-business values to implement is very similar to the choices that still confront executives embarking on JIT. e-Business functions that focus squarely on *cost reduction* are clearly the easiest to implement and can be seen directly on the firm's balance sheet. However, technology applications that are directed at simply automating cost out of processes constitute a purely tactical response to value generation and their impact provides only temporary advantage. While a degree more difficult to achieve, e-business applications enabling *time reductions* are similarly focused on improving operating efficiencies and bottom line results. For example, while an e-procurement portal may succeed in reducing maintenance, repair, and operation supplies (MRO) order processing and delivery times, the application simply is supercharging components of the already existing purchasing process.

In contrast, e-business initiatives surrounding *revenue expansion* and *relationship enhancement* provide companies with *strategic* opportunities to generate not just bottom-line efficiencies but also radically new regions of competitive advantage. Similar to a JIT project that requires not only tactical but also strategic changes to the way a company relates to their supply chain trading partners, e-business tools enable companies to deconstruct and reassemble exciting new value chains or even reinvent whole industries. For example, when DuPont Performance Coatings, the $3.8 billion-a-year automotive paint arm of DuPont & Co., contemplated e-business tools, it was decided to implement a Web site where customers could place orders. However, the system was not fully integrated with their enterprise business system (EBS) with the result that orders had to be manually entered from the Web front-end to the EBS. Nine months after the implementation, management decided this tactical e-business initiative had to be completely reinvented. To begin with, the Web site was restructured to permit direct input of customer data into and out of the EBS. In addition, customers could now visit the Web site to access information on their accounts, place and change orders, and check order status. The Web site could also be customized so that customers using Web services would have a personalized experience. DuPont now is looking to extend the strategic reach of its e-business architecture by tightening relationships with its customers' customers—the body shops that actually buy and use the paints [3].

Viewing e-business, as DuPont initially did, as simply a piece of automation software misses the real potential of Internet commerce. JIT can not realize the opportunities it holds for both cost reduction and radical competitive enhancement until executives finally understand that JIT is also a management philosophy guiding not just individual companies but entire supply chains. In a similar fashion, e-business strategies must be seen not merely as a component of a company's technology suite, but rather as a business chain model that requires entire supply channels to be transformed into *value networks*. What are the characteristics of these value networks? They seek to utilize e-business architectures to achieve both superior customer satisfaction as well as company profitability. They provide collaborative links matching customer demand with flexible, agile manufacturing product design and delivery. They provide for the real-time transfer and synchronization of plans and information enabling channels to bypass costly distribution intermediaries. They leverage digital technology to provide a seamless channel structure with network suppliers. They permit entire supply chain systems to adapt quickly to constant change.

As a prelude to discussing in the next section how to effectively create an Internet-driven *value network* strategy, it might be valuable to conclude with a definition of what e-business is *not* [4].

■ *e-Business is not the synonymous with the "new economy."* In the early days of the dot-com revolution it was said that the Internet was such a potent new force for business that it marked the end of the old "industrial economy" and inaugurated a new "digital economy." In retrospect, such claims were obviously overinflated. While it is true that e-business does provide today's enterprise with radically new tools for data transfer, information analysis, transaction management, value delivery, and collaboration, it does not obsolete but rather acts as an enhancement to traditional businesses. Similar to the JIT revolution that shook the economic environment during the last decades of the twentieth century, e-business has changed the nature of product and service development and delivery, enabled the pursuit of often radically new competitive values, engendered new business infrastructures and, in some sectors, whole new businesses.

In today's business environment, almost every business has implemented some facet of Web-enabled tools and many executives, whether consciously or not, are moving their enterprises incrementally toward the use of e-business to link their core competencies closer to customers and suppliers. In the previous edition of this book, I stated that as the decade of the 2000s moved forward, the ubiquitous *"e"* that once preceded just about every application of the Internet would disappear. As we move into the second decade, that statement has rung true—with the tremendous advancements in computer communications and networking as well as general acceptance by the public in the private and public lives, the ability to conduct business over the Internet is the norm in processes management. In fact, "e-business" has become passé in today's world of social networking and cloud computing.

■ *e-Business is not just about technology.* As companies began to implement Internet applications, managers first considered Web-enabling tools purely as discrete software components that could be bolted-on to the business architecture in much the same fashion as EDI. Today, companies have become aware that the more Internet-based applications are used, the more the lines separating the technology from existing infrastructure, operations, and strategic functions are becoming proportionately more blurred. As has been pointed out, as Internet tools migrate from tactical to strategic, companies also find that their internal organizations, as well as their relations with customers, suppliers, and business partners, have been transformed from being businesses to becoming e-businesses.

Such a metamorphosis requires that enterprises constantly reevaluate their strategies, their infrastructures, and their supply chains. Similar to the adoption of the JIT concept, firms implementing Internet applications are beginning to understand that it is not the technology but the opportunity to deconstruct rigidly held principles and management silos and then to reassemble the components into whole new environments, organizational behaviors, cultures, and competitive values capable of leveraging radically new ways of working with the value chain that constitutes the real importance of e-business. At bottom, e-business is more about changing infrastructure and channel management than it is about managing the technical aspects of Web applications.

■ *e-Business management is not the responsibility of a company department.* Many companies have made the mistake of treating their e-business initiatives either as a computer project and, therefore, the responsibility of Information Services, or as the responsibility of an e-business department. Such operational decisions perceive much too narrowly the scope of e-business. Again, consider the example of JIT. It is virtually impossible to localize a JIT implementation or make a specific group responsible for the project: everyone in the company and in the supply chain, for that matter, must be in pursuit of JIT goals. e-Business, like JIT, cuts horizontally across the enterprise and its supply partners and requires buy in at all dimensions.

Consider the implementation of a Web-based ordering system. Initially, the project is designed to provide customers with a real-time view of company products and services, facilitate the ordering process, reduce order inaccuracies, and accelerate deliveries. Pursuing all of these objectives, however, means that not only sales, but just about every functional department will be impacted. Depending on how closely the order process is connected to business partners, the implementation could also reverberate throughout channel inventory replenishment, logistics, and outsourced functions as well. e-Business requires process architectures uninhibited by internal or external infrastructures that can constrain the flow of data and information necessary for full e-business functional application.

■ *e-Business is not a close-ended project.* Earlier in this chapter, it was said that all competitive advantage is *temporary.* This means that all management strategies, as well as product and service life cycles, have been continuously shrinking, and the prognosis is that the process will only accelerate in the years to come. To ensure that they are not caught behind the change management power curve, managers today have been countering the pressure by attempting to utilize e-business tools to remain competitive. In fact, much like in the heyday of ERP, companies are gradually moving into an e-business arms race era. Companies are aware that neither their legacy technologies nor their past Internet investments guarantee them a source of marketplace differentiation tomorrow. Web enablers, such as online catalogs and the ability to build order management front-ends that once produced so much wonder, are now commonplace for most companies. Considering the disruptive power of the Internet, a technology enhancement implemented by one business chain can have the power to destabilize the equilibrium and throw an entire industry into chaos.

In such a volatile environment, firms constantly need to be reviewing and planning how Internet technologies can be utilized in the continuous search for competitive advantage. Defining a purposeful Internet technology architecture is not simply a matter of checking off a list of software applications. Rather, it is about intelligently choosing e-business alternatives to migrate, say, from using the Web for information and performing trading partner transactions to enabling full channel network collaboration. What can be said for sure is that very few companies that have begun the journey into e-business have decided to abandon the project as providing little or no competitive advantage. The goal for managers is to be vigilant in fashioning the business strategies that merge e-business enablers, core competencies, and supply partners capabilities with the continuous fracturing of what currently constitutes competitive marketplace value.

Preliminary Steps in SCM Technology Deployment Strategy

Perhaps the fundamental principle of SCM can be described as the capability of supply channel network partners to integrate and synchronize the flow of goods and information through the supply chain. SCM requires the establishment of business networks characterized by flexibility, agility, collaboration, and a focus on cross-enterprise processes. With the application of Internet technologies, the connectivity of SCM is greatly enhanced to permit instantaneous visibility to market conditions and to optimize process management. Whole supply chains are now freed from the tyranny of the *chain pairs of relationship model* and are capable of architecting networks where the impact of information, transactions, and collaborative productivities can be broadcast in real-time to all trading partners unencumbered by rigid channel tiers.

Assembling technology-enabled SCM value networks is also about dramatic shifts in the mindset of companies. While leveraging core competencies and pursuing brand and service leadership remain central, the SCM concept requires a migration of business strategies from the traditional focus on competition to collaboration, from considering information as proprietary to information as a shared resource, from company-centric data to the pursuit of open interenterprise and network thinking where close trading partner coordination dominates. Pursuing the activation of SCM requires that executives possess the capability to continuously reinvent their businesses and the nature of their supply chain networks in light of the tremendous changes occurring in the marketplace and in integrative information technologies. In this section, the basics of fashioning an effective technology-enabled SCM network strategy will be discussed.

Opening Issues in SCM Technology Strategy Development

For today's top enterprises it is *speed* that conveys the essence of their business strategies. Speed means the difference between satisfying customers and spiraling into extinction, between meeting marketplace demand and being stuck with unproductive assets, between success and failure. Speed, however, also means that companies must be constantly vigilant to ensure their businesses pursue optimal combinations of agility, visibility, intelligence, and technology to counter a marketplace characterized by increasing uncertainty, continued compression of business cycles, and expanding supply chain complexity. Responding to such a rapidly mutating business environment requires that companies more than ever have a coherent technology-enabled SCM strategy in place. The good news is that companies capable of meeting the challenge of change management with effective technology-enabled networking strategies can reap fabulous rewards; the bad news is that those that do not are destined to vanish as viable competitors.

Designing an effective technology-driven SCM strategy can seem a daunting task. Immediately a multitude of questions is before the management team. For example, when contemplating an e-business project should the project focus on B2C or B2B or both? How should working with independent trading exchanges (ITXs) or private trading exchanges (PTXs) be determined? What will it take to rebuild the current corporate/divisional strategy around e-business functions? What is the ROI and is it feasible? Should the existing ERP system be upgraded, a best-of-breed approach be followed, or a customized solution undertaken? What impact will the new SCM strategy have on customers, suppliers, and the internal business infrastructure? Will the implementation of new SCM technologies transform the competitive landscape and the company's position in it? These and a host of other questions must be posed, prioritized, and integrated into the new business strategy.

Once such questions are sorted out, the crafted decision may take several directions depending on circumstances and the company's desired objectives. Some firms may start by building onto their initial I-Marketing initiatives by implementing a customer relationship management (CRM) system that will provide customers with the ability to enter, revise, and track their orders. Another company, who may even be in the same industry, may start on the supply side by forming a PTX or joining a constortia trading exchanges (CTX). Still other enterprises might decide to embark upon a number of e-business initiatives simultaneously. Whatever the direction, the strategy must be designed to support both the firm's customer value proposition and supply chain effectiveness.

Preliminary Steps

The task of establishing a purposeful SCM technology strategy requires a number of preliminary steps. The goal of these first steps is to focus the enterprise on the impact of what the contemplated information technologies will mean to everyone both within the organization and to trading partners out in the supply channel network. The critical starting point will be for the executive team to ensure that everyone involved, from employees to channel partners, understands that the new SCM technologies will transform traditional roles for all involved. It is only through the detailed analysis of the current and anticipated interactions with customers and suppliers that a comprehensive technology strategy supporting a value proposition that provides for ongoing competitive differentiation can be created. Achieving this point in e-SCM strategy development involves a five-step approach (Figure 9.3).

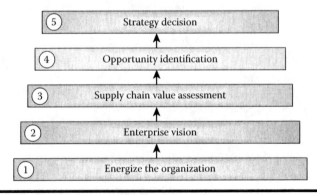

Figure 9.3   Initial SCM technology strategy steps.

Step 1: Energize the Organization

Preparing the organization for new SCM technologies is absolutely critical before a comprehensive business strategy can be articulated. Over the past decade executives have become aware that information technologies by themselves are rarely a source of sustainable competitive advantage. Successful technologies can be easily copied or supplied by software vendors, and Internet technologies are no exception. Rather, it is when technologies are utilized in support of well-conceived business strategies by being focused on an individual firm's employees, customers, and suppliers that a company can gain significant, and potentially sustainable, competitive advantage. Preparing the organization for a SCM technology project requires two major human resources initiatives: getting top management on board to spearhead the effort and energizing and integrating the company's people organization into the selected SCM technology. The following steps should be followed to inform and activate the top management team:

- *Educate about SCM and Technology.* The executive team should be educated on the basics of SCM and the targeted technologies to be implemented. In the past, CEOs could depend on fundamental principles in making decisions about products, markets, investments, and resources. Today, SCM and integrative networking technologies have challenged virtually all of those assumptions. Executives will need to understand what SCM technology means, not just in theory, but how it can be used to respond to their business's current competitive challenges.
- *Act as a Sponsor.* Once education is complete, the executive team must act as champion or sponsor of the SCM technology effort. Such an effort is doomed without a major commitment and the involvement of senior management.
- *Develop a SCM Technology Strategy.* The development of a SCM technology strategy often means both redesigning the supply chain and integrating in new capabilities now available through the new technologies. CEOs need to make sure their supply chain and SCM strategies are an integral part of the big picture. Business vision, software applications, Internet e-commerce objectives, supply chain, and customer management strategies all have to be aligned.
- *Develop the Firm's Human Resources.* Designing infrastructure around cross-functional collaboration is a significant challenge for most companies. Many firms are still organized around the departmental silos of the past and have yet to be restructured around SCM thinking.
- *Invest in Supply Chain Improvement.* Ongoing effectiveness of SCM technology strategies requires a supply chain improvement budget. Executives need to establish the means by which constant improvements in technology, external resources, and business processes can be continually funded.

Once the new SCM technology strategy has gained the support of top management, the second part of Step 1 is energizing the company's people organization. According to Manheim [5], there are six major "thrusts" that can be used to properly integrate the SCM technology strategy and the company's human resources. The first thrust serves as the overarching theme of the business strategy; the next five are supportive thrusts, each of which reinforces and amplifies the first thrust.

Thrust 1: *Enhance the Ways in Which People Work.* The implementation of a new SCM technology strategy will fundamentally alter the ways in which people, the organization, and supply chain partners work. Besides top management, the organization as a whole, and often trading partners, must be educated in the new technologies and the SCM philosophy. Similar to initiatives such as ERP or JIT, unless the people who will run the daily business do not fully buy into the technology, the project will achieve only a fraction of its potential.

Thrust 2: *Build Powerful Multienterprise Processes with Appropriate IT Support.* A critical opening step is identifying across the supply chain critical business processes impacted by the technology strategy and designing or reconstructing them as necessary. Besides rendering processes more efficient and productive, this thrust also requires network partners to implement channel-integrating technologies that permit the management of processes within the enterprise and between enterprises in the supply chain.

Thrust 3: *Balance the Roles of People and Technology.* The adoption of a new SCM technology strategy heightens the need for integration between people and operating the technology. Three critical issues come to mind. To begin with, SCM-based business is about relationships. Trading partner perceptions, particularly Internet-based businesses, are heavily influenced by trust and the uniqueness of customer–supplier relationships. Second, the dynamic nature of today's business environment requires that companies be prepared to reinvent supply channel models in response to changing competitive conditions, customer preferences, financial conditions, and supplier offerings. Finally, companies pursuing a new SCM technology must be adept at quickly turning channel data into knowledge useful for effective management decision-making.

Thrust 4: *Manage Multienterprise Processes Flexibly and Dynamically.* The growth of Web-based applications and protocols has made it possible to tightly integrate the critical work flows occurring between network partners. The ability to perform transactions, store and retrieve documents, and pass information across different organizational units and supply chain partners must be determined before a new SCM technology strategy can be deployed.

Thrust 5: *Manage Knowledge Strategically.* The effectiveness of technology-based SCM rests on the concept of gaining competitive advantage through people, enhanced by Internet networking technologies. Being able to standardize the enormous amount of human knowledge, work flows, checklists, process rules, and best practices of the workforces to be found in individual companies and supply network systems is essential in managing and directing intellectual capital to continuously enrich e-SCM strategy development.

Thrust 6: *Enhance Individual Effectiveness.* The ability to increase the effectiveness and productivity of individuals at any node in the supply chain is the fundamental objective of a SCM technology strategy. The philosophies and technologies associated with today's SCM can provide powerful enhancements to the way people think and act. SCM concepts enable people to deepen their relationships with channel partners, while computerized tools permit them to more effectively transfer and store data, communicate with their work partners, and facilitate change and decision management.

Step 2: Enterprise Vision

Visioning the competitive power of the business is the next step on the journey to building an effective SCM technology strategy. This step is about defining the nature of the competitive competencies possessed within the current infrastructure and outside in the supply chain network. In defining the enterprise vision executive teams need to think about such factors as these:

- What is the historical nature of the firm?
- How has it traditionally approached the marketplace?
- What processes add the most value to customers?
- How have relationships with suppliers grown through time?
- What is the nature of the internal organization?
- What are the strengths and weaknesses of business partners?
- What capabilities are the most important in creating and sustaining competitive advantage?

The goal of this process is to ensure a deep degree of awareness on the part of executives concerning just what the new SCM technology strategy means to the company, the steps necessary to build an effective SCM model and strategy, and how a new integrative networking technology-driven value proposition would translate into specific processes. This is not to say that other factors should be excluded from the visioning process. Current market conditions, existing channels of supply, product characteristics, competitive pressures, and legacy technology infrastructure should also be included to broaden the proposed competitive vision.

For example, Fresenius Medical Care, a manufacturer of dialysis equipment, has visioned itself in the position of a value network integrator in its industry. The company is successfully developing a total therapy concept based on its chain of over 1100 clinics. These clinics are arranged into a network of therapy centers providing a uniquely integrated value proposition linking equipment development and service to fit patient needs. To realize this strategic vision, the management system will use the Internet to link patients, nurses, physicians, researchers, pharmaceutical companies, and health insurance companies into a treatment, therapy, and development process monitored and administered over the Web [6].

Step 3: Supply Chain Value Assessment

The decision to implement SCM technology applications must be driven by a thorough understanding of which critical business processes will be impacted. For example, a move to Internet-based functions will require a close review of what company activities would be moved to e-business. Which processes to focus on should be those closely linked to the enterprise visioning activity detailed in Step 2 above. Basically, companies should seek to convert to e-business those processes that deliver the most competitive advantage. Noncrucial processes should be left untouched, except in cases where they are impacted by an associated e-business initiative.

Perhaps the most effective method to begin matching new technology applications, business processes, and strategic visioning is to perform a supply chain value assessment (SCVA). The object of this activity is to identify and then prioritize which technology initiatives should be undertaken that will provide the greatest enterprise and trading partner benefit. Far from being an isolated activity, an effective SCVA requires a collaborative effort where internal value assessment teams are closely integrated with analogous teams appointed by supply chain partners. The ultimate objective is to determine whether the networking technology vision and the impact it will have on the supply chain will be *evolutionary* or *revolutionary*.

Technology initiatives that are *evolutionary* are normally focused on improving core business functions and sustaining the competitive advantages they drive. Typically, these initiatives tend to center on process automation, are usually low-risk and low-return, tend to be inward oriented, and focus on short-term bottom-line return. Examples would include selling products online through a catalog with standard prices or automating information on shipments and inventories. In contrast, *revolutionary* SCM technologies attempt to create radically new supply chain network architectures that can actually transform internal core processes as well as those possessed by trading partners. Typically, these initiatives seek to create new value propositions, customers, and revenue streams and are by their very nature high-risk and long-term, supply channel oriented, and focused on capital investment. An example would be the implementation of a full SCM business software suite capable of helping businesses achieve order-of-magnitude competitiveness and profitability by enabling them to better predict demand trends, negotiate the best possible deals with suppliers, optimize inventory levels, and better coordinate distribution channels.

Determining whether the SCM technology vision is evolutionary or revolutionary is the central outcome of the SCVA process. Unfortunately, popular assumptions are that all technology initiatives are revolutionary and, likewise, that all revolutionary initiatives can be implemented through the same efforts as evolutionary initiatives. Obviously there is a qualitative gap between the two. Companies seeking revolutionary SCM technology change will soon find that the approach requires a massive transformation in infrastructure, culture, competencies, learning capabilities, funding, people motivation, and, of course, technology. It will also normally destabilize existing channel configurations, time proven business conventions, and long-standing partner relations.

Performing an effective SCVA can be distilled into three fundamental steps.

1. A collaborative team consisting of company and supply chain partners is formed. The operating basis of the team is to integrate supply chain, business process, and technology knowledge. It is the responsibility of the team to identify company and supply chain business issues, prepare an as-is model of competitive processes, and begin detailing the implications of evolutionary versus revolutionary approaches to utilizing SCM technologies for competitive advantage.
2. In the second step, the SCVA team breaks their findings down into key performance indicators (KPIs) and supply network opportunities. As the broad outlines of possible SCM technology applications become apparent, the team begins to investigate and detail solution approaches and concerns, obstacles and risks, and benchmarks to validate future performance.
3. In the third step, the SCVA team begins to match KPIs with proposed SCM software applications to determine such decision points as objective of the initiative, risk/return profile, major risk factors, outcome metrics, value-adding processes impacted, competencies required, and overall impact on the organization and the supply chain. When the exercise is completed, both the firm and its supporting supply chain partners should be left with a detailed portfolio of possible SCM technology alternatives to select from. It is this list that will be then used in the prioritization process to come.

Step 4: Opportunity Identification

The SCVA exercise should provide the collaborative SCM technologies team with a map of possible choices that will provide solutions to the KPIs identified in the previous step. Perhaps the first activity is to prioritize the possible SCM technology solutions. Accomplishing this task will require the SCVA team to divide initiatives into those that are *evolutionary* and those that are *revolutionary* in nature (see Figure 9.4). This map will then enable the firm to begin the process of determining just what kind of SCM implementation they wish to embark on, the range of competitive opportunities made available, and the approximate costs both to the enterprise and to supply chain partners.

As SCVA teams begin detailing and prioritizing possible solutions, several issues need to be kept to the forefront. To begin with, teams must understand what the proposed technologies do and do not do. During the dot-com craze, some futurists proclaimed that the Web signified the end of business as we then knew it.

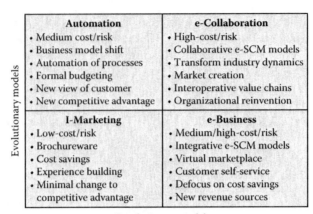

	Automation	e-Collaboration
Evolutionary models	• Medium cost/risk • Business model shift • Automation of processes • Formal budgeting • New view of customer • New competitive advantage	• High-cost/risk • Collaborative e-SCM models • Transform industry dynamics • Market creation • Interoperative value chains • Organizational reinvention
	1-Marketing	**e-Business**
	• Low-cost/risk • Brochureware • Cost savings • Experience building • Minimal change to competitive advantage	• Medium/high-cost/risk • Integrative e-SCM models • Virtual marketplace • Customer self-service • Defocus on cost savings • New revenue sources

Revolutionary models

Figure 9.4 SCVA for possible e-business implementation.

In reality, instead of creating a "new economy" where traditional transactional processes and relationship mechanics were simply replaced by radically new ways of doing business, the introduction of the e-business enabled new opportunities for efficiency and effectiveness around business processes that had always existed. Until a futuristic time when some fabulous machine can instantly transform thoughts into products, economies will have to contend with solving the age-old problems of production, demand, and distribution regardless of how automated the process may become.

Another critical dimension of prioritizing the results of the SCVA revolves around executives understanding the expanding degree of involvement required of supply chain partners as the SCM technology solution moves from evolutionary to revolutionary. The true value of network technology is that it enables companies to move beyond viewing the marketplace purely from *inside* the business. Even small-impact Internet automation projects soon force companies to shift their attention from managing internal to architecting interbusiness processes that are synchronized and intermeshed. Such *systemic* thinking will prevent SCM technology projects from failing prey to the deficiencies that plagued dot-com strategies at the beginning of the century, where the elegance and apparent simplicity of the technology were not in sync with the processes whereby products and services were acquired, stored, and finally delivered to the customer. Business, in reality, has always been about the interconnections that exist between supply chain partners; it has only been with the rise of networking technologies, however, that those linkages have not only become visible but not hold out the possibility of radically new mechanisms of delivering value to the customer.

Finally, SCVA teams must be aware that as the level of networking in the supply chain increases, the pressures on traditionally structured organizations will grow incrementally. In fact the growing requirements for collaboration and the

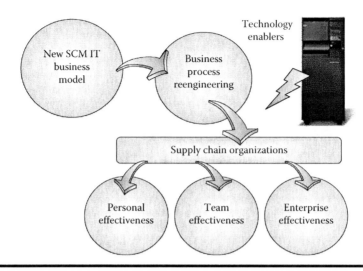

Figure 9.5   SCM organizational reengineering.

cross-functional nature of Internet-driven applications tend to undermine hierarchical organizations. No matter in which of the four quadrants (see Figure 9.4) a company may wish to begin, business processes must be engineered to support the channel network model (Figure 9.5). For example, a B2B initiative would require changes to purchasing, planning, sourcing and negotiations, supplier relationships, delivery and receiving functions, and accounts payable. As the pace of technology changes increases, this process of technology—organizational realignment will be constantly repeated to accommodate changes to network models and new opportunities.

Step 5: Strategy Decision

Once the SCM technology solutions map has been completed, company executives can then begin the process of planning a for the acquisition of software solutions and the development of a concise implementation plan. Regardless of whether the proposed solution involves a cautious evolutionary tactic or a dramatic Web-based strategy, the decision should focus on expected advantages. Whether the SCM technology initiative is focused on automating and integrating processes, reducing costs and increasing the flow of information through the supply chain, or engendering whole new businesses and forms of customer value is not important. What is critical is the understanding by the executive team that by itself the technology accomplishes nothing and that the real objective of the SCM project is to utilize the power of trading partners to amplify existing marketplace advantages or realize radically new ways of providing value to the customer.

At this point, the preliminary steps necessary for the selection of a SCM technology strategy have been completed. In the next section, executive planners turn their attention to selecting the corporate SCM strategy and setting in train the mechanisms for continuous strategy review.

Developing the SCM Technology Strategy

Earlier in this chapter, it was stated that the sum total output of a company and the way its infrastructure and value-enhancing resources are organized to meet its marketplace demand constitutes the firm's *business architecture*. To stay competitive, today's companies are being required to seriously reconstruct and, in some cases, reinvent their business architectures by including new supply chain networking technologies. Undertaking such a task of enterprise deconstruction and rebirth can be a serious project, especially as firms move toward more *revolutionary* SCM models. In any case, once the preliminary analysis of what technology solutions are available to the company has been completed, strategic planners must begin the task of transforming the enterprise, and often its supply chains, to realize the decided upon SCM value network strategy.

To assist planning executives create business architectures that enable them to successfully leverage the technology models they would like to pursue, a design framework diagram has been created and is illustrated in Figure 9.6. While it is impossible to design a strategic planning model that would be applicable to all businesses in all industries, the framework at least can offer a broad brush landscape to assist planners to effectively begin their implementation plans. As can be seen, the diagram, first of all, is portrayed as a never-ending cycle where

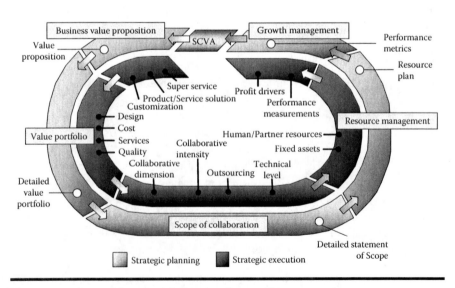

Figure 9.6 Structuring the SCM technology architecture strategy.

enterprise SCM architectures and marketplace objectives are constructed, operated, reviewed for performance, and then reconstructed as business and technologies change through time. Also, the diagram shows that the design framework consists of two interconnected flows, one focused on continuously driving innovative strategic thinking and the other operational execution. To be effective, an SCM value network must be constructed utilizing each segment of the framework. Examining each of these segments is the subject of the remainder of this section.

Constructing the Business Value Proposition

The purpose of performing the preliminary steps to SCM technology strategy development discussed above is to have available a map of possible technology solutions, what impact the application of any one of the solutions will have on business infrastructure and competitive positioning, and what changes will have to be made in supply chain arrangements. The process of actually selecting the technology solution to be implemented is the focus of this first step in SCM technology strategy development.

At the very core of strategy selection is the *business value proposition.* Companies exist to satisfy a particular need or want of their customers. In the past, firms could be half-hearted in listening to the "voice of the customer." Today, in this age of "temporary advantage" companies must be ever vigilant in ensuring that their organizations and their product and service offerings are synchronized to provide total value to the customer. Companies, however, must be careful to view a value proposition in its widest sense. While the end-customer has been the main focus, the term *customer* can be expanded to include suppliers, channel members, partners, or even competitors.

In defining a SCM technology value proposition planners are essentially concerned with the performance of two major activities. In the first, the customer segments to be served by an initiative, such as an e-commerce Web site, are identified. Here the goal is to look for mismatches between the expected results of the technology to be implemented and the value of the targeted marketing segment. For example, in order for McKessonHBOC, one of the world's largest providers of health care products, to be able respond to catalog information, pricing, real-time ordering, and delivery tracking needs of a wide range of customer segments from drug stores to mass merchandisers and hospitals, the company established a self-service Web site. When the site was launched, more than 6000 customers immediately began to use it in their dealing with McKesson. While traditional customer services remain, this experiment in e-business helped the company fill the gaps in what had been a fragmented view of servicing a widely differentiated group of customer segments.

Once Web technologies are matched to intended customer segments, the next step is to ensure that the technologies to be implemented will fulfill the service values expected by the customer. According to Bovet and Martha [7] an effective value proposition must be ready to respond to three possible service values:

- *Super Service.* The ability to provide superior service enhances both the value of the product/services mix presented to the customer and the competitive differentiation of the provider. The two primary attributes constituting super service are *speed* and *reliable delivery.* Speed refers to the velocity of response desired and offered by the provider; reliable delivery refers to the receipt by the customer of orders that meet expectations for capacity to absorb changes, accuracy, completeness, timeliness, fitness-for-use, and location.
- *Product/Service Solutions.* What constitutes "value" to the customer varies by the nature of the type of solution desired. Products and services that are commodity in nature have easily identifiable values, such as ownership, availability, low-cost, convenience of acquisition, and a recognized level of quality. On the other hand, noncommodity-type products are surrounded by more complex customer values, such as possession, service performance completion, or unique product–information–service combinations that permit customers to enrich their own competitive strategies. Dell Computer, for example, provides not just computers, but configurable information technology solutions designed to provide customers with value beyond mere possession of the product.
- *Customization.* As today's customers increasingly look toward *solutions* instead of just *products* and *services*, the capability for providers to offer configurable, customized choices that fit the precise need of the customer is growing as a prime attribute of a competitive strategy. Such a strategy can be realized by following an assemble-to-order (ATO), make-to-order (MTO), or through various forms of supply chain postponement.

The technology value proposition is not a separate part of the overall business strategy: it and other value initiatives *are* the *business value proposition.* Aligning a company's technology initiatives with its overall strategic framework sounds like basic advice, but for most firms this is the exception rather than the rule. Often technology initiatives do little to support a company's core marketing strategy or long-term customer satisfaction goals. Failure to view the business value proposition as an integrated part of the business strategy can not only endanger the technology initiative, it can also endanger the firm's existence.

Defining the Value Portfolio

Today's Internet networking technologies have dramatically accelerated the changes occurring to the nature of product and service offerings. In the past, companies competed by selling relatively stable product lines consisting of standardized, mass-produced product/service offerings. Marketing concentrated on persuading customers to purchase products whose value was fixed in the form of standard pricing.

Table 9.1 Changes in Products and Services

Marketing Function	Past Market Values	Internet Market Values
Form Utility	Standardized product and services offerings	Configurable products providing customer choices
Time Utility	As available; customers willing to wait.	When wanted; customers want immediate availability and delivery
Place Utility	Available were most sales occur	Available everywhere, physically and electronically
Quality Utility	Acceptable level of product quality	Total quality that exceeds customer expectations
Price Utility	Focus on offering the lowest standard price	Pricing based on value of the customer solution
Services Utility	Minimal service offerings	Complex packages of product and services
Information Utility	Product/service contains minimal to no information	Product/service contains rich information

The purchasing transaction was considered as the culmination of the sales process, after which neither seller nor buyer seldom expected ongoing opportunities for increased value-added to the products or services purchased. Today, as illustrated in Table 9.1, past views of product and service value have yielded to new market-place requirements that stimulate the development of deep, sustainable relation-ships among producers, distributors, and customers. The central theme of the table is to illustrate the radical shift in the portfolio of what constitutes product and service value as it migrates from standardized to more customized offerings, always available, delivered instantaneously, and accessible through electronic mediums.

To leverage the enabling power of the Internet, companies need to closely align their SCM technology strategy with their operations capabilities to continuously provide the product/service wrap that satisfies the unique needs of the customer. The following process developments need to be structured to effectively support the *business value proposition.*

- *Design.* Products and services have been dramatically impacted by continu-ously shrinking life cycles and the acceleration of new product/service intro-duction. For example, the life cycle of any of Panasonic's line of consumer electronic products, such as CD players, TVs, and VCRs is just 90 days.

The market for personal computers is so volatile and dynamic that a product can be obsolete while still in production. The ever-shrinking window of opportunity directly impacts accompanying service and delivery functions. Companies must continuously seek to uncover ever new opportunities to wrap intelligent services around products and activate logistics functions to speed product movement through the distribution pipeline.

■ *Cost.* Effective cost management requires companies not only to design product/service offerings with an eye toward continuous process improvement and cost reduction, but also to be able to squeeze the time it takes from concept idea to sales. Achieving such objectives means that companies must leverage the capabilities of supply chain partners. Cook and Tyndall [8] relate a story of how Dell was able to steal time, cost, and leadership from a competitor through the interactive connection of design processes with its suppliers. Because of its real-time demand management channel connectivities, when a sudden surge in two-gigabyte disk drives called for a switch from one-gigabyte drives, it quickly changed its ordering with its supplier. Because of a conventional 6-week demand forecast, a Dell competitor missed the trend and continued to build the one-gigabyte PCs. By the time the competitor got the product to market, nobody wanted one-gigabyte drives. As a result, the competitor lost market share and had to take a sizeable write-off.

■ *Services.* Customers today, especially those utilizing Web technologies, expect their products to be accompanied with a matrix of value-added services. For many products the associated service package is often more important to the customer than the product itself.

■ *Quality.* Over the past years, the concept of quality has moved from a concern with the standard dimensions of performance, reliability, conformance, and so on, to the capability of *choosing* between a multiplicity of products and services, to today's Web-driven requirement for product and service *individualization.* Customers now expect suppliers to have the ability to assist them in selecting the right combination of product and/or service offerings, and then configure the purchase to meet unique requirements. For example, IBM's customer order system contains a sophisticated configurator translator that converts model and features decisions made by customers into a buildable product, and then hands the configuration to production, The entire process is designed to remove redundancies, reduce cycle times, and ensure the delivery of a quality product to the customer.

Structuring the Scope of Collaboration

Once the *business value proposition* and the *value portfolio* have been formulated, strategic planners must then determine the scope of trading partner collaboration. In this step companies need to decide what will be the scope of the firm's processes and activities and correspondingly what will be the level of collaboration with trading partners necessary to supply missing resources and competencies. Today, everyone accepts the idea that a company can not and should not try to do everything and that collaboration with supply network partners, suppliers, and customers is a necessity. But while the idea of collaboration is not new, drafting a collaboration agenda has become in today's marketplace a complex affair. Similar to other management concepts such as Lean and TQM, most of the ambiguity stems from the fact that collaboration is a multifaceted philosophy of business that can be approached from several angles. Simple forms of collaboration occur when companies exchange information periodically. In contrast, collaboration can take the form of highly integrated, Web-enabled intermediaries, like Covisint/Compuware, that are focused on the real-time transfer of information and complex multiyear product development and marketing projects. The following points will have to be reviewed in determining the *scope of collaboration* when architecting the SCM technology value network strategy.

■ *Determining the Collaborative Dimension.* The supply chain enables companies to leverage the competencies and resources to be found in their trading partners to assist in the sourcing, creation, and delivery of determined value portfolios. According to Sawhney and Zabin [9], strategists can view this collaboration as having a *vertical* as well as a *horizontal* dimension. The *vertical* dimension can be said to consist of the matrix of network partners that assist in sourcing a business's inputs (suppliers) and delivering its outputs (channel intermediaries). This network can be said to constitute the supply and demand chains, respectively, of a given business. In contrast, the *horizontal* collaborative dimension consists of channel partners that enhance or reinforce a firm's value portfolio and customer relationships. A vertical partner contributes resources and competencies directly to the value portfolio, while horizontal partners utilize a company's value portfolio to perform value-added enhancements. For example, the thousands of software packages that run on Microsoft's operating system have dramatically expanded the corporation's reach into the marketplace and protect its competitive advantage.

Corporate strategists must be keenly aware of how dependent their value propositions are on the channel network. Often a new value proposition that significantly expands the depth and breath of the demand channel, while requiring greater supply channel cooperation, will also require a new collaborative value network. To cut cost and expedite customer service, Owens & Minor Inc., a health care products distributor, implemented an Internet order-fulfillment system with the product catalog system of key supplier Kimberly-Clark. When a customer clicks for more information on a Kimberly-Clark product, there is a connection to Kimberly-Clark's own product information, such as safety data sheets, videos on proper usage, product substitution notices, and FDA announcements. The results are that Kimberly-Clark gets a direct link to the customer, Owens & Minor does not have to maintain an online catalog, and both are assured the information the customer is accessing is current [10].

■ *Collaborative Intensity.* Regardless of the dimension of collaboration needed, strategists must determine the intensity of the collaboration with trading partners necessary to realize the value proposition. As mentioned above, collaboration can be pursued on many levels. Low-levels of collaboration, which focus on making information available to trading partners so they can be more efficient in their support, are relatively easy to achieve and require little in organizational change and technology expense. Conversely, high levels of collaboration, which require an increasing symbiosis of fundamental processes and shared goals, are complex and expensive to implement.

According to Prahalad and Ramaswamy [11], there are four levels of collaboration intensity that can be pursued by strategic design teams.

1. *Arms-Length Relationships.* Normally this is the level of collaboration pursued by companies seeking to drive market-based transactions across network boundaries. Often such a strategy can effectively utilize a Web portal site but will not require any greater level of sophistication. The target value of the collaboration effort is to increase the number of participants in the system.

2. *Information Sharing.* This level of collaboration is pursued by trading partners seeks to share a wide variety of information ranging from sales and order data to forecasts and stocking levels. Such efforts will require systems that will network the transaction systems of supply chain members. The target value of the effort is to improve business processes through more real-time work flows. Accordingly, as the number of participants increase, the more effective the information becomes for decision-making.

3. *Sharing and Creating Knowledge.* In this level of collaboration strategists seek to utilize and integrate the competencies of network partners in value proposition and/or value portfolio development. Pursuit of this level will require systems that enable online collaboration networking and unified information access. The target value is the capability to leverage knowledge from anywhere in the network to import needed competencies as well as reduce functional redundancies.

4. *Sharing and Creating New Insights.* At the highest level of collaboration, networked trading partners feel that they share common business value propositions and are willing to jointly leverage competencies and resources. Such a high level of collaboration requires common network information access, collaboration tools, and capacity for rapid knowledge creation and insight building. The value of this level of collaboration is to be found in the capability of companies across a network to devise and share a common vision regarding opportunities that reveal whole new competitive space and the capability of structuring the technical and social architectures necessary to achieve those visions.

As strategic planners begin the process of selecting the level of collaborative intensity, they will need to determine answers to a number of key questions. To begin, at what level of collaborative intensity is the company currently positioned? This will reveal the gap between the existing and expected level. It should also reveal the major hurdles. Second, the analysis should reveal what prerequisites need to be in place, the administrative costs, and the impact of existing cultures. Third, the viability of the existing information technology infrastructure to support the targeted level of collaboration should be made visible. Finally, strategists must access the real and hidden risks and costs of the collaboration initiative both within the organization and with trading partners.

■ *Technical Level.* One of the critical elements in establishing the targeted level of collaborative intensity is the information technology capabilities required of both individual companies and the network. From the start, however, it must be stated that not all collaborative efforts require an Internet technology solution. For decades companies have established effective modes of collaboration with network partners utilizing EDI, fax, and even the phone. Simply being Web-enabled does not mean that a company has achieved a high level of business collaboration. Determining just what the level of technical assistance is required is a matter of finding answers to three questions.

 – What is the level of collaborative intensity required?
 – Based on the level of intensity, do the competitive values available meet the requirements of the value proposition?
 – Based on the answers, what then should be the level of technology necessary to support the scope of collaboration?

A critical mistake is assuming that a single technology solution will be sufficient to meet the requirements of the scope of collaboration. In reality, the actual collaborative partnerships a company has may require quite different technology responses. Smaller trading partners who are working quite well with fax or phone connectivity will be resistant to high-tech solutions such as B2B sites. Strategic planners should be prepared to create a portfolio of technical solutions to meet the possible needs of their collaborative partners.

According to Treachy and Dobrin [12], there are four possible technical responses to meet connectivity needs to support collaboration strategies.

1. *Non-Internet Technologies.* Many companies utilize basic technology tools to connect to their trading partners. Devices such as EDI, fax, and the telephone fall within this area. Normally, these tools are used by companies seeking to pass basic transactions and market information across the channel network. The benefit of this level of technical connectivity is its low-cost and ease of operation.

2. *Visibility.* This technical strategy seeks to provide an open systems approach whereby either information, such as schedules, forecasts, or orders, is broadcast to the channel network or trading partners are

provided with the capability to access system data. For the most part, basic Web-based tools are used to achieve this level of functionality. The benefit of this level of technical connectivity is an increase in the speed and accuracy of information and ability of trading partners to more effectively coordinate channel business activities.

3. *Server-to-Server.* This technical solution is used by trading partners requiring that data physically reside in the systems of trading partners to support the large-scale transmission of information. This solution utilizes concepts such as e-hubs and transmission standardization tools like RosettaNet and CPFR standards. The benefit of this level of technical connectivity is channel information scalability, permitting each trading partner to use their own systems without manual transformation.

4. *Process Management.* This is the most challenging level of technical connectivity and is the focus of channel networks seeking to integrate intercompany processes at the applications level. The goal is to configure Web solutions that provide for real-time work-flow sharing. The benefit of this level of technical connectivity is the capacity for channel partners to accommodate changes and support business process management in the pursuit of radically new competitive space.

In developing an effective technology strategy, planners will need to determine answers to a number of key questions. To begin, what is the scope of collaboration and what level of technology does the firm plan to use to tap into the network of trading partner relationships. Second, have clear business benefits been identified for each technical initiative? The goal here is to prioritize collaborative opportunities and devise simple, low-cost technology tools to achieve the necessary level of collaboration intensity. Third, have strategists designed a useable portfolio of technology connective approaches? Fourth, do the technology solutions chosen represent the simplest and least costly methods or are they characterized by overelegance and complexity?

■ *Outsourcing.* A critical part of the scope of collaboration is the decision to outsource functions currently performed by the firm. In the past, companies sought to control through vertically integrated corporations all aspects of supply and delivery. Today, businesses are increasingly moving toward outsourcing, contract manufacturing, and third party logistics, as short product life cycles, spiraling operations costs, and global competition tighten profit margins. By divesting themselves of labor- and capital-intensive assets not central to their businesses, companies can focus on their core competencies to improve competitive positioning.

When formulating outsourcing initiatives, strategic planners must be careful to understand that outsourcing is not the strategy but rather a vehicle to strategy activation. According to Tompkins, Simonson, Tompkins, and

Upchurch [13], there are a number of advantages to outsourcing business functions.

- *Focus on Core Competency.* A company that outsources functions no longer needs all the resources it had been using to make a product, inventory it, and deliver it to the customer. As a result, companies can redirect all of their efforts to improving core competencies while eliminating those peripheral to the essential processes of the business.
- *Return on Assets.* By reducing costly assets like personnel, warehouses, noncore manufacturing, materials handling, transportation, information technology, and others, return on current assets and capital expenditure can be significantly enhanced.
- *Personnel Productivity.* Non-core functions can be eliminated from company processes, thereby increasing employee productivities.
- *Flexibility.* Outsourcing permits companies to access new markets without initially shouldering the associated costs. As markets change and new products are developed, firms need to be as flexible as possible to manage service requirements, ordering methods, and competitive offerings.
- *Customer Service.* Today's mantra of total customer satisfaction has greatly increased the importance of logistics. To assist in realizing this critical channel value, logistics providers have gradually been emerging into businesses that offer specialized services beyond the capability of most firms to achieve.
- *Information Technology.* The increased demands for new information systems and resources can often be far more efficiently met through outsourcing. Whether leveraging logistics providers or contracting a Software-as-a-Service (SaaS) provider, firms can realize ERP, EDI, or Internet capabilities without the need to acquire or develop in-house resources.

In designing effective outsourcing approaches, strategists must keep several key principles in mind. To begin with, companies should never outsource core functions. The loss of core competencies results in a hollowing-out of a corporation that often sets off a downward-spiraling effect that is difficult to reverse. Second, firms should never outsource functions internal personnel do not comprehend. This will result in trying to link the provider to performance metrics that have not been fully defined, communicated, or understood. Third, the channel value assessment team in charge of the initiative should be diligent in defining the precise objectives. Reasons should go beyond a simple focus on cost and consider network collaborative issues as well. Fourth, the exact nature of outsourced functions should be detailed. This step requires identifying expectations, productivity metrics, gain sharing incentives, service requirements, precise operations to be performed, level of electronic connectivity, and partnership boundaries. Finally, it is critical that the outsourcing project have the support of senior management and be integrated into the corporate plan.

Ensuring Effective **Resource Management**

While the business value proposition will determine how the enterprise will approach the marketplace in terms of the products, services, and informational and delivery techniques it offers, the effective application of its assets and management of its internal and supply chain infrastructures are the drivers of growth and profits. Part of the role of plans in this segment of the SCM technology business strategy is to reengineer all business processes that are inefficient and rigorously eliminate all non-value-added activities. Cost containment and optimization, however, is only part of this step. Of far greater importance is the capacity to continuously reconfigure the resources found both within the organization and outside in the supply channel in the pursuit of order-of-magnitude profit growth and competitive advantage. The ultimate goal is to construct business architectures that offer customers what they want: convenience in ordering, solutions to complex needs that often require "killer" services to be encapsulated with the product, speed of delivery, and ease of payment.

The content of an enterprise's resources consists of its assets and core competencies. In general, these resources can be divided into three major areas: the value that resides in human knowledge of processes, customers, suppliers, and business partners; the capital invested in physical assets; and the value to be found in the business's physical assets.

■ *Human Knowledge.* In today's hypercompetitive environment, businesses have been migrating from a departmental focus on *Human Resources* to a far more strategic and expansive focus on Human Capital Management (HCM). HCM can be defined as the repository of human knowledge and skills found within an organization that result in the creation of products, technologies, systems, processes, and relationships. Human resources are intangible and tacit in nature, yet without them neither physical assets nor supply networking resources can be actualized and purposefully directed.

Today's best strategic planners continue to aggressively invest in management initiatives and knowledge expansion solutions that will enable their firms to leverage existing human capital. Key to HCM development has been a growing focus on technology to accelerate recruiting, optimize learning and skills development, measure and appraise performance, and more productivity arrange the structure of infrastructure tasks. Benefits to effective HCM are as follows [13]:

1. *Stronger Growth, Productivity, Performance, and Profitability.* The acceleration in the level of human knowledge application will become more important as the complexity of networking methodologies and information technologies drive increased investment in performance management tools and corresponding opportunities for dynamic growth.

2. *Enhanced Learning and Development.* The tremendous growth in e-learning tools, from CBTs to online learning, promise to raise the productivity,

quality, and personalization of knowledge transfer while significantly cutting costs.

3. *Enhanced Recruitment and Retention.* Leading-edge recruitment functions that automate the work flow and communications associated with recruitment are enabling companies to save significant costs. Such tools are accelerating the time-to-hire while improving the quality of hires.

4. *Personalizing Employee Relationships.* Increased personalization of human work and relationships encourages employee loyalty and motivates the kind of knowledge transfer, personal agility, and entrepreneurship necessary for today's competitive environment. It also provides managers to continuously match knowledge and skills to high-velocity projects.

■ *Physical Assets.* A business's *physical assets* are perhaps the easiest to understand and manipulate. Warehouses, offices, information systems, production and transportation equipment, patents, and inventories are examples of hard, tangible assets. Physical assets provide the mechanisms by which the firm transforms the value portfolio into competitive products and services.

The application of information technology to physical assets has a direct impact on cost and value producing attributes. In fact, there is a direct correlation of the level of information available about demand and supply and the level of capital investment in physical assets. Historically, disconnects in demand information and supply capabilities required businesses along the chain of supply to increase the level of physical assets to counter the infamous bullwhip effect. Increases in unproductive assets to support a lack of knowledge regarding the actual impact of the demand-pull through the supply network simply acts as a drain on cash and profits.

To counter this spiraling upward of channel costs and downward of competitive advantage, strategists need to determine which cost and process improvements can produce real advantage, which service enhancements customers will value, and how to utilize information technologies to integrate and synchronize internal operations into those of customers and suppliers. Networked channel leaders such as Intel see beyond their own operations and linkages with tier-one suppliers to a supply network, or "ecosystem," in which the actions of each member have a direct bearing on every other member. In detail, corporate planners should examine the entire supply chain in an effort to achieving the following process values:

1. *Replacing Physical Assets with Real-Time Information.* Information here refers to gathering accurate customer demand and enabling visibility to inventory and other productive assets as they exist at customer touch points in the supply chain.

2. *Reducing Process Complexity.* According to the Pareto principle, the vast majority of processes are employed to satisfy easily managed customer needs. Disproportionately, it is the complex processes that consume time and money. Eliminating complexity permits trading partners to remove

excess assets that add very little value either to supply chain organizations or to the customer.

3. *Reducing Product Complexity.* Product complexity also increases physical assets. Solutions in this area can be found in greater use of ATO and MTO manufacturing or closely examining the possibilities for deploying product postponement in the supply chain and as close to the end-customer as possible.

4. *Reducing Partner Supply Variability.* The inability of suppliers to receive complete demand information and provide timely delivery also adds to the volume of capital assets in the supply network. Strategists can work on the creation of closer supplier relationships and the application of connectivity technologies to move as close as possible to real-time information transfer.

■ *Business Network Resource Management.* Network trading partners contribute competitive advantage by providing two critical resources: physical assets, such as plant and inventories, and core competencies, such as design or process skills. Partners can provide critical values simply by performing asset-intensive activities that enable partners to leverage their scale, experience, and financial resources. Increasingly, HCM is being expanded to encompass the networks of knowledge and skills that lie beyond the boundaries of the business. Developing and implementing new technology solutions requires planners to explore ways to manage and capitalize not only on the hard assets but also on the competencies of contractors, suppliers, partners, customers, and even, in some cases, of competitors.

Some of the critical dimensions involved in leveraging network trading partner resources are as follows:

1. *Synchronized Delivery and Production.* Tightly integrated, connectivity between trading partners can dramatically enhance marketplace value surrounding capabilities that enhance speed, reliability, convenience, and efficiency. As the level of collaboration increases, network partners can change and better respond to market requirements as they ripple through supply chain asset requirements, priorities, schedules, and optimization and substitution decisions.

2. *Outsourcing.* The capability to utilize the physical and knowledge assets of partners to achieve breakthroughs in competitive advantages is fundamental to a new technology deployment strategy. Outsourcing permits opportunities for increased marketplace value without often premature and massive expenditure yet enabling firms to hold tight the information, knowledge, and vision of the initiative.

3. *Creating Collaborative Solutions.* True collaboration occurs when everyone in the network arrangement receives agreed upon value. Collaboration enables better product development, service design, inventory management, marketing, selling, ordering, and service. Internet tools that enable closer collaboration should be at the forefront of strategic planners' agenda.

Pursuing Growth Management

Perhaps one of the most important components of SCM technology network strategy development is structuring a set of meaningful and focused performance measurements that will allow corporate planners to gauge the effectiveness of their supply chain solutions. Being able to determine the impact of a business strategy on profits and growth has always formed the only real cornerstone of competitive measurement, but with the advent of Internet technologies the requirements for clear focus on supply chain direction and targeted metrics has never been more important. No company today would reject the principle that supply chain trading partners provide the potential for enormous competitive advantage and that exploring ways to gain greater connectivity and synchronization with the supply network is critical. However, determining just what the depth of partnership and the degree of integration should be requires a well-formulated plan and significant ongoing analysis. Today's supply chain strategist would agree that applications such as ERP, CRM, SRM, collaboration demand planning and forecasting, and Internet-enabled procurement, offer tremendous opportunities for making the right decisions regarding trade-offs resulting in greater cost savings and higher levels of productivity. The real issue is that not all companies will opt for sophisticated Internet solutions to run their supply chains and, even if they do, how should they be measured?

Clearly, one of the crucial problems in determining the level of supply chain collaboration is that it requires companies to rethink traditional measurements. In fact, it is being argued that a fundamental overhaul of measurement models has become imperative in the new economy. Reliance on time-worn operations research formulas, linear programs, and spreadsheets are inadequate to provide the flexible levels of information for an Internet economy. In addition, many of the macromeasurements, such as ROI, that are centered on individual organizational performance need to be modified in favor of externally oriented statistics that measure the performance of the entire supply chain. Many companies in the post-dot-com age have spent millions on Internet tools that are out of sync with their value propositions and value portfolios and have failed to architect the promised connectivity with network trading partners. In fleshing out the final component of the technology-enabled SCM strategic process, corporate managers will have to consider the following three areas: supply chain cost, supply chain value, and detailing effective metrics.

Focus on Supply Chain Costs

For over three decades executives have sought to increase corporate value by applying technologies and management methods such as ERP, Lean, TQM, Six Sigma, business intelligence (BI), and others in the pursuit of supply chain cost controls and improved operating performance. The impact of supply chain costs can be dramatic. In some industries, supply chain costs can equal 50% or more of a company's revenues. According to A. T. Kearney, supply inefficiencies can waste up to 25% of a firm's operating costs and, faced with razor-thin profit margins of 3–4%, even a small increase in supply chain efficiencies can double profitability.

Many companies have developed strategies that focus purely on the cost reduction opportunities to be found in their supply chains. According to Kavanaugh and Matthews [14], cost-centered supply chain strategies can be considered as anchored on three models (see Figure 9.7).

■ *Basic Model—Stable Supply Chains.* This model is the least strategic of the three and is normally applied to supply chains that have significant historical stability, such as table salt manufacturers where demand and supply are in equilibrium. Because of long-product life cycles, commodity-oriented processes that utilize scale production, and dedicated capital assets, supply chains using this method are heavily focused on execution with close attention paid to efficiencies and cost performance. Connectivity technologies are normally very simple with little need for sophisticated real-time enablers such as CPFR and complex collaborative synchronization.

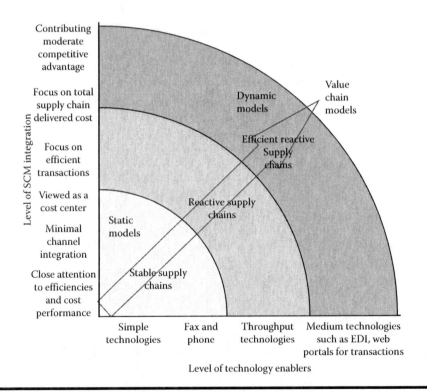

Figure 9.7   Supply chain cost models.

- *Model 2—Reactive Supply Chains.* Companies using this model still have minimal expectations of the value of their supply networks. Typically, supply chains in this model primarily act to fulfill demand by responding to and supporting trading partners' sales and marketing strategies. It is perceived by others and perceives itself as a cost center. Since network nodes are heavily focused on efficiency and cost management, minimal effort is spent on connectivity technologies or capital assets except to support the latest sales and marketing needs. The mantra of a reactive supply chain is to ensure that throughput continues at any cost.
- *Model 3—Efficient Reactive Supply Chains.* In this final model, the supply chain is still perceived as contributing minimal, or at best, moderate competitive positioning to trading partners. While still considered primarily as an operations function, the supply network assists competitive positioning by acting as an efficient, low-cost, and integrated unit. Efficiencies and cost management transcends local departments or company focus and is centered on the total delivered cost of finished goods. Connectivity technology and new equipment dramatically increase in importance as channel partners search to automate functions with a view to reduce labor costs and improve capacity and throughput. As an example, contract manufacturer Flash Electronics subscribes to a hosted Web-based system for managing RFQs. The system links the company to nearly all of its suppliers. The results have been to date an almost 50% reduction in overall quote-process time and up to 65% in quote-processing costs, along with the ability to respond faster to its OEM customers.

The three above strategic performance models perceive the supply chain from a localized point of view as a cross-company cost center. The metrics utilized are based on traditional channel network performance models, which are centered on assessing the ROI of individual trading partners and not on the potential for collaborative payback and joint strategic value. To leverage the competitive strength inherent in supply chains characterized by highly integrated and synchronized connectivity, network partners must begin the process of migrating from a *cost* focus to a *value* focus.

Focus on Supply Chain Value

A SCM technology strategy that focuses on cost management and optimizing channel functions will lead to less waste, higher productivity, greater market share and earnings, and greater competitive positioning. However, the benefits attained by just concentrating on process improvement are destined to be short-lived and will rapidly be neutralized as competitors copy the solution and install the technologies. In order to create competitive advantage that is truly sustainable, strategic planners must look to developing SCM technology strategies that go beyond cost reduction and optimization and actually leverage the resources and competencies of trading partners to *support* if not *facilitate* the generation of value.

In today's environment of rapidly changing economic and technological changes that rapidly destroy competitive advantage, strategies for maintaining sustainable profitability are increasingly factoring in the capabilities of supply chain partners as critical components driving marketplace advantage. While it is true that companies look to local decisions regarding customers and product/service offerings to produce growth, the level of participation and connectivity a business enjoys with its trading partners can dramatically assist in supporting current advantage as well as generating radically new sources of competitive value. Creating e-SCM strategies that will leverage the supply network to generate *value* for the firm will require a dramatically higher level of commitment, collaboration, integration, and synchronization than what is characteristic of channel relationships focused purely on cost management. There are two models of supply chain at this level [15].

■ *Efficient Proactive Supply Chains.* The goal of proactive supply chains is to leverage total network partner resources to actively drive demand and supply requirements through the supply channel in order to support, if not expand, the profit and growth engines of the entire network ecosystem. Perhaps the most distinguishing characteristic of supply networks at this level is the dramatically increased integration of sales and marketing. In this strategic model, entire supply chains *proactively* pursue the management of total channel demand, not only to reduce costs, complexity, and efficiency, but also to drive the creation of new sources of value generation by suggesting product design or service changes. Efficient proactive supply chains invest in Internet technologies to integrate enterprise business systems across the network and enable real-time visibility and synchronization regarding critical sales and marketing data as well as share ideas.

■ *Revenue and Profit Driver Supply Chains.* At the highest level of supply chain performance, strategists perceive the supply network not only as a critical demand and supply integrator, but also as an active contributor to the continuous generation of new forms of customer value. Through the application of advanced technology tools that provide real-time connectivity, network business nodes seek to explore the use of trading exchanges, collaborative product development and information sharing, and synchronization of resources and competencies that enable radically new opportunities for competitive advantage and profit growth. Forecasting, planning, and replenishment processes are fully integrated across the supply network; performance measurements focused on total supply chain revenue, cost, ROI, and profitability are developed.

The ability of today's enterprise to increase profitability and growth is in some degree or another dependent on their supply chain partners, whether they be OEMs, contract manufacturers, distributors, or suppliers. Supply chains are truly evolutionary ecosystems, where the role of each participant is intertwined in a complex matrix of business strategies, trading networks, and levels of collaboration. Simply making the ecosystem more efficient does not guarantee its survival—interdependency enables the entire system to evolve to achieve levels of performance impossible to attain by companies working independently on their own. Similarly, as in the case of the high-tech industry, relationships between network partners must be inherently interdependent if the innovative products demanded by the marketplace are to be created. Assembling supply chain systems that enable order-of-magnitude breakthroughs to profitability and growth are the results of finding answers to the following statements:

- Who are the pivotal supply chain network partners?
- What are the collaborative supply chain strategies and technologies they have in place or are planning to implement?
- What do direct customers and customers of trading partners want the most from their supply chain providers?
- What are the weak links in the supply chain process that have a direct impact on delivering customer value? If customers consider speed to be critical, what channel areas increase non-value-added time to the fulfillment process?
- Once the trouble points are identified, what steps need to be taken to work with channel partners to determine what information could be shared among channel participants that would most improve the network's effectiveness in delivering value to the customer?
- Finally, what would be necessary to map and integrate the value-enhancing steps into the SCM business and technology plan?

Design an Effective Performance Measurements Program

The ability to create effective performance measurements has always been a critical function of a successful enterprise. Performance measurements enable companies to determine the efficiency and effectiveness of business processes and provide overall metrics regarding profitability and growth. Over the past decade, companies have sought to extend the reach of their ability to manage cost, processes, and profitability by focusing on their supply chains. However, while much time and effort has been expended on the physical management of supply channels, there remains an immense gap in the ability of supply network partners to actually measure supply chain performance. While most executives realize that performance metrics are critical to achieve the type of intercompany collaboration necessary to satisfy customers, they are faced with a number of critical issues when it comes to the actual definition of SCM performance measurements.

Perhaps the fundamental problem is that SCM measurements transcend the performance of individual companies and seek to determine how well an allied group of businesses perform in regard to overall costs and profitability. According to Brewer and Speh [16], SCM performance measurements require trading partners to transform traditional performance philosophies in three important ways. First, to provide meaningful information for decision-making, the performance measurements must be designed around true intercompany collaboration. Second, individual companies and their management and staffs must work in collaboration with channel partners. This means SCM metrics must be structured that provide incentives for collaborative behavior. Finally, each network business partner, no matter the position they occupy in the supply chain constellation, must focus on performance that promotes the satisfaction and ultimate cost of servicing the customer from the perspective of the entire channel network.

Architecting a SCM infrastructure that promotes such channelwide metrics is no easy task. Gaps in performance goals can easily occur when local measurements are pitted against global objectives. Even seemingly relevant measurements designed to provide short-term successes can have unintended consequences. Take for instance a company seeking to accelerate the speed of inventory through the supply pipeline. While the effort might provide initial cost reduction for one channel node, in the long run all that happens is that the channel system has to absorb additional inventories. In the end, the cost for holding this inventory comes back to the company in the form of higher prices.

One school of thinking holds that extending performance metrics to a supply chain is a massive task that boarders almost on the impossible. One consultant feels that such an undertaking "is more of an idealistic concept because rarely do the interests of numerous trading partners align together. Most companies have a big enough challenge to meet performance measurements within their own four walls" [17]. Such sentiments must be taken seriously. Companies can become easily discouraged just by trying to map out the number of relationships that constitute their supply chains. Combined with the fact that each company has its own business systems and parochial measurements, the task of constructing a common set of performance metrics does indeed seem to contain almost insurmountable obstacles.

Still, no matter the degree of difficulty, there can be no doubt that there is inherent value in striving for supply chain performance measurements. Since companies can not escape the growing dependence they have on trading partners, it is critical that they develop forms of business performance statistics exchange, supported by analytical tools, which can be utilized from both an operational and strategic perspective. Lapide and others [18] have identified six possible measurement approaches that can be utilized by strategic planners. A short description of each model is as follows:

- *Cash Velocity.* The ability to cycle assets and cash to generate growth is directly dependent on how quickly value can be passed through the supply channel. Cash velocity in the supply chain is best considered as a component of value rather than the value itself and is affected by inventory turnover, transaction costs, current liabilities turnover, growth rate, net profit margin, and the tax rate. Where assets build at various points in the supply network, cash turns to cost. No better example can be seen than in the high-tech sector where companies like Dell face short product life cycles that require rapid flow-through of assets from suppliers to outsourced manufacturers to the customer measured in days. Optimizing cash velocity requires aligning supply network partner processes and resources with channel customers, products, and services to achieve the quickest return. Models to deploy to increase cash velocity include optimal asset utilization (OAU), activity-based costing (ABC), event-driven costing, and cash velocity levers, such as receivable and inventory turnover.

▪ *The Balanced Scorecard.* Originally developed by Kaplan and Norton [19], this model, while not directly created for SCM, can be easily modified to generate an excellent method for managing supply chain performance. Briefly, the balanced scorecard refers to a performance strategy that seeks to achieve a balance between financial and non-financial performance across short-term and long-term time horizons. The model focuses on performance from four different perspectives: *financial results* (cash-to-cash cycle), the *customer* (viability of the value proposition), *business processes* (outputs measured in terms of quality, time, flexibility, and cost), and *innovation and learning* (capability of organizations to learn and grow). Table 9.2 illustrates an example of applying the balanced scorecard to proposed SCM objectives and accompanying metrics.

▪ *SCOR Model.* Originally codeveloped by Pittiglio, Rabin, Todd & McGrath and AMR Research, the Supply Chain Operations Reference (SCOR) model is a tool for translating strategy into supply chain performance goals. The SCOR model divides a supply chain into five distinct management processes: *Plan* (cycle time metrics associated with demand/supply planning and management), *Source* (cost metrics associated with sourcing, unit costs, lead times, and inventories), *Make* (asset metrics associated with production, quality, changeover, capacity utilization), *Deliver* (service metrics associated with on-time shipment, order fulfillment order, warehousing, and transportation), and *Return* (returns and defective products). The goal of SCOR (Figure 9.8) is to decompose each of these processes into detailed metrics focused around the following performance attributes: reliability, responsiveness, flexibility, cost, and assets [20].

▪ *The Logistics Scoreboard.* Developed by Logistics Resources International, a logistics consulting firm, this model recommends the use of an integrated group of performance measurements consisting of four general logistics performance categories: *financial* (e.g., costs and ROA), *productivity* (e.g., orders shipped per hour and number of deliveries made per day), *quality* (e.g., orders shipped without error and percentage of damaged goods received), and *cycle time* (e.g., order fulfillment lead time and delivery lead time). To assist in

Table 9.2   SCM Balanced Scorecard

	Strategic Theme	Strategic Objectives	Strategic Measures
Financial	Financial growth	Channel cost reduction Increased profit margins Revenue growth High return on assets	Cash flow Channel inventory costs Transportation costs Days of open AR & AP
Customer	Increased satisfaction Increased value	Provide strategic solutions View customer as unique Alignment customer service needs and priorities High-velocity delivery	Quality management Timeliness of delivery Flexibility and agility of the supply channel Ability to deliver customized solutions
Business Processes	Cost reduction Flexible response Closer collaboration	Innovative products and services Increased synchronization Increased communication More scalable supply chains Fast flow of inventories Real-time digitization of internal and partner processes and information	Waste reduction Time compression Unit cost reduction Time-to-market reduction Order cycle time reduction Inventory acquisition costs Forecast accuracy Reduced communications time
Innovation And Learning	Motivated and prepared supply channel workforce	Product/process innovation Partnership management Increasing core competencies Motivating workers Skilling workers Staffing the e-business team	Employee survey Personal balanced scorecard Total supply chain competency available Total supply chain information available

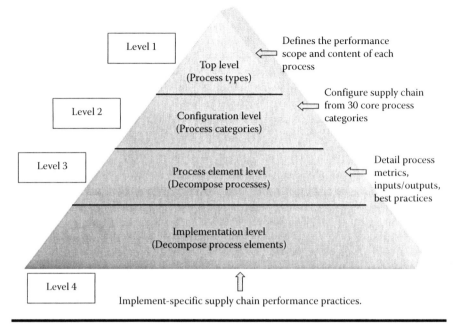

Figure 9.8   SCOR process detail levels.

tracking these logistics costs, Logistics Resources markets a spreadsheet-based tool—*The Logistics Scoreboard*—that firms can employ to measure supply chain performance metrics. The model and tool, however, have a shortcoming in that they are focused on logistics metrics and have limited usefulness in measuring total supply chain activities.

■ *Activity-Based Costing* (ABC). ABC methods were originally developed to overcome the defects in tradition absorption-based cost models. The goal of the method is to ascertain the true cost of processes or products by breaking down the activities necessary to perform them into individual tasks or cost drivers that could then be used to calculate the actual cost necessary to execute each tasks. These detail costs can then be rolled up to provide the total actual time or cost expended. ABC methods are extremely useful in the compilation of supply chain costs. For example, the cost of transporting products through each node in the supply chain can be used as a driver to assist in the compilation of total supply chain logistics costs.

■ *Economic Value-Added* (EVA). One of the criticisms of traditional accounting methods is that they tend to favor short-term profits and revenues while neglecting the long-term economic well-being and potential profitability of the enterprise. To remedy this shortcoming, some financial planners advocate assessing a company's performance based on its return on capital or economic value-add. The EVA model attempts to quantify the value created by taking the after-tax operating profit of a company and subtracting the annual cost of all the capital the firm uses. Companies can also apply the model to measure their value-added contribution to total supply network profit. A defect of the model is that, while useful in assessing earnings above the cost of capital to be included in the executive portion of a balanced scorecard, it is less useful in structuring detailed supply chain metrics.

Supply chain networks strategists must perform a significant amount of work to be able to structure meaningful channel performance measurement programs. While the models outlined above provide excellent examples of where to begin, there is no one recommended approach or definitive set of supply chain measurements. However, such statements should not dissuade companies from taking up the challenge—there is just too much to be gained by harnessing the productive power of the various SCM initiatives that are today being implemented! Whether it is by beginning with a high-level executive performance scorecard or utilizing information technology to automate the capture of agreed upon metrics, it is critical that some plan be inaugurated.

Perhaps the first place to start is to form cross-enterprise performance design teams. To begin with, these teams will need to move beyond a concern with local function-based measurements, which tend to splinter the performance development effort and focus on cross-functional processes and accompanying metrics that will crystallize objectives designed to increase cross-network channel integration. The goal is not to eliminate function-based measurements, but rather to broaden their effectiveness by integrating them with supply chain level metrics that will reveal how well each network business node is individually working toward goals that will improve not only their own performance but also the overall performance of the entire supply chain. In addition, teams must be strong enough to tackle several other critical problems inherent in determining supply chain metrics. The measures decided upon must be in synchronization with individual company and total supply chain strategies. The tendency to capture too many measurements must also be avoided. Participating companies must be encouraged to provide meaningful information on their performance. And finally, supply chain measurements can be beset by problems in defining basic terminology necessary to ensure common understanding of performance standards.

The following key steps should be taken when implementing a supply chain performance program.

- Begin by establishing a multibusiness network performance measurement team. The role of this team is to prepare the program's overall strategy, detail current and future performance metric contents, set priorities, and ensure ongoing progress.
- Ensure that the measurements detailed are in synch with individual company and overall supply chain strategies. Trading partner executives should

articulate the supply chain vision and how the channel performance program will assist in strategic realization.

■ The channel measurements defined must truly support customer satisfaction. The metrics must be focused on adding actual value to the customer and should not be abstract. If customers value receiving goods on the date requested, a metric focused on tracking on-time shipment will not result in a meaningful performance target.

■ Once executive level metrics have been defined, they should be decomposed into tactical and operational measurements. The goal is to track whether actual performance at each business node in the supply network is in overall alignment with executive objectives.

■ Focus only on key supply chain measurements. While literally hundreds of metrics could be applied, select the ones that best track the process measurements focused on the critical time, cost, and quality criteria most important to each channel partner.

■ Use information technologies to gather, process, and analyze the information received. While disparities in data and computer architectures may make this task difficult, today's Internet-driven supply chain management systems, ERP software, and data warehousing tools will continue to make this effort easier in the future.

While these steps will be helpful in getting started, performance design teams must be ever vigilant in the pursuit of new technologies, greater performance collaboration, and the launch of new measurements programs.

Summary

The rise of new forms of supply chain connectivity caused by the application of new forms of Web-based technologies has caused a virtual revolution in the development and implementation of business strategy. In the past, enterprises sought to create corporate marketing, product and service, cost management, and profitability plans that were inward focused. Over the past decade, however, the realization that the resources and competencies contributed by their supply chain trading partners were at least as critical to competitive survival as their own internal capabilities has driven companies to critically re-examine the place of SCM in the drafting of business strategy. Today, the tremendous integrative power of Web-based technologies has elevated this concern with leveraging the supply channel network to a new level of awareness. All executives are keenly aware that success in tomorrow's marketplace will go to those enterprises that focus on the strategic capabilities of their value chain networks, rather than on the temporary ascendancy of company-centric products, services, and infrastructures.

Structuring an effective *supply chain business architecture* requires strategic planners to view the supply chain as consisting of three interdependent dimensions.

To begin with, planners must understand the supply chain from the *physical* point of view. This includes mapping out the terrain of the channel system and the level of value contributed by each channel business node. Second, planners must establish a map of trading partner competencies. The goal is to determine the key capabilities each member contributes to individual company and total supply chain value. The final dimension seeks to detail the type and robustness of the connecting links integrating each channel node. By identifying the technology tools in use in the supply chain, planners can determine how easily and effectively data and process information can traverse the channel network landscape. Viewing the supply chain from a three-dimensional perspective enables strategic planners to architect exciting and radical channel structures providing for a host of possible collaborative and synchronized opportunities to build and enhance market leadership.

Deciding on the proper blend of collaborative technologies and channel capabilities depends on how individual firms want to utilize their network relationships. Some companies may choose to implement simple analog methods to drive cost reduction and facilitate supply channel throughput. Still other companies will want to explore the application of Internet solutions that tightly link trading partners in the search for opportunities for revenue expansion and relationship enhancement. The more technology-driven connectivity strategies are deployed, the more the supply chain model migrates from purely mechanism for moving goods at least cost to becoming a true *value network* capable of generating collective competitive advantage far beyond the capacities of individual companies working on their own.

Structuring effective SCM technology strategic initiatives requires a two-step process. To begin with, corporate planners will need to ensure that certain preliminary steps have been completed. First, individual businesses will need to be energized through education and training from the executive level down to all employees. A detailed vision of the business and the role of channel partners will have to be devised and broadcast. Following, channel strategic teams need to perform a value assessment (SCVA) that will identify which supply chain initiatives are *evolutionary* and which are *revolutionary* and which channel processes need to be moved to the Internet. Finally, the SCVA should provide strategists with a map of possible technology choices that can then be prioritized in preparation for supply chain implementation.

The second part of SCM technology strategy development is the structuring of the actual supply chain strategy. The chapter suggests a possible five-stage planning process. The model begins by requiring planners to construct a *business value proposition*. Activities in this stage center on determining how customer value is going to be pursued with the opportunities provided by the technology alternatives identified in the preliminary step. In the second stage, companies will need to match the value proposition with the products and services constituting the *value portfolio*. Structuring the *scope of collaboration* is the subject of the third stage. Here companies will decide what will be the content of the firm's processes and activities and correspondingly what will be the scope of trading partner collaboration. In stage four, firms will decide how internal and supply chain resources will be used in support of the technology strategy. Basically, this stage seeks to determine how physical assets, human capital, and trading partner competencies can most effectively be utilized. Finally, before the SCM technology strategy can be considered as complete, planners must design a set of meaningful and focused performance measurements that will provide metrics detailing supply chain operational effectiveness and provide for continuous improvement and growth.

Notes

1. Charles H. Fine, *Clockspeed: Winning Industry Control in the Age of Temporary Advantage* (Reading, MA: Perseus Books, 1998), pp. 11–23.
2. Mohan Sawhney and Jeff Zabin *The Seven Steps to Nirvana: Strategic Insights into e-Business Transformation* (New York: McGraw-Hill, 2001), pp. 25–26.
3. Eileen Colkin, "DuPont Jumps Out of Dark Ages into E-Commerce," *Information Week,* December 10, 2001, p. 80.
4. Sawhney and Zabin, *Seven Steps to Nirvana,* pp. 26–34, have been most helpful in compiling this short section.
5. Marvin L. Manheim, "Integrating People and Technology for Supply-Chain Advantage," in *Achieving Supply Chain Excellence Through Technology,* 1, David L. Anderson, ed. (San Francisco: Montgomery Research, 1999), pp. 304–313.
6. Paul G. Lowe and William J. Markham, "Perspectives on Operations Excellence," *Supply Chain Management Review* 5, no. 6 (2001), p. 60.
7. David Bovet and Joseph Martha, *Value Nets: Breaking the Supply Chain to Unlock Hidden Profits* (New York: John Wiley & Sons, 2000), pp. 37–53.
8. Miles Cook and Rob Tyndall, "Lessons from the Leaders," *Supply Chain Management Review* 5, no. 6 (2001), p. 30.
9. Sawhney and Zabin, *Seven Steps to Nirvana,* p. 100.
10. Marianne Kolbasuk McGee and Chis Murphy, "25 Innovators in Collaboration," *Information Week,* December 10, 2001.
11. C. K. Prahalad and Venkatram Ramaswamy, "The Collaboration Continuum," *Optimize Magazine,* November 2001, pp. 31–39.
12. Michael Treachy and Michael Dobrin, "Make Progress in Small Steps," *Optimize Magazine,* December 2001, pp. 53–60.
13. James A. Tompkins, Steven W. Simonson, Bruce W. Tompkins, and Brian E. Upchurch, *Logistics and Manufacturing Outsourcing* (Raleigh, NC: Tompkins Press, 2005), pp. 28–36.
14. Kevin Kavanaugh and Paul Matthews, "Maximizing Supply Chain Value," in *Achieving Supply Cain Excellence Through Technology,* 1, David L. Anderson, ed. (San Francisco, CA: Montgomery Research, 1999), pp. 278–281.
15. Ibid.
16. Peter C. Brewer and Thomas W. Speh, "Adapting the Balanced Scorecard to Supply Chain Management," *Supply Chain Management Review* 5, no. 2 (2001), pp. 48–56.
17. Karen Abramic Dilger, "Say Good-bye to the Weakest Link with Supply Chain Metrics," *Global Logistics and Supply Chain Strategies* 5, no. 6 (2001), pp. 34–40.

18. Larry Lapide, "What About Measuring Supply Chain Performance?," in *Achieving Supply Cain Excellence Through Technology,* 1, David L. Anderson, ed. (San Francisco, CA: Montgomery Research, 1999), pp. 287–297; John Grabski, "Valuation Methods for the New Supply Chain," in *Achieving Supply Cain Excellence Through Technology,* 3, David L. Anderson, ed. (San Francisco, CA: Montgomery Research, 2001), pp. 254–255; and Robert J. Bowman, "From Cash to Cash: The Ultimate Supply Chain Measurement Tool," *Global Logistics and Supply Chain Strategies* 5, no. 6 (2001), pp. 42–48.

19. Robert S. Kaplan and David P. Norton, *The Balanced Scorecard* Boston: Harvard Business School Press, 1996) and *The Strategy Focused Organization* (Boston, MA: Harvard Business School Press, , 2001). See also the excellent article by Brewer and Speh, "Adapting the Balanced Scorecard to Supply Chain Management."

20. For a summary of the SCOR method, see SCOR version 10.0, found at accessed January 15, 2010, http://www.supply-chain.org.

Afterword

It is almost impossible to conceive of the concept and practical application of *supply chain management* without understanding the enabling role of today's information and communications technologies. Up until the advent of computerized systems permitting enterprises to link their plans and processes with those of their customers and suppliers, businesses had nothing more than informal methods by phone, mail, fax, or word of mouth to communicate their needs and capabilities to the marketplace at large. This isolation was fortified by the management techniques of the day, which viewed organizations as hierarchical silos dominated by a command and compliance style. Companies jealously guarded production and distribution processes, safeguarded information regarding product, planning, and price, and were ready for combat with anyone, even close suppliers, who interfered with hard-fought, entrenched positions of market share and customer segment dominance.

The growth of computerized information technologies over the past half-century enabled businesses to slowly tear down these communications and management silos and provided the mechanism for an ever-widening circle of integration, collaboration, and networking. The first computerized applications provided organizations with the ability to streamline internal operations, dramatically increase productivity, and cut costs by automating processes that were highly repetitive, prone to human error, and generated inherent variation. At the same time, businesses systems like *manufacturing resource planning* (MRP II) and then *enterprise resource planning* (ERP) provided the essential linkages by which organizations could connect what had often been isolated departments into a single integrated, mutually supporting whole. While inward focused, the rationalization and standardization implicit in these early applications were essential in enabling businesses to escape from their narrow internal focus.

The first real steps of a nascent supply chain occurred in the 1980s as computer system architecture began to allow businesses to integrate their systems with those of customers and suppliers. For the most part, this integration was one-way: companies could use tools like *electronic data interchange* (EDI) to pass data such as forecasts, orders, and financial information electronically, thereby escaping from

383

oral and paper-based communication. In addition, as connectivity became more sophisticated, businesses began to look at their customers and suppliers increasingly as business partners, each performing a critical role in a continuously evolving supply chain ecosystem of which each were an integral part.

The next leap in the linkage of technology and supply chain management occurred with the establishment of the Internet. In a way that was previously impossible, the Internet permitted companies not only to integrate and network remote parts of the organization on a single database and software system, they were also able to connect with business partners and their systems, regardless of particular hardware and software, in a two-way conversation where data and ideas could be easily and instantaneously shared. Networking transformed work from a sequential to a concurrent process; work between business partners became a dialog where individual business visions and knowledge could be fused to inform and inspire others to act as a single, virtual team focused on activating processes that generated value that transcended individual company objectives.

There can be little doubt that the ability of organizations to leverage human networking teams that not only cross company boundaries but also global space and time is at the core of today's concept of *supply chain management*. Technology-enabled networking enables companies to shed the inflexible organizations of the past in favor of the creation of virtual teams composed of the best resources from across the supply chain who can now link their skills, talents, and technologies to work in an iterative and parallel manner to achieve common objectives. Today's Internet-based technologies have rendered obsolete the concept of the supply chain as simply a pipeline for the efficient management of the flow of goods and information; supply chain management is now a management philosophy and a mechanism enabling the continuous regeneration of networks of businesses integrated together and empowered to execute superlative, customer-winning value at the lowest cost through the digital, real-time synchronization of products and services, vital marketplace information, and logistics delivery capabilities with demand priorities.

This view of supply chain management can be seen in IBM's concept of the supply chain of the future [1]. The authors of the white paper see today's supply chain managers struggling with five critical challenges: continued cost containment, requirements for greater visibility to supply channel events, navigating through increased business risk, gaining closer customer intimacy, and coping with increased globalization. Responding to these challenges will require companies to deploy technologies capable of converging the physical and digital infrastructures found in today's integrative technologies. On the transaction level, sensor technologies will make it increasingly possible for devices to communicate and network without human intervention. On the strategic side, Internet technologies will enable supply chain systems to interactively network with other systems including transportation and warehousing, financial markets, and the social network to provide even deeper levels of collaboration.

IBM's vision of the supply chain of the future has singled out three essential technology-based drivers:

- *Instrumented.* Technologies in this area seek to automate and informate transaction processes. Supply chain information currently managed by people will increasingly be generated by sensors, *radio frequency identification* (RFID) tags, meters, actuators, *global positioning systems* (GPS), and other devices. Not only will these technologies squeeze risk and cost out of processes, they will provide visibility to possible disruptive events as well as day-to-day information as to the status of shipments, the location of containers, and the position of components and finished goods with timely planning and real-time execution.

- *Interconnected.* Smarter supply chains will utilize the tremendous networking power of Internet technologies to drive complex levels of collaboration not only between supply chain partners and their systems but also with smart devices as they transmit information on supply chain processes on a real-time basis. Enablers like social networking will expand customer intimacy and link channel teams with marketplaces who will increasingly drive product innovation, performance, and branding dimensions. Supply chains will be capable of collective planning and decision-making on a global basis.

- *Intelligent.* Smart supply chain managers will be capable of more effectively countering risk and selecting optimal cost-responsive channel trade-off through the deployment of intelligent systems that will enable them to assess constraints and opportunities and to test choices through simulation. Intelligent technologies might be able to make automatic decisions, such as automatically reconfiguring a supply chain when a significant constraint appears or even requisition supplier assets like production facilities and transportation fleets on demand through virtual exchanges. Using sophisticated modeling and simulation capabilities, intelligent supply chain technologies will enable channel managers to "move past sense-and-respond to predict-and-act" [2].

Enabling this vision of the technology-enabled supply chain will require companies to embrace management styles that are more global, flexible, responsive, and networked through increased instrumentation, interconnection, and intelligence. When the first edition of this book was written, such technologies were only beginning to form on the horizon. Everything was about the interconnective and new business opportunities brought about by the breakthroughs occurring in the Internet world. Today, globalization, deepening supply chain interdependence, and rising risk and volatility are driving companies to move far beyond the Internet Revolution to espouse the transformational principles to be found in collaborative networking through the continuous development and application of supply chain information technologies. It should be interesting indeed to see where the changes in the business climate of the 2010s will drive the direction of technology and supply chain management.

Notes

1. IBM Global Services, *The Smarter Supply Chain of the Future.* (Somers, NY: IBM Corporation, 2009).
2. Ibid., p. 35.

Index